To So Few

Deflection

Books by Cap Parlier:

—

Anod series
The Phoenix Seduction (1995)
Anod's Seduction (2004) [reprint of The Phoenix Seduction]
Anod's Redemption (2004)

—

Sacrifice (2000)
The Clarity of Hindsight (2016)

—

To So Few series
To So Few – In the Beginning (2014)
To So Few – The Prelude (2014)
To So Few – Explosion (2015)
To So Few – The Trial (2016)
To So Few – The Verdict (2017)
To So Few – Frustration (2018)
To So Few – Deflection (2019)

—

and with **Kevin E. Ready:**
TWA 800 - Accident or Incident? (1998)

—

These and other great books available from Saint Gaudens Press
Post Office Box 405
Solvang, CA 93463-0405
URL: http://www.SaintGaudensPress.com
Visit Cap Parlier's Web Site at: http://www.Parlier.com

—

To So Few

Deflection

by
Cap Parlier

SAINT GAUDENS PRESS
Phoenix, Arizona & Santa Barbara, California

Saint Gaudens Press
Post Office Box 405
Solvang, CA 93464-0405

Http://www.SaintGaudensPress.com

Saint Gaudens, Saint Gaudens Press
and the Winged Liberty colophon
are trademarks of Saint Gaudens Press

Print edition ISBN: 978-0-943039-47-3

Library of Congress Catalog Number - 2019900302

Printed in the United States of America

The TO SO FEW series books are works of fiction. Any reference to real people, objects, events, organizations, or locales is intended only to give the fiction a sense of reality and authenticity. Other names, characters and incidents are the products of the author's imagination and bear no relationship to past events, or persons living or deceased.

Dedication

—

This volume of the To So Few series
is dedicated all those patriots
who have served this Grand Republic.

—

Acknowledgments

—

Words cannot express my gratitude to John Richard for his critical and constructive review of the manuscript. His care for and interest in this story have made it better in multitudinous ways. Thank you so much for giving so generously of your time.

I would be remiss if I did not convey my sincerest appreciation for the courage and attention to detail of the staff at Saint Gaudens Press for their continuing encouragement and support. They are truly a blessing.

Most importantly, I must publicly thank my wife Jeanne for tolerating my dedication to this story and taking such good care of me. She is a saint.

—

Prologue

—

The war in Europe had been raging since September 1939, after the unprovoked German invasion of Poland. The infamous Blitz—the broad, nightly, indiscriminate bombing of British cities by the German air force—continued through the winter and into the spring of 1941. By the end of October 1940, the Germans abandoned their nearly successful air superiority campaign when their leader decided to postpone the cross-Channel invasion of England in favor of the terror bombing of British cities. No one in Great Britain was invulnerable.

Concomitantly, the devastating successes of the German submarine fleet in the Battle of the Atlantic were choking the United Kingdom of vital supplies not just for defense of the realm but also for their very survival.

From the capitulation of France in June 1940, Great Britain stood alone against German domination of Europe. The British colonial dominions had begun to mobilize in support of the motherland.

Prime Minister Winston Leonard Spencer Churchill, CH, TD, Member of Parliament for Epping, assumed the charge of King George VI and formed a coalition government. The War Cabinet served as the executive committee of His Majesty's Government and was comprised of the leaders of each of the major political parties. Soon after they came to power, Churchill and the War Cabinet took extraordinary actions to bolster their defenses. Among those actions was the unilateral transfer of secret, vital, military technology to the United States by the so-called Tizard Mission—the British Technical and Scientific Mission to the United States of America—led by Sir Henry Thomas Tizard FRS, KCB. The Prime Minister also made a personal, private appeal to his friend and colleague across the Atlantic Ocean for any assistance that could be provided.

President of the United States Franklin Delano Roosevelt had been communicating privately with Churchill since before the war. The President recognized how deeply unprepared for war his nation was as well as the essential need to sustain Great Britain against German aggression. He took serious political risk to approve crucial arms shipments as well as the "lending" of 50 "surplus" destroyers to Great Britain to bolster the British efforts against German submarine attacks on the life-blood supply convoys. In those turbulent and uncertain days, Roosevelt also had been reelected to an unprecedented third term as president. Shortly thereafter, the President broadcast his "Arsenal of Democracy" speech in the 16th edition of his fireside chats and repeated the message a week later in his state of the union message to Congress. Four days after his remarkable speech and in collaboration with the Prime Minister, President Roosevelt created, submitted to Congress, and coaxed the legislature to pass in two months the historic bill. He signed into law the monumental Lend-Lease Program, to resolve the desperate British financial situation, and supply them with the necessary military hardware to rearm and strengthen themselves for stopping, and then defeating the Germans.

By the end of March 1941, the situation of the British people had changed dramatically. After the desperate fight during the summer for their very existence and enduring the brutality of the Blitz night bombing, the Royal Air Force began offensive air operations against German occupation forces in Western Europe and into Germany itself.

—

Beyond the political and military leaders, their strategic decisions, and the nation-states and armies in conflict, individual citizens stood to serve a greater purpose, lived their lives as best they could in the circumstances around them, and collectively carried the weight of the conflict. It is through their eyes that history comes alive for future generations.

Flying Officer Brian Arthur 'Hunter' Drummond, CBE, MC, DFC, of Wichita, Kansas, approached his 20th birthday in April of 1941. Brian Drummond stood out physically among his brethren, standing 6 feet, 2 inches, with an athletic, 185-pound body, a distinctively chiseled, unblemished face, light brown, wavy hair, blue-grey eyes, and a fair complexion. He looked older than he was. His 19 aerial victories and his unofficial designation as a triple ace had come with a price. Four aircraft had been shot out from around him; the last one on Friday, 27.September.1940, had nearly killed him and left him on extended recuperative leave. As he returned to flight status, Brian was transferred from No.609 Squadron to the all-American volunteer No.71 'Eagle' Squadron.

Charlotte Grace Palmer née Tamerlin had saved Brian's life once, when she risked her own life to pull the unconscious pilot from her farm pond and out from under the tangle of the sinking parachute. King George VI awarded her the George Cross – the highest civilian award for heroic action – at the same ceremony where Brian had received his DFC from the King. Charlotte was a strikingly attractive, 27-year-old, relatively tall woman with porcelain skin, prematurely gray hair, blue-gray eyes, and distinctive features. She was a strong, independent and confident person, who ran the 190-acre, Hampshire, Standing Oak Farm that had been in the Tamerlin family for more than two centuries. Charlotte and Brian married in December 1940, and their first child was due in June 1941.

Brian's closest brother in arms, Flying Officer Jonathan Andrew Xavier 'Harness' Kensington of Newcastle had been Brian's classmate during their flight training and fortuitously gained assignment to the same squadron, seven weeks after Brian. Jonathan cut quite the figure of virile British manhood – curly blond hair, ice blue eyes, and half a foot shorter and 20 pounds lighter than his buddy, but also an accomplished Spitfire pilot. Jonathan held the distinction as one of the few 'line' fighter pilots to fly captured enemy aircraft with the exploitation team at the Royal Aeronautical Establishment, Farnborough.

Brian's benefactor and protector in the Royal Air Force – Air Commodore John Henry Randolph Spencer, CMG, DFC – had been squadron mates with the Great War, American volunteer, fighter pilot and ace Malcolm Bainbridge. Malcolm had been Brian's flight instructor since the young man had been nine years old. John was now 43 years old, of moderate stature – 5 feet 9 inches tall and 155 pounds at his last check – green eyes, and dark brown hair, streaked with grey, now limited to a laurel band just above his ears. He luckily managed to marry the beautiful, out-going and eight years younger Mary Elizabeth Ann Spencer née Armstrong 12 years ago. John's commitment, energy and expertise garnered him the promotion and assignment as Chief Controller, No.11 Group, at Uxbridge, the air defenders of Southeast England, who had borne the brunt of the German air assault last summer. Although he did not and never would brandish his family connection, John Spencer was also a nephew of Prime Minister Churchill, to whom he had introduced Brian before the war and after the American's arrival in England.

Squadron Leader Lord Jeremy Robert Kenneth 'Mud' Morrison, now commanding officer of No.32 Squadron and the younger brother of the 8th Duke of Cottingstone, had been Brian Drummond's first RAF flight instructor and had become friends with the American.

Trevor Thomas Andersen graduated from Cambridge University with a degree in European history in 1926, and he already had a job. During his college years, he attracted the attention of an influential man, Director of Naval Intelligence Vice Admiral Sir Geoffrey Ian 'Jumper' Pike, KCB, DSC. Trevor's frankly rather ordinary appearance, long-ish light brown hair, blue eyes and medium build attracted little notice. Trevor's fluency in French, German and Polish, along with the unusual ability to quickly switch to one of several dialects, made him nearly ideal for intelligence fieldwork. After several apprenticeship missions, Trevor was given a code name – 'Diamond.' He also picked up several alias persona, including that of Robert Henry Stone Johnston, a leather goods salesman. After the formation of the Special Operations Executive (SOE) – tasked by Churchill to set German-occupied Europe ablaze – Andersen transferred from the Admiralty to SOE.

And so, here begins our story.

—

Chapter 1

Bearing their birthrights proudly on their backs,
to make a hazard of new fortunes here.
-- William Shakespeare, King John, II, i, 70

Wednesday, 2.April.1941
Standing Oak Farm
Winchester, Hampshire, England
United Kingdom
16:50 hours

The farm erupted in the glorious colors of spring and the weather matched the scene. Brian wanted to release the taxi at the turnout at the last ridgeline, to enjoy the idyllic view for a few minutes, and then walk the rest of the distance to the house. However, he knew it was milking time for the farm's cows. Charlotte and her hired hands could always use the extra help. The driver responded professionally to Brian's direction to continue. No one appeared to greet him, which seemed to confirm the location of everyone.

Brian paid the driver and thanked him for his service. He went inside, and removed his hat, uniform tunic and necktie, laying them across the nearest chair. He proceeded directly to the barn and traded his uniform shoes for his muck boots and donned his water-repellent work apron. Sure enough, Charlotte was in the barn along with her long-term hired hands Lionel Bridges and Horace Morgan as well as their new hand, 13-year-old Jacob Holden, hired just last month.

Horace was the first to notice Brian. "Our hero has returned."

Other greetings came and were acknowledged. Charlotte looked around the far cow she was tending to as Brian walked toward her. She did not miss a beat, rhythmically squeezing and pulling two of the udder's four teats and squirting successive nice streams of milk into the pail positioned below the cow's udder. Brian patted the cow's rump and leaned over to kiss his wife. She silently mouthed I love you, as she continued her rhythmic chore. Brian returned the gesture, and then turned to gather a stool and pail near the next cow. He positioned and secured the cow, and began the process of relieving the cow's distended udder.

The afternoon milking session took another hour to complete including the preparation of the product as well as the clean up of their equipment and barn. Horace, Lionel and Jacob finished loading the truck for the journey into Winchester after supper. Charlotte finally embraced her husband, hugged him tightly despite her amply pregnant abdomen, and kissed him passionately.

"How long do I have you this time?' she asked.

"Forty-eight hour pass and I've consumed six of those hours already." Brian smiled.

"Then, we had best get us all fed, so we can get the lads on their way and I can welcome my lord and master home properly."

"What is this lord and master nonsense?"

"Just traditional paternalism . . . not worth the words." They reached the house. Brian opened the door and allowed Charlotte to enter first. "Why don't you get comfortable while I prepare our meal." Brian nodded his head, retrieved his tunic, hat and tie, and went to the bedroom to change out of the remainder of his uniform. He hung his uniform in the closet and returned to Charlotte.

"How have things been since I returned from the U.S.?"

"Spring is here," Charlotte answered without looking up from her chopping task.

"I noticed as I crested the ridge. This is my first spring with you and I am amazed at the beauty." Brian wrapped his arms around her from behind and below her late pregnancy belly. He could see her smiling.

"I am glad you are here, Brian. I only wish it was permanent, and we could live in peace."

"We will get there, my love. It will just take time."

"And blood."

"I shall do my best to avoid further donations."

"You'd better. I will never forgive you . . . putting me through that process, again."

"Under . . . ," Brian interrupted his response as Lionel, Horace and Jacob entered.

"Supper will be ready in 30 minutes," Charlotte announced loudly, and then turned to Brian. "Why don't you get a bottle of wine or two from the pantry."

Charlotte had accumulated an ample supply of wine before the war that sustained her farm during the sparse months after the invasion of France. Surprisingly, Charlotte had connected with a few British vintners in Southern England and Southern Wales, and managed a quid pro quo barter process to maintain her wine supply. Brian selected two bottles of a 1932 Beaujolais. Everyone hoped wine would eventually resume flowing at least from the South of France.

Brian opened the first bottle and tested a small amount as Anne Booth had taught him back during the Phony War last year. The red wine was comparatively light and in good form. As he poured the first glasses, he asked Jacob, "Can you have wine?"

"Of course, Mister Drummond," the boy responded. "My parents have served it occasionally."

"One glass," Charlotte added without turning from her task.

Brian split the first bottle among five glasses, handed a glass to everyone including Jacob, and then raised his glass for a toast. Charlotte stopped her effort to join them. "Here is to the maid of the manor, to Standing Oak Farm, and to blessing of the bounty before us."

"Hear, hear," the others said in unison. Everyone took a sip of wine.

Charlotte sat with them at the table while the food finished cooking.

"How is your flying, Mister Drummond?" asked Jacob.

"Jacob!" Lionel objected, followed immediately by . . .

"No!" Charlotte protested. "We will not discuss the war at this table." No one objected. Silence occupied the space between them. "My apologies. I did not mean to react so sharply."

"It's OK, sweetheart. We understand."

Charlotte nodded her head slightly, stood and placed the pot at the center of the table. She ladled the country stew into each bowl. They ate in silence. Not a word was spoken. The hands finished their meal, excused themselves, and departed in the truck with the milk cans for the market. Brian helped Charlotte clean up and put everything away. She opened the second bottle of wine and poured another glass for both of them. Gesturing with her head, Charlotte led Brian to the front living room, sat on the couch and patted the cushion next to her.

"I received an interesting letter two days ago."

"And . . . ?"

Charlotte retrieved a single page, handwritten letter from the upper right drawer of her desk. "Here, read it for yourself," and handed it to Brian.

March 10, 1941

Dear Charlotte,

> *I pray you will not think of me as too forward, but I feel I know you, even though you do not know me. My name is Gertrude Bainbridge. My friends call me 'Gerty.'*

> *Brian Drummond just left Wichita yesterday to return to you and his duty. He is the son my deceased husband and I never had. I love him dearly. He is a very special young man, but I suspect you already know that.*

Brian informed me of your marriage last December and the impending birth of your first child together. I offer my heartfelt and sincerest congratulations as a future 'grandmother.' I look forward to the day I can meet you in person and hold my first grandchild.

I asked Brian and he suggested that I write to you, to open communications between us. While I am not close in proximity to you, I wanted to offer to you my support and a sympathetic ear to listen as you may wish to share. Brian confided in me your tragic family losses as well as your reluctance to marry our mutual fly-boy. I know in part the apprehension you must have felt and probably still do feel. Malcolm loved flying and he conveyed that passion to Brian. We share more than a few things in common. I say this only to commiserate with you while Brian is away.

This is probably already too much, so I will stop here, and await your reply and look forward with eager anticipation to expanding our friendship and solidifying our common bond.
My very best wishes to you, Charlotte.
With great respect,

Gerty

"What did you think?" Brian asked.

"Sounds like you made me out to be some charity case."

"Charlotte . . . really!"

Charlotte stood, walked back and forth a few times, and then sat back down. "I'm sorry, my darling. This is my first pregnancy and it's wearing me down. I have been rather snippy, lately."

"I guess I should not say I understand, because I can't possibly understand. But, I wanted Mrs. Bainbridge to know us . . . to know you. Like she said, she is like a second mom to me. As I explained when I returned from the States, she is going to be in our lives for several reasons."

Charlotte stared into Brian's eyes for several minutes. "Then, Gerty will be my second mom as well."

"Will you write to her?"

Charlotte smiled. "Of course I will write to her. She is my Mom, now, too."

"Thank you. She is a good woman, who cares a lot. I think you will really like her. No, I know you will really like her. I just hope you get to meet her before too long."

"Together we will make it happen. Now, take me to bed and make me a happy woman."

"It's early. Are you sure you're ready for bed?"

"Yes, I'm ready for bed. I'm not ready for sleep, but I enjoy being in bed with you."

Brian smiled broadly. "Then, off to bed we must go."

They lit a single candle and closed the blackout curtains. They made sure everything was in its place, in case they needed to move throughout the house at night. They managed to work around Charlotte's uncomfortable abdomen to more than once find their moments of pleasurable reward.

—

Friday, 4.April.1941
American Eagle Club
No.28 Charing Cross Road
Covent Garden, London, England
United Kingdom
10:45 hours

Most of the remodel construction had been completed a week ago. The contractors were finishing up a few last items. The planned and scheduled dedication would proceed even though the conversion of the four-story office building was not entirely complete. The building was abuzz in anticipation of the arrival of their honored guest—Her Majesty The Queen Consort Elizabeth, the wife of King George VI.

United States Ambassador to the Court of St James's John Gilbert 'Gil' Winant stood toward the back of the simple but nicely appointed lobby, near the staircase leading to the upper floors. With him were four senior military officers from the embassy—all in service uniforms.

-- Special Naval Observer Rear Admiral Robert Lee 'Bob' Ghormley, USN [USNA 1906],

-- Military Attaché Brigadier General Raymond Eliot Lee, USA,

-- Naval Attaché Captain Alan Goodrich Kirk, USN [USNA 1909], and

-- Assistant Naval Attaché Colonel Roy Stanley Geiger, USMC

"Welcome back, Bob," Winant said to Ghormley. "I've not had a chance to congratulate you and ask you how the staff conversations conference went."

The admiral fidgeted a little, clearly uncomfortable with the last part of the ambassador's greeting. "I would suggest, Mister Ambassador, that I brief you more thoroughly in the privacy of your office. However, in summary, I think all of us who attended would report they went exceptionally well." Ghormley looked to Kirk, who nodded his head in concurrence.

"When did you get back?"

"About a week ago. Captain Kirk and I had to leave early when a transportation window opened up. The conference concluded earlier this week."

"Welcome back as well, Alan," the ambassador said and shook hands with Kirk. "Am I informed correctly, are you with us for only a short time more?"

"Yes sir. I will leave at the end of the month. I will temporarily hand off my duties to Colonel Geiger," Kirk responded, nodding to the Marine colonel, "who shall hold down the fort until my successor is named and arrives."

"What is your next assignment, if I may ask?"

"I will become the director of naval intelligence next month."

"N-2 . . . a difficult job in these troubled times."

"Yes sir . . . without a doubt."

Winant turned to Geiger, "The gentleman in green must be our newly arrived Roy Geiger." The distinctive olive green service uniform was a lighter shade of green from General Lee's forest green service tunic. The Marine colonel's uniform displayed the golden wings of a naval aviator above his already impressive stack of ribbons, including the Navy Cross—the second highest award for combat valor within the Department of the Navy.

"Yes sir," Geiger responded, as they shook hands.

"Welcome aboard."

"Thank you, sir."

"Always a pleasure to see you, Raymond," Winant greeted the highly respected general, who stood up to the Ambassador's pro-fascist predecessor.

"Likewise, Mister Ambassador."

"Hopefully, we can dispense with the formalities quickly and get back to work." They all chuckled softly. "We were quite fortunate . . ." Winant paused and gestured to the elegant dressed man approaching them, ". . . there he is, the man of the hour. Gentlemen, may I introduce Mister Charles Sweeny—our benefactor for this fine facility."

". . . one of the benefactors," Sweeny corrected, as the two men shook hands.

Appropriate introductions were made.

"Charles led the way in raising the necessary funds to purchase and renovate this building," explained Winant.

"A home away from home for the rapidly growing number of American servicemen in London." Sweeny added.

"I'm certain our fellow citizens will appreciate the touch of home that a hamburger or hotdog, soda and a movie will bring to them while on duty over here."

"That is precisely the intent," Charles replied. "When is this little ceremony supposed to kick off?"

"The plan says 11 o'clock," answered Winant.

Sweeny glanced at his wristwatch. "Then, we should be starting shortly."

"The Lord Mayor should appear to announce and introduce The Queen," Winant added. "She will sign the register, say a few words, and then be on her way."

"That should work," Charles said.

Several minutes passed in silence. Colonel Geiger leaned toward Sweeny. "I understand you and your uncle were instrumental in forming Seventy-One Squadron."

"I wouldn't say instrumental, but we certainly did advocate an American volunteer squadron since we gathered up more than a squadron's worth of American pilots. Seventy-One was formed in September of last year. We have enough other pilots for one, two or more squadrons."

"Congratulations," Roy said.

"Thank you, we try . . . ," Charles was interrupted by a commotion at the door.

"There he is," announced Winant.

The Lord Mayor of London Lieutenant Colonel Sir John Dawson Laurie, KBE, appeared at the doorway dressed in a conservative, medium gray business suit with the symbolic collar chain of office around his neck and on his shoulders. Absent were the ermine-bordered, ceremonial, red robe and black, furry, tri-corner hat common in pre-war occasions of such as this. He nodded to Ambassador Winant, who returned the gesture.

"Ladies and gentlemen," the Lord Mayor said loudly. He waited to quiet. "We are honored to be here for the dedication of this important facility, and most of all, we are grateful for the blessings of the Crown today. It is my distinct honor to introduce our guest of honor for the dedication ceremony. Ladies and gentlemen . . . Her Majesty The Queen."

Queen Elizabeth—Elizabeth Angela Marguerite Bowes-Lyon—entered the lobby and waved to the small gathering. She walked directly to the lobby desk, as if by a pre-briefed, established plan, and signed the register as the very first guest of the American Eagle Club. The audience clapped modestly and respectfully. She turned at the desk to face the group. "Thank you very much for the warm welcome and the pleasure of dedicating The Eagle Club. Lord Mayor, Mister Ambassador," she said, looking to Winant, "our two English-speaking countries have deep bonds that trace back three centuries. In addition to the diplomatic personnel who have been with us over all of those intervening years, we have American ex-patriots who are interwoven in our society and culture, and now we have a growing contingent of military personnel like your embassy's attachés," she said, nodding to the officers standing with Ambassador Winant, "and, volunteers who have so unselfishly joined our armed forces in this desperate battle for freedom itself. We are deeply indebted to the United States and especially to those Americans who are here with us. The King and I trust this facility will make the extraordinary sacrifices of American citizens in service in our abused capital city a little less lonely. Lastly, to all Americans in this country and serving throughout the world, The King and I offer our sincerest, heartfelt gratitude. Thank you for the honor of joining you for this dedication. On behalf of the Sovereign, I declare the American Eagle Club open for service." The audience applauded enthusiastically. The Queen nodded her head and waved to everyone in the lobby. She departed and was followed by the Lord Mayor.

"And, thus concludes the dedication of our fair club," Sweeny observed.

"They are allowing us to tour the building. Anyone interested?"

"Thank you, Mister Ambassador," answered Ghormley. "I, for one, shall pass at the moment—perhaps another time. I need to return to Grosvenor Square."

"I'll join you, Gil," Sweeny said.

The military officers departed, choosing to walk back to the Embassy, despite the gloomy, overcast, darkening sky. Winant and Sweeny ascended the stairs to the next floor. The officers collectively decided to stop at Claridge's Hotel for lunch before returning to the Embassy.

—

Friday, 4.April.1941
Cabinet War Rooms
New Public Offices
Whitehall, London, England
United Kingdom
17:45 hours

The full War Cabinet had assembled in the conference room, along with the Cabinet secretariat support staff and by invitation, the three service ministers. Churchill had delayed his usual departure on this Friday afternoon for his weekend at Chequers in order to receive the most recent news and more importantly discuss immediate actions.

"Pray tell, what is the latest information from North Africa?" asked the Prime Minister.

Secretary of State for War Henry David Reginald 'David' Margesson, MC, PC, Member of Parliament for Rugby, leaned forward and answered, "General Wavell has ordered the evacuation of Benghazi and is withdrawing. The *Afrikakorps* once again appeared where Sir Archibald least expected them to be. Rommel has proven himself to be quite the magician when it comes to maneuvering armor and infantry forces. Wavell intends to carry out a spirited withdrawal."

General Sir Archibald Percival Wavell, KCB, CMG, MC, PC, had been the senior army officer in Northeast Africa since July 1939. He had been given the larger appointment as Commander-in-Chief Middle East with far broader responsibility from East Africa to Persia. Since the previous February, Wavell's principal adversary in the desert of North Africa was *Generalleutnant* Erwin Rommel—one of Hitler's favorite generals and an accomplished armor commander.

"To where?" snapped Churchill. "To Alexandria, to Cairo, to the Suez, to Palestine, or even the Persian Gulf? Should we just bring the Army of the Nile home now? When or where is he going to stand and fight?"

Total silence filled the room . . . other than the air handlers replenishing fresh air. Margesson wisely waited to gain control of his emotions. He cleared his throat. "We have knowingly depleted his forces to reinforce Greece. He has insufficient forces, stretched precariously thin, to withstand the German and Italian onslaught."

"Nonsense. Balderdash, I say. We are all thin. He should ask the Royal Navy about thin in the cold of the North Atlantic, or ask those intrepid young fighter pilots, who rose against daunting odds many times every day during those harrowing months last summer. Wavell has made his position quite clear. He desires more troops. Unless we are prepared to abandon our allies or weaken the Home Forces in the face of a possible renewed spring or summer invasion threat, we have no more forces to give. Even the Commonwealth countries under threat like Australia and New Zealand are reluctant to send more divisions for fear of weakening their defenses further."

"Be reasonable," begged Margesson.

"There is no reasonable in war, David," snapped the Prime Minister. "There is only success or failure." Churchill paused and looked down at the papers in front of him. Again, silence remained. Winston looked back to David Margesson. "I want you and General Wavell to inform me and the War Cabinet where he intends to draw the line and fight. I want that plan today, or tomorrow at the latest. Is that clear?"

"Yes sir. General Dill and I will confer with General Wavell as soon as we are through here."

"Very well. I anxiously await the plan. I cannot resist recounting my counsel last week as we reviewed the results of the Battle of Cape Matapan. This is precisely why I advocated for the offensive, to keep Rommel on his heels and deny him the initiative. Well, that opportunity has clearly been lost, unless General Wavell can muster up the wherewithal to regain the initiative." Again, Churchill paused for his thoughts. "I do not intend to berate General Wavell further. I await the plan as promptly as humanly possible. Now, David, is there anything else for the War Cabinet from the War Office?"

Margesson cleared his throat. "Actually, I have more bad news from North Africa. Cairo reports that Generals O'Connor and Neame were captured last Thursday as they were attempting to rally their troops. Confirmation took two days. The Germans were moving very fast and overran them both."

"With Wavell on the defensive, we are not likely to recover them. Neither of them is read into the special intelligence program, as I recall our access list."

"No sir," answered Margesson, "neither of them had access to special intelligence."

"Thank goodness for that small blessing. Conditions in Cyrenaica are fluid to say the least. Perhaps, we will get lucky and recover them before they are transported to Italy or Germany. Please inform their families of their capture and that we are doing what we can to return them to the fold."

"Certainly sir."

"I will write personal notes next week, if they have not turned up within a few days. Anything else?"

"No sir."

"Very well, then, let us turn our attention to the Atlantic. Your turn, 'A.V.'"

First Lord of the Admiralty Albert Victor 'A.V.' Alexander, CH, PC, Member of Parliament for Sheffield, Hillsborough Division, shifted in this chair. "Convoy SC Twenty-Six was attacked by a small wolf pack for two days and lost ten ships, 58,000 tons of ships, precious seamen and cargo. One U-boat was reported sunk."

"Not a good ratio by any measure," said Churchill. "We have seen hints the Hun are referring to these days as 'The Happy Time.'"

"No, it is not a good ratio, and yes, we have seen those references. We have been at this battle for not quite two years, now, and we have yet to gain the upper hand, but that day is coming. I would like to report to the War Cabinet our joint radio detection team in America delivered encouraging news last week. The team carried out a successful airborne test with the prototype equipment that combines our cavity magnetron and the American duplex switch to achieve good results against airborne targets. The real surprise of the mission was the team's ability to detect a submarine periscope."

"Excellent. We need that equipment today. When can they deploy that capability? What do we need to do to speed up the production process?"

"They agreed to take the risk of production start-up as they complete development work. Several capable American companies have agreed to accept the risk with surety from the U.S. Government. They are working on the schedules as we speak. Everyone associated with this equipment knows and understands the priority. Archie and I contacted Sir Hugh on his mission to the United States." Air Minister Sir Archibald Sinclair nodded his head in agreement. "He is scheduled to visit the team, to witness the performance of the equipment, later this month – the 29th, I do believe. Sir Hugh knows the importance. He will also directly review the production schedule to get operational units to the field as soon as humanly possible. Archie and I have asked for his assessment promptly after his visit. We have the Aeronautical Establishment working our fitting plan, including which aircraft, provisions for installation to be accomplished in advance of delivery, and even training of ground and air crews to minimize the time to operational effectiveness."

Secretary of State for Air Sir Archibald Henry Macdonald 'Archie' Sinclair, Bart, CMG, PC, Member of Parliament for Caithness and Sutherland, 4th Baronet of Ulbster and Leader of the Liberal Party, had been the ministerial leader of the Royal Air Force since Churchill became Prime Minister in May 1940. Sinclair had also been instrumental in the removal of Air Chief Marshal Sir Hugh Caswall Tremenheere Dowding, GCB, GCVO, CMG, as Air Officer Commanding-in-Chief, Fighter Command, in a highly controversial change of command at the end of last year, after the Battle of Britain had been won by Fighter Command.

"I appreciate the information, A.V., but you did not answer my queries."

The First Lord did not hesitate. "We expect the first units to be deployed for operations in a matter of months, not years. I do not believe, and I will let Sir Archie respond for himself, that there is anything additional we can do. I believe President Roosevelt shares your concern and interest. The Americans have applied extraordinary pressure to get usable equipment to us as soon as possible. The first units should be acceptable for Coastal Command aircraft, however it will be too large and heavy for the fighters. The team is focused on getting the equipment to operational status, and then they will turn their attention to size reduction for fighters."

"I concur with A.V.'s presentation," Air Minister Sinclair added.

"I will ask my persistent question, once again. Is there anything more we can do to protect these vital convoys, or at least turn our loss ratios positive?"

"You have done your part, Prime Minister," Sinclair responded. "The Americans have transferred as many surplus destroyers as they can spare. With the passage of the Lend-Lease Act, they have engaged and have committed their shipbuilding capacity to merchant ships and destroyers. Our shipyards are at capacity with no identifiable potential for expansion. Sir Hugh has been working the American aircraft production schedules quite effectively, to speed deliveries. Toward the end of summer, we expect to begin receiving our Consolidated B-24 Liberator long-range bombers, configured for Coastal Command and the anti-U-boat role. Sir Hugh has confirmed that those aircraft will have space, weight, power and mounting provisions for the new radar unit, if not already installed, which in turn will hasten installation of the new equipment when it does arrive."

"'Sir Archie's information is spot on," Alexander added.

"And, the acoustic torpedo?" asked Churchill, continuing to press his ministers.

The First Lord answered, "Progress in fits and starts, I'm afraid to say. The hardest task is making the tracker see a specified or desired acoustic signature rather than the loudest sound the seeker detects. They have had success . . . just not reliable success. At the end of the day, the torpedo must find, track and hit the submarine, not a merchant struggling to escape or an escort churning to protect."

"Understood," Churchill said. "Very well, then, is there anything else to be discussed on this dreary Friday afternoon?"

"One last item, if you will permit me," Sir Archie answered. The Prime Minister nodded his consent. "'C's field lads managed to report information our intelligence branch interprets as the first flight of a new developmental aircraft, we believe, is the twin-turbine Heinkel Two Eight Oh – designed to be a high-speed interceptor."

"Another jet, as you call it?" asked Churchill.

"Yes, so it appears."

"This turbine engine, you say, is similar to Whittle's engine . . . accelerated air for thrust . . . propulsion?"

"Yes, as far as we can tell, so far."

"What is our progress to our version of this jet fighter?"

"Not as fast as we would like, given the apparent progress of the Germans."

"I'm sure you will stay on things."

"Yes sir."

"Anything else?"

No one spoke up and they quietly dispersed.

—

Monday, 7.April.1941
Cabinet War Rooms
New Public Offices
Whitehall, London, England
United Kingdom
16:30 hours

The quickly arranged combined meeting of the War Cabinet and the Defense Committee convened with the Prime Minister at the center of the head table. The news from the various engagements with the Germans and Italians continued to worsen, as the negatives seriously outweighed the positives.

Secretary to the War Cabinet Sir Edward Ettingdene Bridges, KCB, MC, opened the meeting. "Good afternoon, gentlemen. Mister Margesson, your report, if you please."

The War Minister did not hesitate. "The situation in North Africa continues to deteriorate as General Wavell and the Army of the Nile conduct a fighting withdrawal to slow the German advance and avoid being overrun."

"When do we see what we are going to do to stop them in North Africa?" asked Churchill.

"As you requested last Friday, General Dill and I are working with General Wavell to number one clarify the situation, number two develop a better intelligence picture of the enemy's condition, and number three to establish a proper defensive and eventually offensive plan. We should have a draft plan in a few days time."

The Prime Minister nodded his head and remained uncharacteristically silent.

Margesson continued. "Complicating the situation in North Africa, the Germans invaded Yugoslavia and Northeast Greece yesterday morning. The size and composition of the enemy force is believed to be far superior to the allied units on the frontier, including our army and air force units in country. Without sizable reinforcement or diversion of the enemy, the outcome is only a matter of time."

"We do not have reinforcements," Churchill observed and nearly muttered calmly.

"No. We do not. With the Army of the Nile fully engaged, we have nothing in reserve. The colonial recruitment, stand up and transport will not be fast enough. Even if we attempted to draw down all our forces in India and the Far East, we cannot possibly move them fast enough to alter the outcome."

Lieutenant General Sir Henry Maitland 'Jumbo' Wilson, GBE, DSO, commanded the recently deployed Commonwealth Expeditionary Force in Greece, often referred to as 'W' Force, and was compromised of an Australian infantry division, a New Zealand infantry division, and a British armored brigade.

"General Wilson shall have to do the best he can," said the Prime Minister, again in a rather somber tone.

"He has deployed 'W' Force to the west of Thessaloniki, on the left flank of the Greek 2nd Army," added Margesson.

"I might interject here," said Chief of the Imperial General Staff General Sir John Greer 'Jack' Dill, KCB, CMG, DSO, "I was gravely disappointed in my recent conversations with General Papagos. Wavell and Wilson concur. He has been wholly unaccommodating and defeatist. His defense plan ignores the fact that Greek troops and artillery are capable of only token resistance, especially given the preponderance of armor the Germans brought to the fight."

Lieutenant General Alexandros Papagos had been Chief of the General Staff for the Hellenic Army since 1936, and was Sir John's counterpart for the Greeks.

"Has he given up already?" asked Clement Richard Attlee, PC, Member of Parliament for Limehouse, Lord Privy Seal and Leader of the Labour Party. Attlee had been an executive member of the War Cabinet since it wsa formed by Churchill in May 1940.

"No sir," Sir John responded. "He has actually inspired a spirited defense. It is just our considered opinion that he has spread his forces too thinly in deference to their frontiers rather than using the mountainous terrain to concentrate his forces to maximum advantage. General Wilson is doing his best to support the Greek defense plan. However, I think we are looking at another evacuation to save as many troops as we are able."

"Then, best get along with the planning," the Prime Minister said. "The Admiralty will need time to gather the necessary transport. What of Yugoslavia?"

"We have less information of a timely nature," responded Margesson. "MI6 is working on communications with their agents in country, to achieve better currency, but I suspect things are a bit fluid."

"Very well," the Prime Minister answered matter of factly.

Sir Edward picked up his cue. "Does the Air Ministry or Admiralty have anything to add on the situation in the Balkans or Aegean?"

Sir Archie raised his hand and did not wait for recognition. "The air situation is comparable to the land condition. We are grossly outnumbered and confined to defensive operations."

"Can we spare any squadrons from Egypt?" asked Arthur Greenwood CH, PC, Member of Parliament for Wakefield, War Cabinet Minister without Portfolio and Deputy Leader of the Labour Party.

"You know the answer to that query, Arthur," barked Churchill.

The Air Minister ignored the Prime Minister's admonition. "No," Sir Archie responded calmly. "We have insufficient fighter aircraft to defend Alexandria, Cairo and the Suez, set aside operations to assist Wavell's withdrawal in Cyrenaica. Even though the day fighters are not engaged in defending the Home Islands, as they were last summer, we have not been relieved of preparations for another invasion attempt by the enemy, and as you know, we have been conducting offensive fighter operations in Northeast France and Belgium. Plus, even if the War Cabinet decided this moment to redeploy 10 or more fighter squadrons to Egypt or Greece, it would take more than a month to move those squadrons . . . by sea, our only means available."

Another long pause added a further chill to the room. Sir Edward spoke to move the meeting along. "Does the Admiralty have anything to contribute to this report?"

"Yes, actually, I'm afraid," responded the First Lord. The Prime Minister nodded with mounting impatience by now. "The German Air Force struck the Greek port of Piraeus early this morning. In short order, they sank six of our military cargo ships that had only just begun to offload their cargo. The worst of it . . . one of those supply ships was the Clan Fraser with 200 tons of high explosives still on board that detonated and destroyed much of the harbor facilities." A.V. Alexander paused. No one reacted. "Pressure effects of the massive explosion were felt in Athens—windows broken and doors blown open, five miles away. Now, we will have to find alternative means to move supplies into Greece. If we are to support a potential withdrawal of 'W' Force from Greece, we need to find other embarkation points. Based on the immediate assessment of the damage, we are not likely to accomplish sufficient repairs or reconstruction to use Piraeus. Without proper air cover, both tasks become infinitely more difficult."

Winston wiped away a tear before it could descend his cheek. "Enough. We have had sufficient bad news for this particular Monday afternoon. I need time to think. We are adjourned. Go about your work and do it well." The Prime Minister did not wait and walked out of the conference room ahead of everyone. He knew there would be much more bad news before they would eventually see the light at the end of the tunnel. He always had hope, but this afternoon, he felt the nibble of the Black Dog and he knew he had to get ahead of his looming depression.

—

Monday, 7.April.1941
Oval Office
The White House
Washington, District of Columbia
United States of America
14:00 hours

President Roosevelt had asked for this private meeting with Secretary of War Henry Lewis Stimson and Secretary of the Navy William Franklin 'Frank' Knox—both senior Republican Party politicians had joined the President's Cabinet last summer and served as the principal defense ministers. Not even Harry Lloyd Hopkins, the President's trusty aide, attended this particular meeting. He wanted it off the record, and for both Republican leaders and cabinet secretaries to feel as uninhibited as he could possibly make them, as they had each other and he had no one.

"Let us begin with the latest situation in the Battle of the Atlantic, if you will, Frank."

"Certainly, Mister President. The British had a rough week. They lost ten loaded merchants and only managed to bag one U-boat in the process."

"Where are they on their counter-measures, developmental work?"

"We have been most impressed with their acoustic detection efforts. The term we have come to utilize is sonar, which stands for sound navigation and ranging . . . essentially underwater listening and when necessary, propagation sound in water similar to how radar uses radio waves. They send out a ping – a sound wave – and they listen for a reflected return. I should also add that the Radiation Lab joint team found particular success last week, during an airborne test of prototype equipment. They repeatedly detected, tracked and homed in on a periscope of a moving, submerged submarine. We, and especially the Brits, are excited by the results. They are also making slow progress on development of an acoustic homing torpedo. There are other developmental programs on-going, but truth is, we need time . . . to get these various equipments into operational form and deployed to the fleet."

"How can we gain them time?"

"The best we can do until new destroyers descend the slipways, and some of these anti-submarine devices and weapons can become operational will be extension of our escort patrols to perhaps the Icelandic longitude, or farther. Doing so would be equivalent to adding escorts to the Royal Navy. We should also initiate secret talks with Iceland to use air bases there for airborne patrols to fill at least part of the gap, mid-Atlantic."

"Risky."

"Yes, indeed. Factions within both parties might well see such escalation on our part as tantamount to joining the British in this war."

"Quite so, but we must do something more to help until our industry can be brought to bear. To be blunt, they are losing the battle. If the losses continue at the rates they have suffered, they will reach a point of non-viability – insufficient fuel and ammunition for the Royal Navy, or the RAF, or food to sustain their people and their army. We are shipping them boatloads of coal just to keep them warm in winter. I do not know how much time they have left, but it is certainly measured in months rather than years. Discuss the patrol extension with Admiral Stark and give me your recommendations as soon as you are able. I shall refrain from further biasing your deliberations with my personal opinion. As quickly as possible, if you please."

"Yes sir, Mister President."

"Now, to the purpose of this meeting. As I know you are aware, I have been troubled by the paucity of strategic intelligence available to the Office of the President. I understand our current intelligence structure. It has served us adequately for several decades, but that system is not serving this office or the government as a whole. We have discussed this before. We agreed last year in July and again in December, to dispatch 'Bill' Donovan to England, and then to unoccupied Europe, North Africa and the Middle East. He has completed his assessment. Both of you have seen his report. I have read his words with great care. I would like to hear your private opinions, shared between the three of us, as a private conversation. We need to hear each other. I am rapidly approaching a decision, out of necessity, and I need the three of us to be in agreement. So, who wants to go first . . . it will not be me?"

Colonel William Joseph 'Bill' Donovan, also known as 'Wild Bill' or 'Big Bill' to some, was the senior and founding partner of the prestigious and highly successful, Wall Street law firm of Donovan Leisure Newton & Irvine. More notably, Donovan was a recipient of the Medal of Honor for combat valor in France during the Great War, along with the Distinguished Service Cross, Silver Star, and three Purple Hearts, among other awards. While he had no direct experience in the intelligence community, his expansive international contacts and impeccable credentials made him an ideal candidate to fulfill the President's vision and needs.

Knox and Stimson looked at each other. Henry nodded and Frank took the cue. "I will go first, Mister President," said Knox. "Henry and I have discussed what we both recognize as an inevitable conversation. For both of us, I thank you for availing us the opportunity to offer our perspectives. We feared a directive, without debate, as you have been quite direct regarding your dissatisfaction." Roosevelt smiled and nodded his acknowledgment. "Hence, I shall speak for myself, as Henry has other relevant experience. The Office of Naval Intelligence evolved in a void. The leaders of the Navy several decades ago recognized their vital need for intelligence. Those sources of intelligence did not exist. They sought to fill the void they faced. As a consequence, naval intelligence has grown to serve the mission of the Navy. ONI has no interest with infringement on the domain of other branches of government. ONI does an incredible job, ensuring the secretary, and the CNO and fleet commanders have the best intelligence possible, but our mission does not involve land armies, or embassies, or diplomacy. Like you, I have carefully studied Bill's assessment and recommendations. I have met with Bill several times since his return as well as CNO and ONI. We accept the need. We are concerned about the path to fulfill the need. We would prefer a joint intelligence committee to handle the broader taskings and analyses."

"So, a separate entity?"

"Of agents and analysts from the four primary intelligence branches."

"So, you would include State and Justice, at least in the form of the FBI?"

"Yes sir, precisely."

"Who would be the decision makers of such a committee?"

"Designated representatives from each branch."

"So, your notional committee would be subordinate to the existing intelligence branches?"

"I would not suggest that."

"But, in reality, if the designated representative reports to the G-2, or the ONI, or Director of the FBI, are they not subordinate to the branch chief, and thus driven by the service chiefs?"

"I see where you're coming from here."

"Do you?"

"I believe so, Mister President."

"Very well," the President said and paused. He looked to Stimson. "Before we go too much farther down this road, I would like to hear your perspective, Henry."

"My pleasure, Mister President. Frank and I share similar views. As you know, when I served as secretary of state during the Coolidge administration, I was very much against the department's 'Black Chamber' signals intelligence operation. As secretary of war with a looming global conflict, I see intelligence in a far different light. Also, as you know, intelligence is a vital, essential element of every military operations plan. The Office of the President depends upon us to provide the best possible plans based on the best possible intelligence we can gather. We are concerned about dilution of the base intelligence. Further, we see no benefit to competing intelligence operations. The Army and Navy have evolved a reasonable, implied, consent agreement to avoid stepping on each other's toes. We do not want to upset the apple cart, so to speak."

"I understand your argument, Henry, I would say you are happy with the performance of the G-2, and you do not feel interference, or infringement as Frank says, from the other intelligence branches."

"There is always room for improvement, but yes, in the main, we are satisfied with the way things work."

"What if I am not happy?"

Stimson and Knox both stared at the President. Roosevelt held Stimson's eyes without the slightest emotion. "That would be a problem," Henry finally answered.

Roosevelt's expression remained cold and distant. "Therein lies the rub, doesn't it. You both are happy with your intelligence branches. I suspect Cordell and Edgar would be happy with theirs. So, I ask you both, where is my intelligence branch?"

"We serve at the pleasure of the President, sir," answered Stimson. "If we are not doing our jobs, then you should find the correct person to do the job you believe needs to be done."

"Yes, well, there is always that, but frankly and candidly, you are both doing a magnificent job. It is just that you are rightly focused on your job," he paused and looked to both men, "and, I still do not have the intelligence I need to make properly informed decisions. I have suffered the consequences of that dreadful gap. I had good intelligence from the Navy on what the Royal Navy, or the *Kriegsmarine*, or the *Marine Nationale*, are doing or going to do. What I did not, and do not have to this day, is what the leaders of the belligerent countries are thinking or what they are likely to do. Fortunately, I have a close, personal relationship with the British prime minister, but that could change overnight with one vote of no confidence in Parliament. That paucity of national, or more appropriately international, intelligence leaves us all vulnerable and prone to mistakes or missteps. I have no intention of altering your intelligence branches. I believe they are working well for the purpose they serve. Simply put, I need a national intelligence service to give me the whole picture, not parts of particular situations."

"Then, you are inclined to accept and implement Bill's recommendations?" asked Knox.

"Short answer, yes. He has done a yeoman's job of what we asked him to do. I believe his recommendations are spot on, as the British say. With all due respect, a joint committee of subordinates will not fill the gap we all suffer. I do believe a joint committee is appropriate and warranted to ensure our intelligence is properly shared and disseminated, but at the bottom, any organization depends upon the people who make it run. If the service intelligence chiefs wish to torpedo a national intelligence organization, they will. I need you both to understand, appreciate and support any proposed change, and more importantly, I need you to ensure the G-2 or ONI make such an organization work for all of us."

"That is a tall order, Mister President," Stimson said.

"Yes, it is, which is why I believe you are the best men in the country for the jobs I have asked you to do for the nation. We do not have much time left before we will be in this fight up to our eyeballs. We were gravely ill prepared when hostilities broke out in the Far East and now in Europe. We will not be allowed to sit this one out. The fight will come to us eventually; it is only a matter of time. We must be as prepared as we possibly can be. That is my number one job under the Constitution."

"What are you going to do, and how may we best assist you?' asked Knox.

"To be frank, Frank," he answered and smiled, "I have not decided, as yet. I recognize the office politics, the turf issues, associated with this question, but I also know the status quo will not help us win the coming war. This change is inevitable. Filling the intelligence gap is simply too damn important. Your two departments are the two most pivotal here. I can handle State, and I believe we have a way forward with the FBI. I want you to take a week or two to think through how we implement Donovan's bureau of information, or whatever we are going to call it, and by that I mean both the intelligence and special operations elements he outlined in his report. If this means replacing people who are incapable of change, then so be it. You are to assume something like what Bill has sketched out for us is going to happen. I want to know how you are going to implement this change for maximum effectiveness. I believe you both see the need, the gap as I call it. We are putting in place the teams we will be going to war with in the not too distant future, and we simply must fill the gap with as little blood loss as possible."

"Yes, Mister President," they said in unison.

"I want your input no later than the end of this month. We have much to do, and I'm afraid we do not have much time left. Also, Frank, please do not forget the Atlantic patrol extension matter."

"Never fear, Mister President."

Both men stood. Roosevelt extended his right hand to each man and shook hands firmly. "Thank you for your time, gentlemen. Have a good day. Please ask 'Missy' to find Harry for me. Thank you."

Marguerite Alice 'Missy' LeHand had been a friend, confidante and administrative assistant to Franklin Roosevelt since his vice presidential campaign in 1920. When Franklin became President, Missy took up her natural post as Principal Private Secretary to the President.

Both men departed and closed the Oval Office door behind them. "Well, this should be interesting," Franklin muttered aloud.

—

Wednesday, 9.April.1941
RAF Martlesham Heath
Woodbridge, Suffolk, England
United Kingdom
10:35 hours

No. 71 Squadron landed safely by sections at their new airfield, after a comparatively short transit flight south from Kirton-in-Lindsey. Ground crewmen directed them to their assigned parking area. The 12 'XR' Hurricane fighters shut down. Some of the squadron ground crews were present. The remainder would arrive later this afternoon. As with their other transfers, a base officer gave them the introduction and orientation to their new home base. For the veterans, the operational procedures and routine of RAF Martlesham Heath were essentially the same as every other RAF fighter base they had served on for the war years, while this was only the third base for the newer pilots.

Once the official welcome was complete, Squadron Leader William Erwin Gibson 'Billy' Taylor, the first American to command an RAF fighter squadron, gathered his pilots in their new Dispersal Hut, without Corporal Harris, who would arrive later in the day with the remainder of the squadron personnel. "OK guys, it will be lunch time shortly. We will adjourn and gather up back here. We will be held in Released status today. I expect we will be returned to full strength by this evening. Our plan at the moment entails flying our area orientation flight tomorrow morning and moving to Available status by late morning, early afternoon, tomorrow. Lastly, I am informed that we will have a high-ranking visitor . . . ." The grumbling and muffled resentment stopped Billy. He waited for the distraction to die out. "As I was saying, we will have a high-ranking visitor from Eleven Group this afternoon. While I share your dissatisfaction with the circus in which we find ourselves, you will restrain yourselves while our visitor is here. Am I clear?"

"Yes sir," they answered in unison.

"Let's go to lunch," Billy said and led them through the orthogonal streets and identical green, clapboard buildings, across the base to their new Officer's Mess.

The noontime fare was virtually identical to other bases as well – cheese or egg sandwiches, some form of potatoes, in this instance a version of potatoes au gratin, and hot tea or water. None of the Eagle Squadron pilots displayed any urgency in returning to what they knew all too well would be another dog-and-pony show. Several pilots took a circuitous route back to explore their new base. Brian Drummond walked by himself directly to Dispersal, found a garden chair, opened it outside, sat and closed his eyes. A nap, however short it might be, seemed appropriate. Brian heard and felt someone duplicate his sitting process next to him. He did not respond in any fashion.

"What do you think of this move, Hunter?" asked Pilot Officer Edward Jonas 'Bulldog' Tolly—Left Wing in Hunter's Green Section.

Brian kept his eyes closed and remained motionless. "We are getting closer to the action," he answered, eventually. "Eleven Group covers the southeast corner of Great Britain. The Group is the closest to the enemy and took the brunt of the German attack during The Battle, last summer." Brian paused, and then added, "We are the northern most base in Eleven Group area, so the farthest away of the Group fighter squadrons."

"What does that mean?"

"My guess, we will probably fly cover for squadrons crossing the Channel."

"We'll get back to France soon enough," added Pilot Officer Paul James 'Dusty' Langford of Los Angeles, California, and veteran of both the Battle of France with le Armée de l'Air and the Battle of Britain with No.151 Squadron. Dusty was Right Wing in Hunter's Green Section.

Brian had not heard Dusty join them and was not sure whether he was sitting or standing. "Spot on," Brian said.

Things quieted. Brian placed his cover over his face and drifted off to slumber. He lost awareness for an unknown period of time, until he heard Pilot Officer Robert Charles 'Sweet' Sweeny Jr. say, "You don't see that every day." Sweeny was the younger son of ex-patriot, American businessman Robert Charles Sweeny and nephew of Colonel Charles Sweeny.

Brian initially resisted looking, but curiosity took hold. He removed his cover, allowed his eyes to adjust to the bright sunlight, and eventually noticed a de Havilland DH-82 Tiger Moth approaching to land. Oddly, the aircraft had no tail designation markings and taxied to the Base Operations building and tower, turned, parked and shut down. A single pilot got out and walked into the building. He was too far away to recognize, other than he was wearing an RAF blue uniform with a broad stripe on each sleeves.

The telephone rang. Everyone perked up in Pavlovian response. Billy Taylor appeared at the door. "Gather up inside, gentlemen. Our guest will be here shortly." They did as they were commanded. Once everyone was inside and seated, he continued, "I remind you to withhold any derogatory remarks, speak only when spoken to, and most of all, be respectful." Billy saw their guest first and commanded, "Attention!"

All the pilots stood and took a position of attention. Brian recognized their guest immediately. One broad stripe replaced the thin stripes of a junior officer.

"Gentlemen, may I present Air Commodore Spencer, Chief Controller and Operations Officer, Eleven Group."

"At ease," John said. "Please be seated." He waited for the shuffling to cease. "Welcome to Eleven Group. We are eager to integrate this squadron in our operations. I needed some flight time, which gave me an excuse to visit the all-American Eagle Squadron . . . well, pardon me, Flight Lieutenant Whittington, all-American except for one remaining Brit." Everyone laughed. "That should be remedied soon enough."

Flight Lieutenant Charles Gordon 'Whitey' Whittington was the last remaining British citizen fighter pilot in No.71 Squadron and was also a veteran of the Battle of Britain.

"No hurry," interjected Whitey. "These are good lads, sir."

"Thank you for that," John Spencer responded. "In addition to welcoming you myself and on behalf of Air Vice Marshal Leigh-Mallory, I thought it might be useful to give you a feel for what lays ahead. You may expect to fly cover patrols along the south coast for squadrons on CIRCUS or rhubarb missions across the Channel. You will also fly escort for bomber raids over France, Belgium and Holland. Eventually, you will take your turn attacking enemy installations across the Channel. We do not perform fighter defense missions these days, as the enemy has confined their operations to nighttime bombing raids, and the night fighters are handling those missions. Our procedures are identical to those you have used in your training and operations in the other groups, so you should fit in quite nicely. Now, are there any questions I may answer for you?" The room remained quiet. "Very well. Again, welcome to Eleven Group. Fly well, as I know you can. Squadron Leader Taylor, with your permission, I would like a word with Flying Officer Drummond."

Air Vice-Marshal Trafford Leigh-Mallory, CB, DSO, had been Air Officer Commanding-in-Chief, Eleven Group, since he relieved the highly respected and renowned Air Vice-Marshal Keith Rodney Park, CB, MC+Bar, DFC, and was currently John Spencer's boss.

"Certainly, sir. You are welcome to use my office."

"Thank you, Billy, but outside will be sufficient," John said and walked out.

Several of the pilots poked and pushed Brian for being singled out by a senior officer. Taylor nodded to Brian for him to go outside.

"Yes sir," Brian said, as he walked several paces to Air Commodore Spencer.

"Brian, I have some other business to tend to here. If you are free, I would be honored if you would join me for dinner."

"It would be my pleasure, sir . . . when and where?"

"Let say half seven. I will pick you up at the Mess. I am given by a reliable source that a public house known as The Unruly Pig in a nearby village has simple but well prepared wartime food and good beer."

"We have not had a chance to visit any pubs, yet, so your source is far better informed than me."

"Excellent, then I shall see you at half seven. Thank you, Brian."

"Thank you, sir."

Air Commodore Spencer turned and walked off toward the Operations Building.

Brian returned to understandable ribbing from the other pilots and the myriad of questions asked in different ways as to why he was singled out. Brian claimed ignorance. Right Wing Red Section Pilot Officer Frank Oscar 'Red' Burns decided to add what he knew from his days with Hunter in No.609 Squadron, last year. Brian neither agreed nor disagreed; he simply chose to feign ignorance, since he really did not want to discuss his relationship with Air Commodore Spencer. Squadron Leader Taylor knew more of the story, but chose not to involve himself in the speculation. Billy also chose not to question Brian.

—

Wednesday, 9.April.1941
The Unruly Pig
Orford Road
Bromeswell, Suffolk, England
United Kingdom
19:55 hours

The RAF staff car, complete with an enlisted driver and Air Commodore Spencer in the back, waiting for Brian, had arrived directly in front of the Officer's Mess precisely at 19:30 hours sharp. Brian joined John in the back. The rather leisurely drive from the air base on the narrow country roads through the farm fields to the north side of Bromeswell village brought them to a two-story, sturdy, standalone, brick, public house. A large caricature of a pig's head dominated the large, classical, hanging sign with old English lettering underneath—The Unruly Pig.

"It appears we have arrived," announced Air Commodore Spencer.

"Yes sir. This is The Unruly Pig pub," the driver responded. "Please call the Operations Duty Officer when you are ready to return."

"Thank you, Aircraftman."

The Unruly Pig was a refurbished but authentic, 16ᵗʰ century, brick, public house with a more contemporary exterior and a classical interior complete with massive oak beams that made Brian duck underneath to avoid clipping his forehead. They passed through the bar and found an empty corner table.

"Thank you for taking the time to have dinner with me," Air Commodore Spencer said.

"My pleasure, sir."

They ordered their pints and meals. The waiter actually offered a small bowl of nuts, an unusual rarity for wartime England. Rationing was taking its toll on the cuisine and selection of everyone in England.

"As I said in my remarks to the squadron, welcome to Eleven Group. We are the pointy end of the sword and it is quite nice to have the Eagle Squadron with us. We have had good success with our cross-Channel operations, and I know you and your mates will do well on those missions. The Germans may not be as overwhelming as they were last year, but they still have sharp teeth and a vicious bite, that you need to be mindful of over there. We are still dealing with The Blitz, but that is in the hands of the night fighters."

The waiter delivered their pints.

"We hear they are doing better."

"You hear correctly. They have made great progress with their new equipment and their hunting tactics. Night intercepts are never easy and always nerve-racking, but the night fighters do have some advantages. Their targets have no fighter escorts and the gunners, when they carry them, have no means of engaging the fighters, at least until they open fire, and that is often too late."

"I can't imagine attacking another aircraft at night, not seeing your target with your eyes."

"Me either. Not easy, certainly. Yet, they climb up to do battle every night."

The waiter brought their meals – fish pie and fish stew – simple but nourishing. They also ordered another round of beers. They ate several bites and took several swallows before John cleared his throat to continue their chat.

"While it is always a pleasure to talk about flying machines, Brian, I actually had another purpose for my visit and our private talk." John paused and stared at Brian, who did not budge. "This is rather awkward, so it is probably best if I jump right on this beast." Again, John paused to see any reaction, which Brian did not give him. "Mary told me everything."

Brian could not contain his expression of shock and serious discomfort. He actually felt nauseous and struggled to suppress the urge to vomit. *Is this the moment my dreams come to a nightmarish end? Dear God above, what on earth do I possibly say?* Brian shifted nervously on his chair, fighting the nausea welling up within him. *What did she tell him? He really has not said anything serious. Why am I acting like this? I am acting guilty. Hell, I am guilty.*

John smiled and held up his hand. "Calm yourself, Brian. I am not here as an angry, cuckolded husband, but rather, I am here as a proud and grateful father." Brian continued his efforts to suppress the twisting, churning sensations within him and the pounding pressure of his racing heart. "I mean what I say, Brian. You were not the first man with whom she has had a dalliance. She told me she pressured, pushed, cornered, coerced and virtually demanded your submission to her will. In her heart, she knew, as most women do, you could give her what she . . . what we both wanted – a child. Mary gave birth two days ago."

"Ah . . . ah . . . congratulations, sir."

"Thank you. We agreed to name him Malcolm Brian in honor of our mutual friend and to recognize your contribution in this . . . this production. Mother and Son are in good health and doing well. To be candid, Brian, I would be lying to you, if I made any attempt to deny or lessen my sense of betrayal, when Mary first told me. Yet, as she is so capable, Mary convinced me to take the larger view, to accept the blessing we have been given, and most importantly, to confide in you, which is precisely why I am here. This could only be done in person, eye to eye. I must convey my true feelings." John paused to give Brian a chance to catch his breath. "Mary and I have tried to produce children since we married 12 years ago, obviously without success. The doctors cannot tell us why; it simply is. We both have wanted children. I am eight years older than her, so I don't know if I'm the problem. She did not tell me what she was doing, but she confessed her desire existed from your very first meeting that evening for dinner, shortly after your arrival in this country, two years ago." John stopped and waited. He ate a few more bites. Brian ate nothing, drank nothing, as he was still nauseous and did not want to vomit.

"Air Commodore Spencer," Brian began, "I have no idea how or what to say. I never meant to hurt you. I never intended to cause you any apprehension. I never wanted to hurt anyone."

John held up his hand, again. "Brian, you have not hurt me." He then raised his hand high and held up two fingers, presumably to signal their waiter for two more beers. "I am not confronting you, Brian. Mary and I agreed that there should be nothing hidden between us. We wanted you to know."

"Yes, Mary did pressure me. She is very skillful and she was relentless. I am truly sorry and remorseful that I was not a stronger man. She did not tell me she wanted a baby. I owe you so much, sir."

John laughed heartily. "Nonsense, Brian, I readily know how persuasive she can be. She is a very attractive woman and she knows her body. You were not the first, and I doubt you will be the last. It is who she is, and I love her for all she is, and I know she loves me. I have only asked for discretion from her. You owe me nothing, Brian. I do what I can to help you because I can and because of my love for Malcolm. He was a very good man, may God rest his immortal soul."

"Yes, she is very attractive. If I may ask, sir, how long have you known?"

"Let me see," he said and lapsed into contemplation. "I suppose three or four months. Yes, that's about right. It was after your Military Cross award and dinner with Uncle Winston and before Christmas, so that would be four months."

"Why did she tell you?"

"Well, good question. I have not asked her. Perhaps one day, you can ask her. I do believe she would say she did not want to carry that burden into childbirth, and I suspect she wanted to give me time to process the information before the birth. She is a very wise woman."

"Yes, she is, and you are a very generous and understanding man."

"Well, thank you," he said and smiled. "Now, if your stomach and nerves have settled down, you should finish your meal before it gets any colder."

"Yes sir." His stomach had calmed sufficiently to allow him to eat. John was nearly finished with his stew, but he stretched it out. The beer tasted even better for some odd reason.

Brian's curiosity worked on him. "So, you knew when you helped me return to Kansas a couple of months ago?"

"Yes, I knew by that time. I was honored to help you after suffering such a terrible tragedy."

Brian nodded his head and finished his meal, as did John. They both slowed down with their beers.

"I trust your journey was successful."

"Yes, thanks to you. I have no siblings. There were no other family members to settle their estate. I really needed to go back. I cannot thank you enough for arranging my transportation."

"How was the flying boat? I've never flown in one."

"Amazing understates the experience. Going out, I found out in flight that President Roosevelt's chief of staff Harry Hopkins was onboard and returning to the United States. It was an honor to talk to him. On the return flight, it was the same crew, and they let me fly the aircraft a bit. The Boeing Model 314 is an enormous aircraft, a very comfortable form of flying, like a hotel in the air, I must say. We departed from Bournemouth, and stopped in Lisbon, Horta, Azores, and Hamilton, Bermuda, before we reached New York City, 25 hours in total. The trip was one helluva lot better than the eight days it took aboard the *Liverpool Lady*, and no rolling seas." They both laughed. "An Army Air Corps C-39 was waiting for me in New York. Did you arrange that aircraft, too?"

"No, I cannot take credit for that segment. That recognition must go to the assistant naval attaché at your embassy in London – Colonel Roy Geiger of the American Marine Corps. He is an exceptional aviator, one of the first Marines to be designated as a pilot, as I understand his history."

"If I ever meet him, I will be sure to thank him."

"Quite appropriate, I should think. Now, I need to take off early tomorrow and get back to Uxbridge. I am fairly certain you will be on duty tomorrow morning as well. Let me pay our bill, and we shall be off."

"Do you have a room at the Mess?"

"I am taken care of, Brian. Thank you. Are you up for a walk back? It is a bit cool, but at least it is dry, and it should be invigorating for us."

"That work's for me, sir. Thank you for dinner."

"Thank you for enduring my little exposé." John Spencer paid their bill. Brian followed the air commodore outside into the cool and fortunately dry night air. John turned to Brian. "The next time you are in London, if you can make time for us, I know Mary would love to introduce our son to you. He won't remember it, but he will eventually."

"I'll do my best, sir. I don't know when my next London visit will be, but I will call you when I do."

"Excellent. We cannot ask for more. Let's walk."

Off they went walking through the village and into the farm fields between Bromeswell and RAF Martlesham Heath. An RAF truck and a staff car stopped to offer a ride for the two officers. Both drivers were clearly surprised to find an air commodore walking along a country road at night in blackout conditions. The moon was two days short of full and two hours short of local zenith, but the moon was high enough to light their way and the sky was crystal clear that night. The two men laughed, talked and even danced in the roadway during their 1.5-mile walk. An air commodore and a flying officer walking up to the armed guards at the closed gate to RAF Martlesham Heath produced a hesitation of suspicion and close scrutiny of their credentials by the corporal of the guard. Satisfied the two officers were in fact who they claimed to be, the guards opened the gate and closed it after the two officers passed. All said and done, Brian felt at peace after a rather tense evening. Mercifully, sleep claimed him quickly.

—

Chapter 2

We should all be concerned about the future
because we will have to spend the rest of our lives there.
-- Charles Franklin Kettering

Tuesday, 15.April.1941
Headquarters, Fighter Command
Bentley Priory
Stanmore, Middlesex, England
United Kingdom
12:30 hours

The office secretary for the Air Officer Commanding-in-Chief, Fighter Command, Air Marshal William Sholto Douglas, CB, MC, DFC knocked and opened the tall office door. "Sir, it appears General Arnold's staff car has arrived," she said.

"Thank you. Please ensure lunch is ready in my dining room."

"It is, sir. Ready when you are."

"Very well. Thank you."

Air Marshal Douglas rose from his desk, retrieved and donned his cover, and departed his office for the reception lobby. Douglas stepped outside as the U.S. Embassy staff car pulled to a stop. The red flag with two white, 5-pointed stars side by side were displayed on the right front fender, signifying the car carried an Army major general. Major General Henry Harley 'Hap' Arnold, USA [USMA 1907] had his position as Chief of the Air Corps since September 1938, yet, leadership of the Army air forces remained ambiguous, especially to the British, who had reorganized the Royal Air Force into a separate, equal service at the end of the Great War. Arnold did not wait for the driver to open the right-rear, passenger door.

Douglas extended his right hand. "General Arnold, I presume."

"Indeed." They shook hands. "Hap Arnold, and you must be Air Marshal Douglas."

"I am. Great to finally meet you, Hap . . . may I call you Hap?"

"By all means, I prefer the familiar."

"Very well, then, William for me, if you please. Lunch is ready, so let's jump right into our discussions while we eat."

Arnold followed Douglas into the old manor house built onto an old medieval monastery, which had been converted into the headquarters building for Fighter Command during the inter-war years under the leadership of Air

Chief Marshal Sir Hugh Dowding. They entered a small dining room adjacent to Douglas's office. The small table was set for two, across a corner of the table. Two male stewards served them promptly and departed.

"The first order of business, if you don't mind," Douglas said and waited for a consent nod from Arnold, "would you be so kind to explain how the American air force is organized? To confess my ignorance, I do not understand the relationship of your Air Corps to the Army proper and perhaps more significantly to the General Headquarters Air Force."

Arnold laughed. "To be candid, William, I'm not sure we understand it either. Rather than spend the time to explain the genesis of the air forces in the United States, let me just say, we are in the process of reorganizing our air forces, and I would rather focus on the future than the past. Secretary of War Stimson and Chief of Staff General Marshall have agreed in principle that the air forces must have a place at the table. In that sense, your air force is more advanced than ours. You are an equal sister service to the Army and Navy. We are not quite there, yet, but I do believe we are headed in that direction. I am here as the chief of our air forces . . . well, at least the Army air forces, since the Navy and Marine Corps have their own aviation components. When the reorganization is complete in a month or so, my position will be equivalent to that of Sir Charles, as Chief of the Air Staff."

Air Chief Marshal Sir Charles Frederick Algernon Portal, KCB, DSO & Bar, MC, became Chief of the Air Staff at the end of October last year, having previously served as Air Officer Commanding-in-Chief, Bomber Command.

"Very well, then, we shall take that direction. Next order of business, I presume you have seen our shopping list."

"Yes. In fact, since the passage of the Lend-Lease Act last month, I have had numerous meetings with Bill Stephenson, Sir Hugh and even a couple of meetings at your Embassy with Lord Halifax and his staff. We have reviewed your purchase list. Our industry is still ramping up, but we will meet your requirements at least in type and quantity, but frankly our greatest concern is delivery schedule. We have serious worries about the shipping situation in the Atlantic. We are looking closely at how to ferry the aircraft to you, where the capability exists, so that we avoid the risks of using ships to move the aircraft to you."

William Samuel 'Intrepid' Stephenson, MC, DFC, had been assigned personally by Prime Minister Churchill, shortly after his ascendancy to the premiership, to be his eyes and ears in the United States, and to perform as a personal and private conduit for the Prime Minister. His cover for the public was running the Passport Control Office. Within select segments of the British and American governments, he was the chief of the British Security Coordination office and occupied two floors of the International Building, Rockefeller Center, Manhattan, New York City. Administratively, Stephenson reported to the Director-General, Secret Intelligence Service; however, practically, he reported to the Prime Minister. He was also a primary contact for Bill Donovan, and in deference to their physical size difference, he was occasionally referred to as 'Little Bill' in contrast to Donovan's informal moniker of 'Big Bill' or 'Wild Bill.'

"Do you think you can improve the delivery schedules?"

"We are working that problem hard, as we speak. Let me ask about the air defense situation and the renewed invasion threat."

"We have not faced daylight bombing attacks since late last summer. The nightly Blitz continues to be problematic. Our night interceptor capability improves by the day, but remains a long way from even minimally acceptable. The joint work at the Rad Lab in Massachusetts has been quite encouraging, both for our night interceptor problem as well as Coastal Command's U-boat hunting difficulty."   Douglas paused as the stewards entered and served their main course of lamb chops, mashed potatoes and gravy. "We have not diminished our vigilance, however, we do not see the signs we saw last summer. Our intelligence apparatus informs us the Germans have thinned their ground and air forces across the Channel. Several months ago, we began flying fighter sweeps into France and Belgium. The German response we encountered appears to validate the intelligence analysis. We suspect the German chancellor has turned his attention to the east."

"The Soviet Union?"

"It appears so."

"That may be a little much for Germany to chew."

"Perhaps . . . yet the German land army has not been defeated since 1918. Their tactics and equipment exceeded our intelligence and the forces we placed before them. Nonetheless, their diversion is welcome relief for us."

"Quite so. With a diminished invasion threat, does that change your fighter aircraft needs?"

"That is a point of continuous debate. The government's position remains . . . aircraft on order are needed and will be pressed into service as soon as they are delivered. Future orders may well change from our current thinking, as the RAF transitions from air defense to bombardment of the German homeland and positions of occupation."

"Which brings me to training. We have issued orders to our Training Command to devote a substantial portion of our existing training capacity to the RAF, and we are working to significantly increase our capacity to support both your pilot needs as well as ours. We are opening and building several new training sites as well as engaging supplemental contract services to fill the gaps in our capacity."

"We can use . . . ," William stopped his sentence when the stewards entered. "Would you care for anything else?" Douglas asked Arnold. The American shook his head in the negative. "Please clear the table and leave us."

"Yes sir."

The two senior officers waited for the two stewards to complete their work. The service door closed.

"As I was saying, we can use all the help we can gather."

"We shall do our best, which brings us to rumors I have been informed of, even before I departed Washington for this mission. I understand you and your staff are not happy with the American volunteer fighter pilots you collected up into . . . what is it, Seventy-One Squadron."

Air Marshal Douglas stared at General Arnold, and then cleared his throat. "I had not wanted to touch on their performance."

"Why not?"

Douglas shifted in his chair. "We are grateful for their contributions to our defense. After all, they are volunteers and have no obligation to be here."

"But . . . ?"

Air Marshal Douglas was clearly uncomfortable on this topic. "As I am sure you are aware, several of the American volunteer fighter pilots survived last summer's epic and pivotal air battle. In fact, one of those Americans is the third highest scoring ace in all of Fighter Command and decorated by The King for his achievements."

"And . . . ?"

"We collected our American volunteers into an all-American squadron. Yes, you are correct. We have designated the American unit as Seventy-One Squadron. The Press has referred to them as the Eagle Squadron. They are very popular in the Press. However, they have not coalesced as a fighter squadron . . . rather unruly and undisciplined lot, to be rather blunt. Yet, I must tell you directly they have made progress. In fact, we just moved them south, into Eleven Group covering the southeast. They will soon fly escort and fighter sweep missions into France. Their progress toward a cohesive fighter squadron has been agonizingly slow and fraught with setbacks."

Arnold retained a stern expression. "If you are not happy with them, send them home. You have no obligation to retain them. You will certainly not hurt our feelings. Combat is serious business, as you well know. Either they perform to your satisfaction, or you should cut them loose. Neither of us can afford to carry along pilots who cannot perform the basic mission."

Douglas was struck by the directness and sharpness of Arnold's response. "Well, I appreciate your understanding. Fortunately, I do not think we are quite to that point, as yet. They are a high spirited bunch, who lacked the necessary discipline for productive fighter operations."

"I have little to no sympathy for such conduct, William. They either perform to your satisfaction, or cut them loose."

"A few months ago, I was rapidly approaching that point, but as I said, they are ready for combat operations alongside their brethren. Considerable credit must go to their commander and experienced pilots, who have brought them together."

"As you wish," Arnold said curtly.

"We have a goodly number of additional American volunteers at various stages of integration. We are likely to constitute at least two more, American volunteer, fighter squadrons in the coming months. I suspect the problems we had with the first lot were predominately due to the lack of prior, basic, military training. Many of them came from the French air force after the collapse. Only one was with us before the shooting started."

"That one must be Drummond."

"Yes, I do believe you are correct. Quite the capable fighter pilot, I must say . . . an ace nearly four times over and decorated by The King himself, twice. Do your plans involve meeting him or the pilots of Seventy-One Squadron?"

"No, I'm afraid not. I have a very full dance card, as it is. I must learn as much as I am able to absorb, to prepare our air force for what most of us believe is our inevitable entry into this conflict. It is only a matter of time, in the opinion of the Chief of Staff."

"General Marshall?"

"Yes. George Marshall . . . good man . . . tough job."

General George Catlett Marshall, Jr. [VMI 1901] had been Chief of Staff of the United States Army since 1.Septermber.1939, and remained the senior military advisor to the President.

"I've not met him, as yet, but I am certain I will in due course. What is the next stop on your agenda, if I may ask?"

"Eleven Group at Uxbridge. I will observe their operations this afternoon and tonight."

"Daylight operations are orders of magnitude less intense than last summer. Night operations remain quite vexatious, as mentioned earlier. You should observe a reasonable dose of how we operate. If I am properly informed, you will be with us to view overall fighter operations."

"Yes . . . and bomber operations . . . that is the plan. Now, I'm afraid I must beg your pardon. I must be off, if I am to remain on schedule."

"By all means."

"Thank you very much, William, for lunch, and your hospitality and generosity. Please review our training support plan, when the Air Ministry releases it to you. We will do our level best to support your needs. It will never be enough, but I can assure you, we will try to meet your expectations."

"Thank you, Hap. I am certain you are correct."

The two aviation leaders shook hands. Douglas walked Arnold to his waiting staff car. Air Marshal Douglas waited for the embassy car to disappear out the guarded gate and behind the stone, perimeter wall before returning to his office. He did not sit down and decided to descend the stairway to the underground operations room for all of Fighter Command.

—

Thursday, 17.April.1941
No.10 Annexe
New Public Offices
Whitehall, London, England
United Kingdom
02:15 hours

Assistant Private Secretary John Rupert 'Jock' Colville placed the telephone handset in its cradle and stood from behind his anteroom desk as Prime Minister Churchill approached, followed closely by General Ismay and Detective-Inspector Thompson, having just come down from the building roof, observing the nightly bombing as he so often did.

Major General Sir Hastings Lionel 'Pug' Ismay, KCB, DSO, held several key positions in the Churchill government—Principal Assistant to the Minister of Defense (Churchill), Secretary of the Imperial Defense Chiefs of Staff Committee, and Deputy Secretary of the War Cabinet—all of them placing him close to Churchill, as the Prime Minister's right-hand man for military affairs.

Special Branch Detective-Inspector Walter Henry Thompson had been Churchill's bodyguard from 1921 until his retirement from service in 1935. He had been recalled to service on 22.August.1939, as the trusted personal protector of Winston Churchill.

"It sounds bad out there, even from in here," Jock said, as Churchill and Ismay entered the reception room for of the Prime Minister's Annexe Apartment.

Jock followed them. Thompson remained outside at his post, when the Prime Minister was in residence.

"Much worse," responded Winston. "The bastards are attempting another fire-bombing attack. They are trying to burn our beloved city down to the ground . . . and even then, such an accomplishment might not be sufficient for them. They are dropping more incendiary bombs than usual, and there must be larger formations up there than we have seen since the Great Fire. The damned attack is still on going. They probably won't stop until dawn."

"I've heard no reports from the Home Office or Emergency Services."

"The fires are not as bad as they were the night of the Great Fire, wouldn't you say Pug?" The night bombing on London that began on the evening of 29.December.1940, and lasted all night, was often referred to as the Second Great Fire in deference to the first massive fire of London in 1666.

"I believe you are correct," Ismay responded.

"The Fire Brigade has managed to keep the pumps running and stay ahead of the fires, for the most part."

"Agreed," added Pug.

"Hopefully, the fires do not reach the Great Fire proportions."

"Indeed."

"Prime Minister," Jock said, "'C' asked me to inform the duty officer once you returned, as they had an important message for you to see before you retire. I did so when you came down from the roof. I am not sure who is coming, but at least a courier is on his way and should be here shortly."

"Very well. Send him in as soon as he arrives." Jock returned to his desk, just outside the apartment. "Would you care for a finger or two of cognac to vanquish the chill, Pug?"

"I think I will join you," Ismay replied and went to the liquor cabinet to pour two snifters of fine Hine cognac. He handed one to Churchill. They clinked glasses, and then took a healthy sip.

"We have got to find a way, Pug," Churchill muttered, seemingly more to himself.

"How so?"

Winston looked directly to Ismay for several seconds before he spoke. "They have been abusing our beautiful capital city virtually every night for nearly eight months, now. We must find the means to give them some of their own medicine . . . as disgusting as that is."

"We need the aircraft and the capacity to conduct such operations into Germany."

"Yes, yes," Churchill responded with some irritation, "but, those aircraft are coming along."

"Talk to Portal and Bomber Command unofficially and quietly. They should be thinking in those terms . . . getting those heavy bombers into operations as soon as practically possible, and prepare several squadrons to conduct fire bombing operations to German cities."

"Yes sir. I'll try to do that later today."

"We must give back as they have given as . . ." A knock at the door interrupted the Prime Minister. ". . . as soon as possible," Churchill finished his thought and turned to the door.

Colville appeared. "Mister Journeyman has arrived from Broadway House." . . . meaning from the headquarters building of the Secret Intelligence Service, otherwise known as MI6.

A rather average, non-descript, middle-aged man dressed in a conservative grey business suit appeared with the Buff Box manacled to his left wrist and a noticeable bulge under the left breast pocket of his suit jacket. Churchill presumed the bulge was a pistol in a shoulder holster since it was not obvious that he had an armed guard with him and he was more than a courier. The man froze when he saw General Ismay. The Prime Minister picked up the hesitation.

"It's alright. General Ismay is on the access list."

The man nodded. "Good evening to you, Prime Minister. 'C' asked me to bring three Boniface messages to you as soon as you returned."

"Not particularly a good evening, I must say . . . with the insult of tonight's bombing still underway. I've not seen you before."

"Excuse me, sir. I should have introduced myself. My name is Stephen Journeyman. I am a deputy to Carl Acton in the Operations Branch. I happen to have the duty tonight."

Churchill extended his right hand to Journeyman and shook hands. "Nice to meet you. Now, what have you . . . to warrant a late night visit."

Journeyman placed the Buff Box on a nearby table. Churchill retrieved his access key and unlocked the case. He withdrew the only two sheets of paper from the red folder. The Prime Minister read the first message.

MOST SECRET - ULTRA

SECRET
DATE: 17 APRIL 1941

```
TO: COMMANDER AIR CORPS 11
FROM: CHIEF ARMED FORCES HIGH COMMAND
OPERATION MERCURY
BREAK
EXECUTE OPERATION BY APPROVED PLAN NO LATER
THAN 20 MAY BREAK UNABLE TO RESUPPLY BY SEA
BREAK ESSENTIAL PLANNING INCLUDE CONTINUOUS AIR
RESUPPLY UNTIL CRETE SECURE BREAK HAIL VICTORY
BREAK HAIL HITLER
END
SECRET
```

MOST SECRET - ULTRA

Churchill handed the first paper to Ismay and looked to Journeyman. "If I recall, Air Corps Eleven is Student."

"Yes sir. The commander is Major General Kurt Student, the highly regarded leader of the German air force's parachute troops. These are the same folks that were so effective in Belgium last spring. Air Corps Eleven is a full corps of several parachute divisions—a formidable group."

"A very capable general."

"Yes sir, that is our opinion."

Churchill glanced at Ismay, who nodded his concurrence. "So, it is Crete."

"Yes sir, it appears so . . . although we must note that it is quite unusual for the Germans to refer to an operational objective by name, even in an encrypted message."

"Misinformation?"

"That would indicate they believe Enigma has been compromised, and we have seen no evidence of that suspicion."

"And, the execution date. Pug please work with CIGS to scrub the source and get the information to Generals Wavell and Wilson. We do not have much time to prepare. We cannot sacrifice Crete. The island is far too close to Egypt and Cyrenica. The message suggests their invasion attempt will be an airborne assault without naval support. Having the date and method of assault should enable the assault to be repelled. It also suggests the Germans are feeling quite confident in their ability to subdue Greece and Yugoslavia"

"I would urge caution," interjected Ismay. "Student's airborne troops are highly trained, experienced, and exceptionally well equipped. They will not be pushovers. And, where do we get the troops for the defense of Crete?"

"At the rate the situation is deteriorating in Greece, I would say Wilson's 'W' Force will be extracted from Greece and moved directly to Crete."

"They are in combat now. They will need rest and refit."

Churchill waved his hand dismissively. "They will have to make do. We have no reserves."

"Should we move them now . . . before we have to do so?"

"We cannot abandon Greece."

"We cannot defend both."

"Indeed! We'll need to get this to the War Cabinet." Churchill turned his attention to reading the second message.

MOST SECRET - ULTRA

```
SECRET
DATE: 14 APRIL 1941
TO: SUPREME COMMANDER NORTH AFRICA XX CORPS
ARMOURED AFRICA GROUP
FROM: COMMANDER AFRICA CORPS
BREAK
SUPPLY SITUATION MUCH IMPROVED BREAK 30 COMBAT
DAYS ON HAND BREAK HAVE BEEN NET POSITIVE FOR
SEVERAL WEEKS NOW BREAK SUPPLY LINE MUST BE
PROTECTED END
SECRET
```

MOST SECRET - ULTRA

"Well," said Churchill, "this is not good news, is it?" The Prime Minister handed the second sheet of paper to Ismay.

"No sir," Journeyman replied.

"I presume Broadway House and Station X are looking for opportunities to alter this situation."

"Yes sir . . . as we are everywhere else."

Churchill stared at Journeyman, but resisted the urge to snap at the intelligence officer. "Can you leave these with me?"

"If you wish, sir; however, our procedures suggest it would be preferable to have a courier return with the messages to support your meeting later this morning."

Again, Churchill stared at Journeyman with a scowl on his face this time. "Very well." Ismay handed both messages back to Journeyman, who in turn placed them in the appropriate folder and locked the case. "It is interesting that Rommel's message is to senior Italian generals with no indication they have gone to the German High Command."

"That is his chain of command . . . at least for now," Journeyman answered.

"Yes, yes, quite so. One left," the Prime Minister said, as he lifted the last message from the folder.

MOST SECRET - ULTRA

```
SECRET
DATE: 14 APRIL 1941
TO: FM ONLY FO
FROM: CHARGE USA
BREAK
NO DISSEMINATION
JAPANESE DIPLOMATIC CODE COMPROMISED BEING
DECIPHERED BY AMERICANS BREAK SOVIET AMBASSADOR
TO U S INFORMED ME DIRECTLY BREAK CONSIDER
SOURCE ABSOLUTELY RELIABLE BREAK RECOMMEND
NOTIFICATION OF JAPANESE GOVERNMENT SOONEST
BREAK HAIL HITLER
END
SECRET
```

MOST SECRET - ULTRA

"Dear God above, this cannot be true!" Churchill exclaimed.

"We have discussed this message inside the Service more than any other I can remember. The implications are profound, broad and incalculable. The Foreign Minister was notified two hours ago. 'C' wants to discuss this matter with you as soon as you are ready this morning before any actions by His Majesty's Government are taken."

The Prime Minister ignored the request for the moment. "Is this Thomsen?" asked Churchill.

"Yes, Hans Thomsen. He has been the senior German diplomat in the U.S. since the recall of Ambassador Dieckhoff, after the Night of Broken Glass in November of 1938. He's an experienced and savvy professional."

". . . to Ribbentrop himself?"

"Yes sir."

"Who is the Soviet ambassador in Washington?"

"Konstantin Aleksandrovich Umansky. He has been the Soviet ambassador since May 1939—a master of language, we understand."

"I've never met him. Back to your request, yes, no action until 'C' and I chat. No one else is to see that message." Churchill wanted to say more of what was on his mind, but he needed time to think. He rechecked the candy-striped folder. It was empty. Winston quickly scanned all three messages, and then placed them back in the cover folder and into its proper place in the Buff Box. Thank you for making the trip, Mister Journeyman. Be safe returning to Broadway House."

"Thank you, Prime Minister. Good night or rather good morning." Journeyman departed.

Churchill and Ismay finished their cognac without a word. Winston secured everyone for what was left of the night and retired to the bedroom of the Annexe apartment. It was going to be a long day before he could travel to Chequers for the weekend. After all, it was a waning half moon that would allow the use of the closer country retreat. Winston instinctively knew he needed sleep, but the multi-layered threat of the third ULTRA message would be difficult to disengage from in these early hours. He had to give it a go.

08:30 hours

Churchill had managed a few hours sleep. He awoke surprisingly early and called for 'C' while he dressed. Those around him every day knew this was not normal and something serious had happened or was about to happen, but no one conveyed their concern. Winston decided they should meet in the security and protection of his small office underground in the Cabinet War Rooms bunker below the New Public Offices building. The two men arrived from different directions at nearly the same time.

The simple letter 'C' was the traditional designation for the Director General, Secret Intelligence Service (SIS or also MI6) in honor of the founder of MI6--Captain Sir George Mansfield Smith-Cumming, KCMG, CB. Colonel Stewart Graham Menzies, DSO, MC, had become the third director general in November 1939, on the passing of Admiral Sir Hugh Sinclair—Cumming's successor and Menzies predecessor.

"Good morning, Prime Minister," greeted Menzies.

"I hardly think so," responded Churchill with a rather grumpy tone and without looking to his chief spy. He had not even looked outside to ascertain the weather. Winston gestured for Stewart to follow him into the office. Menzies did so and closed the door behind him. They sat across from each other at the small, four-place conference table in the Prime Minister's underground office. "Do we know any more?"

"About the Thomsen's disclosure?"

"Yes, yes," he answered, now with an angry tone. Winston calmed his voice. "I cannot imagine anything being more serious than this."

"No sir. We have no new information."

"As I read that message, it indicates the Americans shared their special intelligence with the Soviets or the Russians have a mole at a rather high level inside the American government. We have no direct acknowledgment from the Americans that they have broken the Japanese diplomatic code, do we?"

"No sir. We have hints based on sensitive intelligence the Americans have shared with us, but we have no direct confirmation and certainly no exchange of raw material."

"And, they do not know about ULTRA?"

"No sir . . . not to my knowledge. I am not aware of any decision by you or any other person with access to or knowledge of ULTRA to allow that exchange. Further, we have perceived no hints from the Americans that they might have knowledge or suspicions of ULTRA, or our capability."

"So, we cannot inform the Americans directly." Churchill stared at Menzies while he considered options. "Perhaps I should travel to Washington and meet eye-to-eye with President Roosevelt."

"That is your choice, of course, Prime Minister. You alone have that authority. Short of that, I would strongly urge silence. We cannot risk even the slightest leak, even if the Americans are well intentioned."

"How do we inform the Americans without exposing ULTRA?"

"I have struggled with that very question most of the night. This is far too sensitive and the implications are incalculable. They have not acknowledged their achievement, if we assume Thomsen's message is true and accurate."

"Perhaps it is time to share with the Americans? That day of intelligence exchange is rapidly approaching."

"Mister Prime Minister, the specter of a Soviet mole within high levels of the American government with access to such sensitive intelligence material cannot be ignored . . . or over-estimated. We have been extraordinarily careful in providing ULTRA-derived information to the Soviets once we initiated sensitive communications with them after we deciphered Hitler's Directive 21 activating Operation BARBAROSSA. We have never made a direct reference or even implied such knowledge. If we assume the Americans have achieved similar success with Japanese coded communications, as we have with the German encrypted communications, they certainly would protect that capability as we are protecting ours."

"Yes . . . but, as I am reminded so often, they are not at war. We are!"

"There is that."

"Damn it all to hell!" growled Churchill. "After the Kent affair last year and the convulsions we suffered convincing the War Cabinet to proceed with the Tizard Mission, despite the apprehensions of so many about the security of the U.S. government, now we have this. Damn it all the hell!" he repeated. Menzies recognized the Prime Minister's contemplation and remained silent. Several minutes passed. Churchil grunted and groaned numerous times, as he considered whether to act and if so how. "I will discuss this with Eden and the War Cabinet alone. They deserve to know. For now, I want that message secured—no other dissemination, not even to the approved access list."

Secretary of State for Foreign Affairs Robert Anthony Eden, MC, PC, Member of Parliament for Warwick and Leamington, as Foreign Minister, served as the ministerial oversight for the Secret Intelligence Service (MI6). In his capacity as Foreign Minister, Anthony was also a member of the War Cabinet.

"Understood."

"Unless you hear otherwise from me, and me alone, we will sit on it and see what turns up. As a minimum, this is a cautionary tale of how precarious our golden goose is."

"Yes sir. So be it. With Eden's concurrence, I may send a private message to President Roosevelt . . . perhaps through Lord Halifax . . . cautioning him that U.S. Government communications with the Soviet Union may be compromised."

"I do not need to caution you on the criticality of wording in such a message . . . in any form."

"Thank you for that, Stewart. I will have Anthony coordinate with you before we send anything."

"Thank you, Prime Minister. Is there anything else you would like to discuss?"

"Not that I can think of at the moment. Good day, Stewart."

Menzies departed and left Churchill alone with his thoughts in the perpetual light of the Cabinet War Rooms bunker.

—

Sunday, 20.April.1941
Hatfield Aerodrome
Hatfield, Hertfordshire, England
United Kingdom
13:45 hours

Air Commodore John Spencer eagerly picked up the temporary assignment of escorting American General Arnold, who was on an official tour of RAF units and facilities when his assigned escort staff officer suddenly became ill. He was a stand-in for the next day or two. The quick addition to Arnold's itinerary had come only two days ago, when Geoffrey de Havilland informed the Air Ministry and General Arnold of the afternoon's demonstration flight of the new DH-98 Mosquito for Minister of Aircraft Production Lord Beaverbrook.

William Maxwell 'Max' Aitken, Bart, Kt, PC, ONB, was an ambitious, aggressive, Canadian businessman and newspaper publisher when he moved to England and was elected to Parliament in 1910. He was knighted by King George V in 1911, and served as Minister of Supply responsible for aircraft production during the Great War. In the 1917 New Year Honors List, the King elevated Aitken to be 1st Baron Beaverbrook of Beaverbrook in the Province of New Brunswick in the Dominion of Canada and of Cherkley in the County of Surrey, and he moved from Commons to Lords. Last year, newly installed Prime Minister Winston Churchill selected Lord Beaverbrook to be the cabinet minister solely responsible for aircraft production. Beaverbrook had been the driving force behind the dispersal scheme to protect aircraft production during the worst of The Battle last summer.

John and Hap Arnold had at least passing familiarity, when John attended the court-martial of Colonel William Lendrum 'Billy' Mitchell on behalf of the Air Ministry that was keenly interested in the testimony, evidence and outcome, given Mitchell's outspoken advocacy of air power. Then Major Arnold testified for his mentor Mitchell along with other Air Service notables of the day: Eddie Rickenbacker, Carl Spaatz, Ira Eaker, and Robert Olds. John met the defendant as well as his prominent supporters.

They arrived at Halfield Aerodrome ahead of time. Both Spencer and Arnold had met Geoffrey de Havilland, CBE, AFC, RDI, FRAeS, numerous times between the Great War and the current version. Geoffrey was the founder, owner and chief designer of the De Havilland Aircraft Company, Limited.

Arnold had been a principal in the aircraft design and procurement process for the U.S. Army Air Service that became the Army Air Corps for which Arnold was the current chief. In fact, Arnold had bought more than a few airplanes from De Havilland Aircraft in the 1920s.

"It has been more than a few years since I saw you last," pronounced de Havilland.

"Yep, you got that right. My last visit was in 1928, at your Stag Lane Aerodrome facility in Edware," Arnold responded.

"That does go back a few years. We were ultimately forced to abandon and sell that facility due to residential housing encroachment. Operations out of Stag Lane simply became impossible. Fortunately, we found this land, acquired it, and built the runway and buildings. We moved here in 1933." He paused, looked at his wristwatch, and then continued. "Thank you for adjusting your schedule to attend this demonstration. Lord Beaverbrook asked for the event on rather short notice. We are very proud of our Model 98, so we eagerly took the opportunity, and since you were in this country, I thought it quite appropriate for the Chief of the Air Corps to observe as well."

"Thank you for allowing us to visit, Geoffrey," John said.

"You are quite welcome, John. The Air Ministry is my best customer. I understand you have a very busy itinerary, even on a Sunday, so we are honored that you could stop to visit." De Havilland glanced at his wristwatch. "Now, Lord Beaverbrook will arrive shortly and we need to execute the demonstration as soon as he arrives." He gestured toward his right. "If you will, let us make our way to the flight line. My middle son, our chief test pilot, should be about ready for takeoff." Geoffrey continued talking as they walked along an unusually broad alleyway between several large buildings. "Junior flew the first flight of our Model 98 in November of last year. We are about halfway through our qualification program. We have completed the flight envelope definition portion, so the demonstration Geoffrey will fly this afternoon will give you a good idea of the aircraft's performance."

"Twin Merlin engines?" asked Arnold, more as confirmation rather than query.

"Correct . . . two Merlin 21 powerplants, each rated at nearly 1,500 horsepower for combat and driving our three-bladed, constant speed airscrews."

"And all wood?"

"Correct, again, which gives it low radar detectability."

At the end of the last building, the alleyway stopped at a low fence. Arnold followed de Havilland's look to the northeast. The twin-engined aircraft remained stationary with both propellers turning at idle. Hap followed Geoffrey's glance to his left. A dark blue limousine approached.

Geoffrey looked to Hap and John. "If you will excuse me, this must be Lord Beaverbrook." De Havilland stepped away and waited at a spot where he expected the vehicle to stop. John and Hap moved to the edge of the viewing area. As the round figure of Lord Beaverbrook stepped out of the Rolls Royce limousine, de Havilland welcomed him, although John could not hear the words. They shook hands, and then Geoffrey gestured to the viewing area.

"Hap!" shouted Beaverbrook, and then strode swiftly toward Arnold, leaving de Havilland behind him. As he approached, he extended his right hand. "I heard that you might be here." The two men shook hands.

"Great to see you, Max."

"A pleasure to see you, again, John," Beaverbrook extended his hand to Spencer. "I saw your uncle last Friday, and I must say he is in fine form."

"Likewise, Lord Beaverbrook. Thank you for that observation. I have not seen him since last November"

"I see you know each other," de Havilland interjected, "so with introductions dispensed with, the aircraft is standing ready," announced Geoffrey.

Beaverbrook nodded his head and turned toward the airfield.

De Havilland looked to a young man standing ten yards away by a table with a radio on it. The man raised the hand microphone to his mouth. They could not hear what the man said, but he undoubtedly passed the takeoff instruction to Junior.

They watched and listened. The deep groans of the Merlin engines could be clearly heard at a distance. Within several seconds, the sleek-looking, twin-engined aircraft began its takeoff roll. When the tail rose off the grass, the pilot smoothly advanced the throttles to takeoff power. The full-throated roar of the engines rose in volume as the aircraft broke ground abreast of them. The main wheels retracted immediately and Junior kept the aircraft low, with the propeller blade tips just above the grass to gain speed as quickly as possible. By the far end of the runway, he pulled up into a nearly vertical climb, rolled the aircraft once, and then pitched the nose smoothly out of the climb and into a nearly vertical dive. Junior maneuvered the aircraft through a well-choreographed series of turns, climbs and dives, and high speed passes, with the last run coming directly toward them from just above the treetops to the north. The distinctive element of the head-on approach was the near silence of the aircraft until it was about a quarter of a mile away from them, and then the volume of the engines, huge propellers and special whistle of the aircraft rapidly increased in intensity. As Junior neared them, he pulled up, exposing the aircraft underside, and then performed two victory rolls. The combination of the Merlin engines and the singing of the aircraft at high speed gave the demonstration an almost majestic pleasure to the senses. After dissipating his energy, Junior reconfigured the aircraft for landing. The Merlins popped and spit in protest at being pulled back to idle as the aircraft touched down smoothly without the slightest bounce. Junior taxied the impressive machine to a spot directly in front of them and shut down.

Lord Beaverbrook turned to Geoffrey. "Thank you, Mister de Havilland, for arranging this demonstration for me on such short notice. I must say the Mosquito is a most impressive aircraft, and I must genuinely thank you for persisting against my recommendation to bring this design to fruition. Well done, I must say."

"Thank you, Lord Beaverbrook."

"I can say the Air Ministry, and especially the Chief of Air Staff and Bomber Command are impressed as well. You have exceeded the Air Ministry requirements in every category. I wanted to see the machine fly myself before throwing support to production. I am satisfied. I will call Sir Archibald and Sir Charles this afternoon to move this project along."

"Thank you very much Minister."

Beaverbrook nodded his head and turned to shake hands with John and Hap. "Great to see you, again, Hap. I hope you see everything you wish to see, and your visit is productive. I'm afraid I must run. Enjoy your tour. You are in good hands," he said, nodding to de Havilland.

"Thank you, Max," Hap answered. "I have already achieved my objectives, so the rest is gravy."

"Excellent." Beaverbrook said his good-byes, returned to his limousine and the driver drove away.

When the sound of the automobile's tires crunching on the gravel quieted, Arnold turned to de Havilland, with a broad smile across his face, and said, "We've won the war."

"So, you liked our new DH-98 Mosquito," de Havilland replied.

"Liked it, I loved it!" exclaimed Arnold. "It does not get much better than that. I look forward to the day I can fly one myself."

Geoffrey Junior completed securing the aircraft and dismounted. He walked toward the three men.

The senior de Havilland said, "Well done, Son."

"Thank you, Papa."

"May I introduce Major General Hap Arnold, Chief of the American Army Air Corps, and Air Commodore John Spencer, Operations Officer and Chief Controller for 11 Group."

"Nice to meet you, General." The two men shook hands. "Great to see you, again, John," he said to Spencer, as the two men shook hands, having met several times before.

"That was a helluva demonstration, Geoffrey," Arnold declared.

"Thank you. She is an easy machine to fly hard," Junior responded. "She has quite a bit of capacity. We have envisioned a wide variety of configurations: fighter, night-fighter, light bomber, photographic reconnaissance and even transport."

"There are not many aircraft that are that versatile."

"Precisely. Would you like a tour?"

"Of course, by all means."

Junior led the three men to the aircraft in front of them, with the senior de Havilland, lagging behind to allow his son to conduct the tour of the aircraft. John Spencer had heard of the machine's accomplishments, but had not yet seen one up close, so he was keenly interested as well. Junior performed the cook's tour walk around as he had done many times before, pointing out key features of the design. The four men discussed numerous elements and attributes as well as future potential for the aircraft some were already dubbing the Wooden Wonder.

With the aircraft examination complete, Junior excused himself to prepare for another test flight pending that afternoon. Senior adapted to the remaining time allotted for the visit to provide Hap and John a quick walking tour of the current De Havilland Aircraft Company facilities and offered his vision for the expansion of the Hatfield site. They discussed the continuing threat of The Blitz and the dispersal of manufacturing capacity to avoid catastrophic impact on production. They also discussed the plan and expectation for the Mosquito to enter operational service in the RAF by the end of the year. The company already allocated resources for production of the aircraft even before His Majesty's Government signed the contract for initial production.

By the itinerary and prior arrangement, Geoffrey de Havilland selected and assigned a fairly young, production supervisor who managed the company's dispersed facilities to take over and lead their guests on a quick tour of the surrounding area and a half dozen company-owned manufacturing shops where important, if not vital, machine tools had been moved to lessen bomb damage exposure. Arnold and Spencer thanked de Havilland for his generous time and contributions.

Spencer and Arnold followed de Havilland's limousine and driver with their tour guide. Again, de Havilland's generosity enabled the two air force officers to maintain some independence should a situation or condition develop requiring them to terminate the tour. They would wind through narrow country roads from village-to-village to visit a half dozen small and medium size machine shops and component manufacturing facilities. The company-controlled shops were apparently just a fraction of the total number of dispersed buildings and companies created just to support and sustain one aircraft company during the wartime reality of enemy bombing.

—

Tuesday, 22.April.1941
Embassy of the United States of America
No.1 Grosvenor Square
Mayfair, London, England
United Kingdom
15:40 hours

"May I help you, sir," the Marine guard said, as Brian stepped into the small security lobby of the main embassy building.

"I'm Flying Officer Brian Drummond. I have an appointment with Mister Slaughter at a quarter 'til four."

"Please sign in, and then have a seat, sir. I will let Mister Slaughter know you are here."

Brian did as suggested. The Marine guard raised the telephone handset and said something he could not hear. The Marine was dressed in a sharp blue uniform with dark blue tunic with red piping and medium blue trousers with a wide, white belt and holstered pistol. He was not familiar with the rank insignia on his sleeves, but he had one bright yellow, inverted chevron with crossed rifles underneath.

"Mister Slaughter will be down shortly, sir," announced the Marine.

"Thank you."

Brian was not particularly keen on seeing Arnold Slaughter, again. The last time he had seen the man was a year ago this month, in Drem, Scotland, when the man tried to have him arrested for violating the Neutrality Act. Brian had been disappointed when Slaughter's name came up, when he was making this appointment. He expected this to be a confrontational meeting, but a necessary one, nonetheless.

Slaughter appeared from behind the barrier behind the guard. He actually smiled and extended his right hand well in advance of reaching Brian. "Mister

Drummond, a pleasure to see you, again," he said, as they shook hands. "I was afraid I was not going to see you, again."

Well, that is quite a switch. I'd better keep my mouth shut, to avoid saying something I would regret.

"Please follow me," Slaughter said. He did not wait for a response.

Brian followed Slaughter into the interior of the building, down a corridor, up two flights of stairs, and down another corridor to a closed door with frosted glass upper half and black block letters painted on the glass that read – Deputy Commercial Attaché, and Arnold Slaughter in small letters underneath the title. Slaughter held the door open and closed it behind them. He motioned for Brian to take a seat on the near side of his desk, as he went around and sat in a big leather upholstered, swivel chair.

"First, before we get to your business, I would like to extend my personal apology to you for our last encounter. I was simply following orders under the law. I meant no offense personally."

"Yes sir."

"So, all is forgiven?"

"Yes sir."

"Excellent. Now, how may I help you?"

"I had to return to the United States a few months ago, to settle my parents' estate."

"Oh my! Please accept my condolences for your loss."

"Thank you. Anyway, there was a bit of a misunderstanding with the passport folks when I arrived in New York. After clearing things up, the officer advised me to register my presidential pardon with the embassy and have an appropriate annotation in my passport, to avoid any problems in the future. So, here I am."

"May I see your passport and your pardon letter, if you brought it with you."

Slaughter scrutinized both documents and placed them on the desk in front of him. "This is the first presidential pardon I have ever seen," he said, placing his left hand on the letter. "Please give me a moment to validate the letter." He did not wait for a response. Slaughter took the pardon letter and Brian's passport, and left the office, closing the door behind him.

The office was sparsely appointed. The medium size, wooden desk was the most prominent feature in the room. A single portrait photograph of President Roosevelt hung on the wall behind his desk. The window to the right of his desk looked out on a large tree with fresh, new leaves of springtime. He could see hints of buildings beyond the tree, but not enough to orient himself. The small bookshelf opposite the window was fully populated with books that all looked like official books of one form or another, and a couple of book sets. The door opened and closed, as Arnold Slaughter returned to his office and Brian.

"According to our legal attaché," Slaughter began before sitting, "your letter is an original signature, authentic document. My curiosity has gotten the better of my propriety, I must say. If I may ask, how did you obtain this presidential pardon? Who are you exactly?"

Brian was taken aback somewhat by Slaughter's second question. *What, does he think I am someone other than who I say I am? I had not expected this.* "I am Brian Drummond from Wichita, Kansas."

"Of that, there is no doubt, but who do you know that could get you a presidential pardon?"

"Mister Slaughter, I can assure you that the source of this document is as unknown and strange to me, as it is to you. It simply appeared after you attempted to arrest me last year. I had no idea such things existed, set aside the procedures to obtain a presidential pardon. Nonetheless, I am thankful I have this letter."

"I'll bet. If you ever learn how this letter came about, I would surely like to know, so I can learn something as well. So, here is what we can do. We have entered an endorsement as to the existence of this pardon along with its subject, and the legal attaché and I have counter-signed it to validate the information. You will need only your passport and should not have any more problems entering the United States. I would encourage you to protect this letter as well as your passport. Your RAF identification papers should be sufficient in this country, so there is no need to carry your passport or the letter. Any questions?"

Brian accepted the letter and his passport back from Slaughter. He checked the letter and the passport endorsement. They looked good to him. "I was also told that I should report my marriage and impending birth of our child."

"Well, now, you have been busy, haven't you?" Brian just smiled and did not respond. "Yes, we need to change your status. When did you marry?"

"Last December."

"What date?"

"The 30th of December, last."

"Ah yes, I see you notified the embassy," Arnold responded.

Slaughter made several notations in a ledger of some sort. He also informed Brian regarding the naturalization process for a wife and that their child would be considered a natural born citizen upon birth even though the child would be born in England to an English mother, since Brian was the father whose citizenship was restored by the presidential pardon. Slaughter also informed Brian that he should bring Charlotte to the embassy with her birth certificate and current British passport, if she had one. Arnold also told Brian to keep the embassy informed of any status change like the birth of children, divorce, promotions and any awards that he should receive. Brian learned the embassy was already aware of his promotion and his awards from last summer and fall. They agreed the embassy was current on his status.

"Anything else?" asked Slaughter.

"No sir. Thank you for your assistance."

"You are most welcome. Again, I offer my apologies for last year's confrontation and my sincerest condolences for your loss."

"Thank you."

"Now, Colonel Geiger heard you were in the building and would like a word with you. Do you know Colonel Geiger?"

That's the name John Spencer referred to when we had dinner a couple of weeks ago. "I've heard the name, but I don't know Colonel Geiger, sir."

"He is the assistant naval attaché in this embassy. I'll show you the way." Again, Slaughter did not wait for a response.

Brian followed Arnold out of his office and down the corridor toward the opposite end of the building. They stopped at a similar closed door. This one displayed block lettering – Assistant Naval Attaché over Col. Roy Geiger, USMC. Slaughter knocked and did not wait for a response to the open door. The colonel rose from behind his desk. He was shorter and stockier than Brian with close-cropped blond hair, and wore a forest green tunic and trousers with eagles on his epaulets and shirt collar points over a khaki shirt and necktie. The rows of ribbons above his left breast tunic pocket and under his gold naval aviator wings were not familiar to Brian. Brian saluted the senior officer in British fashion. The colonel did not return the salute and extended his hand to Brian.

"You must be Flying Officer Brian Drummond," Geiger said, as they shook hands. "I'm Roy Geiger."

"He's all yours, Roy," Arnold said.

"Thanks, Arnold. I've got him." Slaughter closed the door as he departed. "Have a seat, Brian. You have developed quite a reputation, young man."

Brian nodded his head. *I have no idea what I am supposed to say to that and no idea what he is talking about.*

"I heard some of your accomplishments before I got this assignment. I was hoping to cross paths with you. Then, Arnold told me you were coming to visit. If my information is correct, your squadron has recently moved to RAF Martlesham Heath, not far from London, as I understand locations."

"Yes sir. Your information is correct."

"When do you need to be back on duty?"

"Tomorrow . . . by noon, sir."

"When do you plan to return?"

"I came into London to get a passport endorsement from Mister Slaughter. Once we are done here, I was just going to head back to Liverpool Street Station for a train back to Ipswich and the base."

"Can you make a few hours for me and a few beers, and perhaps dinner . . . my treat?"

What the hell is this all about? What does a colonel want with me? "If you wish, sir, I would be honored."

"I understand the pilots gather at Shepherds Pub, which is not far from the embassy."

"Yes sir. It's across the northeast corner of Hyde Park . . . a comfortable walk on a dry, spring afternoon."

"Excellent. If you have no objection, let's take a walk. You can show me this place. We can have a few beers and get a bite to eat."

Before they departed, Brian asked if he could call Air Commodore Spencer. Roy assisted Brian in using the embassy switchboard and entering the British Telephone system. Brian guessed that this evening's engagement might be a bit protracted. Now that he knew the rail schedule between Ipswich and London, he could make a date for dinner with the Spencers, perhaps tomorrow evening or another day, after they were released. The telephone connection took several minutes to close, but he eventually reached John Spencer. Brian informed John of what he was doing, and they made a date for Thursday evening.

Colonel Geiger checked out for the rest of the afternoon. The two guards at the entrance snapped to crisp attention and saluted smartly. Colonel Geiger was covered and returned the salute, as did Brian. They casually walked south on Audley Street toward Green Park. They talked about Brian's experience in England and flying for the Royal Air Force and Fighter Command. Brian also learned that Colonel Roy Stanley Geiger, USMC, had flown with No. 5 Group, Royal Naval Air Service, at Dunkirk, during the Great War before heading up a Marine aviation unit in France. Brian also realized that Geiger had been flying longer than he had been alive. Geiger did not know Malcolm Bainbridge or John Spencer. He also learned that it was Squadron Leader Taylor who had contacted Colonel Geiger for assistance in arranging Brian's journey back to Wichita. Brian conveyed his gratitude for the assistance and recounted a little of his adventure, including meeting with and talking to Harry Hopkins, and the President on his return journey. As they approached their destination, Brian told Colonel Geiger about the history of Shepherds Tavern, as he knew it.

—

Tuesday, 22.April.1941
Shepherds Tavern
50 Hertford Street
Mayfair, London, England
United Kingdom
19:30 hours

Roy Geiger and Brian Drummond had finished their small pub meals and were on their third pint each, when Squadron Leader Jeremy Morrison walked to their table.

"Look what the cat dragged in," Jeremy shouted.

Roy stood and Brian followed.

"Hey, Jeremy," Brian said. "I'd like to introduce you to Colonel Roy Geiger. Colonel, this is Squadron Leader Jeremy Morrison, more properly Lord Morrison, but he doesn't like that." Jeremy sneered at Brian, and then smiled at Geiger. The two older men traded cordialities. They also exchanged titles and job assignments.

"If this is not a private conversation, may I join you?"

Brian nodded to Roy, who in turn responded, "By all means."

Jeremy grabbed an empty chair from a nearby table. All three men sat.

"Tell me, Colonel, how did you wind up in the company of this degenerate?" he asked, laughed and punched Brian's shoulder.

"I wanted to meet and talk to the highest scoring American ace since Eddie Rickenbacker in the last war."

"You chose well, Colonel, despite my brotherly jab. He's one outstanding pilot and a helluva killer."

"How did you meet him?" asked Geiger.

"I was his assigned flight instructor in our training command before the war began . . . odd position to be in, actually. He was a better pilot than me before we started. His success during The Battle last summer is factual testament to his skills. His flight callsign is Hunter . . . most appropriately, I must say."

Brian wanted to change the subject. "Is your squadron still in One Three Group?"

Jeremy laughed and turned to wink at Geiger. "He doesn't like us talking about him."

"Just like you do not like being called Lord Morrison."

"*Touché*," Jeremy responded and turned back to the Marine. "Tell me, Colonel, if my limited knowledge is correct, your uniform and insignia indicate you are a pilot in the American Marines."

"You are quite correct. However, the only thing I am flying in this assignment is a desk at the embassy."

"A vital position, I am certain."

"So, they tell me. I would like to hear your impressions of the Germans as aerial adversaries."

Jeremy responded first. "I'll let Hunter handle this one. After all, his body count of dead Germans is significantly higher than mine."

Colonel Geiger looked at Brian, expecting an answer.

"The One Oh Nine is comparable to the Spitfire – better in some aspects, not so in others . . . in the hands of an experienced and capable pilot . . . a formidable adversary. Their pilots have had more combat experience than we have, but we certainly made up for lost ground last summer. The Germans have trained very well. They know what they are doing."

"As I understand it, you have 19 confirmed kills . . . one short of an ace four times over."

"Correct," answered Jeremy for Brian. "He's currently the third highest ace in the whole of the bloody Royal Air Force. And, The King has decorated him twice in his short career with us."

"Which decorations?" asked Geiger, looking at the ribbons above Brian's left breast pocket and below his wings. "I recognize the Distinguished Flying Cross, I believe. What is the other one?"

"The Military Cross, sir."

"And, a CBE, personally from The King, for Christ's sake," Jeremy added.

"What is a CBE, if I may ask?"

"Commander, Order of the British Empire, Colonel," again, Jeremy answered before Brian could, ". . . one step short of knighthood, to we Brits."

"None of that helps me fly any better," Brian mumbled.

"Perhaps not, young man," Roy said, "but, it is indicative of the importance the leadership of the air force and this country place on your contributions to the defense of the country."

"I am grateful."

"Good. You should be." Geiger turned to Lord Morrison. "As a squadron commander, how do you view the tactics Fighter Command uses in aerial combat?"

Jeremy laughed hard. "Oh my, Colonel, where do I begin?" Morrison took a long swig of his beer. "I think Brian will agree with me. When the shooting began, our tactics were abysmal and no match for the Germans. Our formations were all about control and hierarchy. Fortunately and blessedly, we had nearly a year, well eight months, before the Germans turned their attention to us. In that time, we learned. To be frank and blunt, we adopted the German fighter tactics. They call their basic flight formation a schwarm, a swarm, a very appropriate sounding name, essentially a flight of four fighters. We refer to the formation as a finger-tip-four," he said, holding up his hand, "the lead section in left echelon and the second section in right echelon, off the right wing, as an example. More importantly, we spread everyone out, so we could get four sets of eyeballs scanning for the enemy rather than just the leader."

"How can your fighters be improved to give you an edge over their experience?"

"We have plenty of experience, now. Certainly, we are a match for them." Jeremy paused to think. "In short, I would say speed . . . never enough speed. A greater speed differential would give us more control of the fight. . . . and more punch. Our three-oh-three rounds are just no match for their 20-millimeter explosive projectiles. When they hit, they hit hard. We put a lot of rounds in the air with our eight guns, but it usually takes many hits for us to bring them down. All too often, they need only one hit. While I'm at this answer, I would also say more range. We have started crossing the Channel, to take the fight to the enemy, and our time over the target is just too short – the exact same problem Gerry had last summer, fighting us in our home skies. We have to stage a cover squadron behind us, so they have more fuel for a fight, should the Germans come up to play, which they invariably do."

Brian nodded his head in agreement.

"The fighters haven't been back since last fall?' Geiger asked.

"No sir," answered Jeremy. "When they abandoned daylight bombing, the fighters pretty much disappeared as well. We had a month or so, early last fall, when they sent formations of fighter-bombers to tempt and engage us. By that time, we handled the bastards fairly well."

"What about the night raids?"

"I am certain you will experience your share the longer you are here."

"Already have."

"Yes, well, night fighters are not my bailiwick. I know those lads have big brass bollocks to do what they do. Day fighter engagements are sporty enough. I cannot imagine doing what we do at night."

"How many night fighter squadrons do you have?"

"My squadron is day fighters only. The squadron took rather heavy losses during The Battle. I picked up the squadron when they withdrew what was left of the squadron last October. We have been in Acklington since we were withdrawn and slowly regaining our strength. We have only a couple of other beat-up squadrons like mine up there; no night fighters."

"We don't have any either at Martlesham," injected Brian. "We had one – Six Oh Four Squadron – when I was with Six Oh Nine Squadron at Middle Wallop. They were getting the new Beaufighter with a new onboard RDF kit when I was transferred to Seventy-One Squadron."

"The highest scoring night fighter pilot and only night ace – John Cunningham – is with Six Oh Four, isn't he, Brian?"

"Yes sir. Good man. His call sign is 'Cat's Eyes' for a very good reason."

"We hear the night fighters are getting better at what they do, but the bombers still get through." Jeremy looked at his wristwatch, and then scanned the pub. The patrons, mostly in RAF blue, were thinning out markedly. "In fact, I must be on my way before The Blitz continues tonight. The eleven o'clock train should put me back to Acklington by morning."

"It was a pleasure to chat with you, Mister Morrison. Thank you for your time and patience with my questions. Have a safe journey back to your base and good luck with whatever lays ahead. As we say in the nautical services and as borrowed from the Royal Navy, Godspeed and following winds."

"Thank you, sir. It was an honor to meet you. I would suggest you make your way to wherever you shelter. The Germans could appear any time, now." He turned to Brian. "You should get along as well, Brian, unless you plan to spend the night in the city and in a shelter. Fly safe, my friend."

"Thanks, Jeremy."

Squadron Leader Morrison shook hands with both men and departed.

Geiger turned to Brian. "One more thing before you go, if you will permit me, Brian."

"Yes sir. I have time . . . if the Germans don't interrupt us."

"First, thank you so much for changing your plans and spending some time with me. Listening to you and Jeremy Morrison has been most enlightening. Second, if you are ever in London and want a free meal, please do not hesitate to give me a call. I do not know how long this assignment will last, but I would be honored to throw down a few beers and buy you a meal, whenever you can spare the time for me. Lastly, I have no means to predict how this tragic affair is going to turn out for any of us. However, if you ever find yourself in need of a real flying job, please give me a call or send me a message wherever we are in the world. The Marine Corps needs pilots like you. The Army tends to be possessed by regulations. The Marine Corps is a pretty wild group of eager pilots. I think you would fit in quite well. We'll teach you some new real flying . . . taking traps on a carrier. Great stuff!"

"Thank you very much, sir. I will keep that in mind, and I will certainly not forget this night. I must say, I am happy doing what I'm doing. The RAF has been good to me."

"Sure, I would not ask you to break the loyalty you feel. I just wanted to offer my assistance, should you ever need a new home."

"Thank you, again, sir."

Colonel Geiger paid their bill. Outside, past the black-out curtains, as their eyes adjusted to the darkened streets of London, the two pilots said good-bye and shook hands. Roy Geiger headed back north toward the embassy. Brian headed east. He entered the Underground at Green Park Station. Fortunately, the Germans were not early on this night. He transferred at Oxford Circus Station and arrived at Liverpool Street Station in plenty of time to board the night train to Ipswich and RAF Martlesham Heath. Brian knew this had been a good afternoon and evening, and his gut told him Roy Geiger was a name he should remember.

—

Thursday, 24.April.1941
No.10 Downing Street
Whitehall, London, England
United Kingdom
16:00 hours

"Prime Minister, your four o'clock appointments have arrived," announced Assistant Private Secretary John Martin.

"Very well. Show them into the small conference room. I shall be there forthwith."

"Yes sir."

Churchill had a few minutes reading to finish the latest shipping report. There was still little good news, but at least the bad news seemed to be diminishing in frequency, so thought the King's first minister. Their loss rate remained precariously just above the sustainable threshold, which meant they were mercifully positive in the net yield from the vital supply convoys braving the snapping jaws of the German U-boats and wolfpacks.

Three men were still standing when Prime Minister Churchill entered the small conference room next to his study. Home Secretary Herbert Stanley Morrison, PC, Member of Parliament for Hackney South; Brigadier Oswald Allen 'Jasper' Harker, CBE; and David Petrie, CIE, CVO, CBE, KPM. The subject of the meeting was the leadership of the Security Service, often referred to by its original designation MI5—the organization charged with counter-intelligence operations and internal security. Harker had been acting director of MI5 since late last year to allow sufficient time for Petrie to accomplish a top to bottom organizational review of the Security Service. Morrison was the ministerial supervision of the Security Service and a leader in the Labour Party.

"Good afternoon, gentlemen," Churchill said. They replied in unison. "Herbert, Jasper, nice to see you both, again. I have not had the pleasure of meeting you and therefore you must be Mister Petrie." Churchill extended his right hand to Petrie. They shook hands.

"It is quite an honor, sir."

"Nonsense. Please be seated, gentlemen." They took four of the six available chairs at the table. Churchill looked directly at Petrie. "I have read with considerable interest your assessment of the 13th. I want to discuss your report with the other principals here. Given that starting point, pray tell me your assessment summary."

"First, please allow me to convey my gratitude to you and the government for being allowed this time to do a thorough assessment of the Security Service." Churchill simply nodded his head in acknowledgment. "MI5 had 30 officers prior to Munich in '38. The pressures of war have dramatically increased the need for field officers to deal with the German threat."

"A most serious threat it is," interjected Churchill.

"Yes sir, quite so. The resultant recruitment and hiring has been haphazard and spotty at best. As a consequence, we have field agents who have keen instincts and others who, shall we say, are less adaptable to the security environment we face at the moment and will face for the foreseeable future. This reality goes beyond the field agents and into the supervisory ranks. This rather odd quilt has caused serious confusion, hesitation and lack of aggressiveness in the chain of command. Morale within the Service is not good. There is a general malaise, perhaps a sense of being overwhelmed by the size of the problem and being encumbered by the rather ragged performance of some supervisors as well as field agents. The word bungling would not be inappropriate given recent incidents."

"Then, what do you propose we do about these personnel problems?"

"If you will permit me to illuminate one additional serious vulnerability or flaw in our current operating system, and then I would like to return to your question in the frame of a whole recommendation." Petrie waited for the prime minister's consent gesture. "MI5 has been operating under an antiquated card-file catalogue system since the founding of the Service in 1909. Those cards held virtually all of the leads, clues, networks and connections in our evidentiary collection. Fortunately, General Kell had the foresight to see the vulnerability and ordered the duplication of the card file – a project that took several years to complete. Our original card file was essentially destroyed in the German fire bombing of Wormwood Scrubs Prison in Shepherd's Bush, last September. Since then, we have been operating with our duplicate and only remaining copy."

"I read all of this in your report," Churchill responded with a hint of impatience in his tone.

"Yes, sorry sir. I have just been at this for a long time. To your query, we must implement a performance measurement system to weed out non-productive agents and supervisors. Mistakes or missteps in this line of work are dangerous to our citizens and the kingdom. I have drafted selection criteria for recruitment and hiring of field agents. We need a very specific type of person. Not everyone is suited to do what we do and what we must do, or deal with the pressures and risks of this work. I have also discussed a reorganization proposal within the Service with both Minister Morrison and Brigadier Harker, and I believe they are in agreement," Petrie said and looked to Morrison and Harker. Both men nodded their heads. "We need stronger, more autonomous regional offices that have clear operating rules and do not have to rely upon HQ to provide specific direction. Sometimes, events in the field require swift action without waiting for approval."

"All of that sounds reasonable to me. Do you agree, Herbert?"

"Yes sir."

"Now, explain to me what you wish to do with the card file."

"Of course. First, we need another duplicate file, located in a different, secure and protected site."

"That is easy. Do it."

"Thank you, sir. Further, we need to devise a different, better method to capture and retrieve information. We call it connecting the dots to see the picture that emerges from the morass of disassociated information. We gather a plethora of dots. We are sometimes too slow to paint the picture and take action. We must have a quick, consistent and aggressive triage system to swiftly assess the threat, urgency and expanse. Minister Morrison has pointed to Colonel Menzies and our sister service. They are apparently developing new methods to collect, retrieve and correlate information at Broadway House and Bletchley Park."

Churchill shot a sharp, stern glance at Morrison. The Home Secretary was the only other person in the room beyond the Prime Minister, who had access to ULTRA information. Morrison almost imperceptibly shook his head to indicate no one had disclosed the existence of ULTRA to either Petrie or Harker.

"Yes, well, perhaps there is some benefit there," responded Churchill. "Please coordinate with 'C' to absorb what you are able. Herbert, perhaps you can sponsor an intelligence services symposium to review how we handle information. I would include Dalton's SOE as well."

"A fine suggestion, Prime Minister. David has proposed a small internal commission of experienced Security Service supervisors and a few field agents to devise improved methods."

"That sounds like a worthy proposal. However, I must remind you, some of the means and methods beyond MI5 are quite tightly controlled. I expect the analysis methods can be discussed without compromising sources or content."

"That is the expectation," Morrison responded.

"Excellent. Anything else, Mister Petrie."

"No sir. That is a reasonable summary of my findings. The details simply amplify the broader findings."

"Excellent. I am impressed with the thoroughness of your assessment, David. I can state and assure you that you have the full support of His Majesty's Government to implement the changes you have outlined, which in turn means we have met your precondition to assume the directorship."

"Yes sir, that is correct."

"Very well, then, on behalf of the War Cabinet and The King, I offer you the position of director general of the Security Service. I also understand you have accepted the continued service of Jasper here as your deputy."

"Correct as well, sir."

"Jasper, also on behalf of The King, I thank you for your service as acting director, to allow David time to complete his evaluation. You are a loyal and noble servant of the Crown. I trust you will continue to serve the Crown well."

"Thank you very much, Prime Minister. I shall do my utmost to support the Director General and the Security Service."

"Excellent. Now, one last question for now, have either of you discussed your assessment and recommendations with Sir Philip?"

"Yes sir," responded Petrie immediately. "All three of us met with the Commissioner two days ago."

"Air Vice Marshal Game fully supports the proposed changes," interjected Morrison. "The Metropolitan Police and the Commissioner have contributed to David's study."

Air Vice-Marshal Sir Philip Woolcott Game, GCVO, GBE, KCB, KCMG, DSO, had served The Crown as Commissioner of Police of the Metropolis since 1935. The Metropolitan Police performed as the execution arm of the Security Service, in essence the front or public face when MI5 needed an arrest or collection of individuals of interest.

"Very well, then. Welcome to the team, David. Now, if you will excuse me, I would like a private word with the Home Secretary."

Both men thanked the Prime Minister in unison, rose and departed the conference room. Harker closed the door behind them both.

As soon as the door latched, Morrison said, "I . . ." He stopped when Churchill raised his hand.

"No need, Herbert. You know quite well how sensitive I am about protecting ULTRA. My expression was a reactionary response to a potential disclosure beyond the closed group. I do not believe you have compromised even the existence of ULTRA. However, I must say, upon further contemplation, perhaps David and possibly Jasper should have access, at least to applicable information derived from ULTRA."

"That would be a wise move, in my opinion. Internal security information gathered by ULTRA has been circulated within MI5 without illuminating or even hinting at the source. I think we can both agree that knowing how reliable certain information is does sway the analysis, and I dare say there may come a time before this dreadful affair is concluded that we will need to move as quickly as humanly possible. Adding time to sanitize critical information might well be the difference between success and failure."

"My thoughts precisely. Please discuss this with Anthony and Stewart. If all of you agree, please come back to me with your joint recommendation. Also, if you both agree to give him access to the raw data, then I want a personal, private discussion with you and David, so that I can intimately convey the sensitivity of what he is about to gain access to in ULTRA. He is not to be given even a hint of ULTRA until I am assured he understands the gravity."

"As you command, Prime Minister."

The two senior ministers traded cordialities about family and friends before the Home Secretary departed. Winston truly appreciated and remained grateful for the excellent working relationship he maintained with senior Labour Party leaders like Clement Attlee and Herbert Morrison. They often made the difference, especially in these early days of his premiership when the two chief appeasers – Neville Chamberlain and Edward Wood – were full-fledged members of the War Cabinet and leaders of his own Conservative Party.

—

Thursday, 24.April.1941
No.417 Sudbury Hill
Harrow, London, England
United Kingdom
19:15 hours

The persistent drizzle all day had yielded an early release from Available status for No.71 Squadron. Brian had taken his time, showering, changing uniforms and heading into London. The journey aboard the London and North Eastern Railway line from Ipswich into Liverpool Street Station had been surprisingly quiet and uneventful, except for a handful of appreciative words and gestures from other citizens. The transit under the city had also been easy enough – Metropolitan Line to Rayners Lane, and then the Piccadilly Line to Sudbury Hill Underground Station, which oddly was above ground. The rain had persisted all day and into the evening, and convinced Brian to engage a taxi for the short ride to the Spencer residence. The last and only time he had been there was a year ago, in late February. Brian was thankful there was no snow on the ground, as there had been a year ago.

Brian recognized the house before the cab stopped. He paid the cabbie and bolted to the covered porch to avoid as much of the rain as he could. He quickly brushed off the drops on his shoulders and framed cover. As he turned to use the polished brass knocker below the large, brass numerals '417' on the black painted oak door, it opened. There was Mary with her infant child cradled in her right arm. She wore charcoal gray slacks and a loose white blouse with her luxurious brown hair pulled back into a tight bun.

"Come in out of the damp, Brian."

Mary stood aside as he entered, and then shut the door behind him. Brian turned to face her. She threw her left arm over his shoulder, pulled him toward her with the baby between them, and kissed him passionately. Brian was a bit surprised, did not pull away, and kissed her back. After what seemed like minutes to Brian, she released him.

"Now, that is the Brian Drummond I remember. I have missed you, my darling. Please, come in and sit with me." Mary led the way into the living room and motioned for him to sit. "Would you care for a beer or cocktail before dinner?"

"No, thank you."

"John called before you arrived. He is running a little late as usual, but he said he should be here in time. Blessedly, it will give us a little time to chat. It has been far too long since I last saw you – dinner with Uncle Winston and Aunt Clemmie at the Annex last November, as I recall. How is Charlotte? How is her pregnancy progressing?"

"Quite well, thank you for asking. She is due in June, they tell us."

"Does she know you are here?"

"Yes. I talked to her yesterday."

"Is she fine with this?"

"Yes, of course. She knows how important you and the air commodore are to me."

"Excellent. Would you like to hold our young son?" she asked, but did not wait for an answer. She stood in front of Brian and lowered the child into his arms. Brian could not take his eyes off the swaddled infant staring back at him. Mary returned to her over-stuffed chair opposite him. "He is gorgeous, isn't he?"

"Yes, he is."

"He has your eyes, Brian." She laughed. "John and I will have a time of it, explaining that little fact."

Oh wow! I had not thought of that. What on earth will they say? They both have brown eyes.

Mary did not wait for a response. "John told me you two had a nice pub meal and chat a few weeks back."

"Yes, we did. Why did you have to tell him? I was so shocked when he told me you had confessed to him."

Mary stared at Brian, as she considered her response. "Simply put, I love him and I did not want that between us. I am glad I did, even though it was uncomfortable for you, and I had told you our secret was safe with me. Virtually the first thing he said when he held Malcolm the first time was, 'Look at those blue eyes.' The baby has brought us closer. He has made it home more often since the birth."

"You could have warned me."

"John asked me not to do so, since he believed you would look for any reason not to see him. I hope you understand, Brian, and forgive me. I want us to remain friends. By the way, you didn't answer, how is Charlotte's pregnancy progressing?"

"From my perspective, quite well. I'm not sure if she would say the same."

Mary chuckled. "Quite understandable."

"She is still managing the farm and has never complained to me."

"When are you going to make her an honest woman?"

"She is," protested Brian.

Mary laughed. "I mean when are you two going to get married?"

Brian smiled. "I understood the first question."

"I didn't know you and Charlotte married. Did you tell John that news?"

"No, it never came up in our conversation."

"When?"

"Last Christmas . . . the 30th to be precise."

"She changed her mind about marrying a fighter pilot serving in war time?"

"I guess so, yes."

"Well, Charlotte knows about us."

"Yes, she does."

"And, she does not hate me."

"No, she doesn't. She has a very big heart, which is one of many reasons I love her."

The baby began to cry. Brian turned his attention to Baby Malcolm and began to gently rock him in his effort to sooth the child. By the time he looked up, Mary was standing in front of him with her right breast exposed. Brian immediately noticed the white drops on her nipple.

"He needs to be fed," she said and reached for him.

Brian raised the child to his mother. Mary purposely brushed her breast across Brian's hand as she took the baby. She positioned the child at her breast before she returned to her seat. Another woman, who Brian did not know, appeared from another room.

"I heard his cries, ma'am. Would you like me to feed him?"

"No, thank you, Grace. He is at breast and suckling quite well. If you don't mind, I will call you when he is finished, so you can change him and put him to bed."

"Yes, ma'am. I shall be close by."

"Thank you, Grace." Brian gestured to the other woman. "My apologies. I should have introduced you. I will when she comes back. We have hired a nanny to help me with Malcolm, in addition to our chef and housekeeper."

"Wow!" Brian was fascinated by her breastfeeding and remained amazed at her unabashed immodesty. *Hold her eyes! Hold her eyes!* Brian wanted to watch, but he instinctively knew he should not.

"Where were we?" Mary asked. Brian did not know where to begin. "Oh damn!" she exclaimed. With her free, left hand, she pulled the left side of her blouse back, exposing the left side of her nursing brassiere. A darkened spot was quite noticeable. Mary unbuttoned the flap and exposed her left breast.

She grabbed a nearby cloth and mopped up the dripping from her left nipple. "Would you help me?"

"What do you need me to do?"

"You could suckle this breast, so I don't leak," she said and laughed softly.

"I can't do that."

"It's quite alright, Brian. John does it for me when he can."

"He can do it when he gets home, then."

"Don't be such a prude, Brian. It does not suit you."

"I'm sorry, Mary. I can't help you that way."

"Suit yourself." Mary continued to periodically dab her left breast. After several minutes, she repositioned Baby Malcolm to her left breast, leaving her right breast uncovered.

Brian heard the front door open. He assumed it was John Spencer arriving home and stood to face the entryway. Air Commodore Spencer appeared. Brian saluted. John returned the younger man's courtesy gesture.

"Welcome, Brian. My apologies for being late." The two men shook hands. John went to his wife. "Good evening, my sweet," he said and kissed her, and then kissed the head of his son still at her breast.

He does not seem to care that her breasts are exposed.

"He is just about done, John. Please tell Harriett we will sit for supper in ten minutes, and also tell Grace she can take him shortly."

John nodded his head and left the room. Mary smiled at Brian and winked, her right breast appearing somewhat deflated. John returned. "Done." John removed his tunic and hung it on a coat hanger on a standing rack. "Wouldn't you be more comfortable without your tunic, Brian."

Flying Officer Drummond said, "Yes sir," and removed his tunic as well. John took it and hung it on another hanger on the opposite side of the rack.

Baby Malcolm was apparently sated and finished. He lay motionless, cradled in Mary's left arm. Both her breasts were now exposed. Grace appeared, lifted the baby from Mary's arms and went upstairs. Mary dried her breasts, re-buttoned both flaps on her brassiere, and then re-buttoned her blouse, as if nothing had happened.

Mary stood and gestured toward the dining room. "Shall we?"

The dining table sat eight, however, this evening, it was set for three. John sat Mary at the end on one side and motioned for Brian to take the place opposite her. Brian waited for John to sit, and then took his assigned seat. Almost on cue, another young woman appeared from the kitchen with a tray of soup bowls. Brian surmised she was the housekeeper, performing as server for the meal. She placed a bowl of fairly thick potato soup in front of each of them.

"I shudder to ask, John, but how was your day?"

"About the same, I'm afraid. We are prepared. After eight months of this nightly obscenity, it has become rather routine for all of us, except perhaps for the night fighter crews."

"Will they attack in this dreadful weather? How do they know where they are dropping their disgusting bombs, and how do they avoid running into each other?"

"They have not failed to bomb us in past episodes of foul weather. So yes, we expect them later this evening. We need to get Brian through the city before they come, as I am certain he has duty in the morning."

"Yes sir, quite so. The last train to Ipswich departs at 23:10 hours from Liverpool Street Station."

"We should have you out of here in plenty of time."

"You are welcome to spend the night with us," Mary added.

"Thank you, but no. I really must get back."

As the meal progressed, it was clear the Spencers were doing slightly better than everyone else with rationing, but the paucity of common items like pepper and spices was unmistakable. Despite the scarcities of rationing, it was a delightful meal. As they ate, John asked about Brian's meeting with Colonel Geiger. The two senior officers had met several times and knew each other. Brian recounted some of the conversation with Jeremy Morrison as well. Just from his choice of words, Brian recognized that John Spencer thought very highly of Roy Geiger.

Mary looked to her husband. "Our young charge informed me that he and Charlotte married last Christmas, or rather nearly New Year's Eve. They are also expecting their first child in early June."

"Well, that is some good news. Congratulations to both of you, Brian. Well done! She certainly impresses us as a very good woman."

"Thank you, and yes, she is."

"In addition to catching up with you, Brian," John said calmly, "we wanted to float the idea of inviting you and Charlotte to dinner here in a few months after your child is born and stabilized. The children are too young, yet, to know each other, but we would certainly like to know Charlotte better. Perhaps the three of you could spend the weekend with us."

"That is a very generous offer," Brian responded. "I must talk to Charlotte, when I can see her next, and God only knows when that will be."

"Yes, by all means. That conversation is best done eye-to-eye," John said. "We can wait until you have had your conversation with Charlotte. Once you give us your consent, we will send a proper invitation to Standing Oak Farm."

The meal was complete, including dessert. The after-dinner conversation wandered from the mundane to the ordinary, the improving spring weather, and trains operating on time for the most part.

Brian felt the advance of time and checked his wristwatch – 21:47. "I do not want to seem rude or ungrateful, but it is getting late," Brian said, feeling the need for fresh air. "I really must get across town, if I am to make it back tonight."

"Again, you are welcome to spend the night here," John offered.

"No, thank you, sir. I appreciate your generosity, but I do need to get back to Martlesham. Please excuse me." Brian stood.

Mary and John joined Brian in the anteroom, as he donned his tunic.

"Do you have an overcoat or a brolly?" asked Mary.

"No. I shall be fine. It is not far to the station."

Mary retrieved one of several umbrellas from the entryway stand. "Here, take this one. Don't worry about returning it."

Brian hesitated. She insisted. He took it. John embraced Brian and thanked him for coming. Mary embraced him tightly, and then kissed him on the cheek. They both told him to be safe.

John went through the blackout curtain with Brian and opened the front door. The cool air felt so good. The rain had stopped earlier, since a few dry spots could be seen on the walkway. Brian handed the umbrella back to John and set off down the hill toward the Underground station.

The night air still felt damp, but it was not raining and the coolness felt refreshing. Brian chose to stand outside on the platform rather than in the warmth of the brick station house. The desired train arrived five minutes later than scheduled and departed on time. Brian reversed his route back to his base in Suffolk. He remained quiet, withdrawn and pensive during the entire journey.

—

Wednesday, 30.April.1941
Oval Office
The White House
Washington, District of Columbia
United States of America
14:15 hours

"**W**hat is the latest from the Rad Lab?" asked President Roosevelt.

The President wanted another private meeting with Frank Knox and Henry Stimson to follow-up their similar meeting three weeks ago. He knew Harry Hopkins was close by, if he needed him, but he anticipated this would be a particularly sensitive exchange, and he wanted both leaders to feel at ease. He needed them to be as open and frank as possible with him and each other.

The Secretary of War smiled. "A month ago, their test equipment detected another aircraft at better than two miles, well beyond gun range, and allowed them to maneuver into a firing position, regardless of the initial aspect angle of the engagement. They have since improved that detection range. More importantly, they have engaged the Massachusetts company Raytheon to build an operational production unit based on the Rad Lab developmental design. The company claims they will have a test unit of the production design ready by the end of summer, and if everything works as expected, they can start producing field units by the end of the year. The joint development team with Raytheon have agreed on the form, fit and interface requirements, so that aircraft can be modified for quick installation when the production units are ready. The British are ahead of us in that compatibility effort."

"They need that equipment more than we do, at this point," Roosevelt said.

"Yes sir. All new night fighter aircraft under orders for the British already have the provisions."

"Excellent."

"The Germans have hardly missed a night since last September in their bombing campaign of London and other major cities . . . well, except for fitful and spurts now and then, as they endured two weeks ago. The British and our observers have reported that the bombing intensity has diminished since the Second Great Fire at the end of last year. The British engineers and pilots, who have worked with the new equipment, believe it will be a significant improvement over their existing units."

"I can't imagine what they have endured for the last year, but God bless them, they have stood to the line and kept the Germans at bay. Yet, spring is here. The weather conditions for a renewed invasion threat have returned."

"Yes sir, however the intelligence folks on both sides of the Atlantic believe the Germans have turned their attention to the east."

"Russia."

"It appears so. They do not have the amphibious shipping staged in the Channel area, or the necessary ground and air forces gathered for an invasion attempt. Thanks to our supplies, the British are nowhere near as vulnerable as they were last summer. It appears the Germans may have missed their opportunity."

"Hitler has always been more interested in communists than the English," Roosevelt observed.

"Quite so." Stimson paused. "One more bit of technical news of interest. The development team at Cambridge decided they needed a separate unit with slightly different characteristics for the anti-submarine forces. They were able to detect a submerged submarine with an exposed periscope; however, they decided to optimize the signal characteristics of their airborne radar for that task. They also have more power available on the larger anti-submarine aircraft, so the radar units will be bigger and have more range."

"Is there anything we need to do to help them get these new units to the field squadrons?"

"We repeatedly ask that question, and the answer is consistently no, and they will surely ask, if they run across any obstacles."

"Well, then, we should not interfere with success." Roosevelt paused, and then added, "I believe I am scheduled to meet Sir Hugh Dowding and Lord Halifax next month. I will be asking him the same questions."

"Good man," Stimson said. "None of us are particularly pleased with how he was treated last year. If we had to pick one man who was singularly responsible for the entire air defense system that maintained air superiority over England last year and prevented what appeared to be the inevitable German invasion, it would be Dowding; and then, to be summarily dismissed and shuffled off to the colonies. We are honored to have him with us, but still we have not received a worthy explanation for such a radical dismissal, but it was not our call."

"Agreed. I shall ask him about that as well, if the opportunity presents. So, I don't forget, have either of you heard Hoover's assessment of the America First Committee rally in New York City?"

"No sir," Knox and Stimson answered in unison.

"Edgar did not take kindly to Lindbergh's speech during a rather raucous gathering of Firsters at the Manhattan Center on West 34th Street. According to Edgar, 'Lucky Lindy' declared that England has already lost the war and our support is wasted."

"I wonder if the British know that?" Knox said. The three men laughed.

"Frivolity aside, this America First group has substantial support—Charles Lindbergh being perhaps the most prominent. The group includes the accomplished novelist Kathleen Norris, and a half dozen senators for Christ's sake. Anyway, Hoover is concerned they may be coordinating with the American Bund and through them with the Germans."

"Serious stuff," Stimson said.

The President nodded his head and turned to the table beside his wheelchair to pick up a folder containing a quarter inch stack of papers. He pulled the top document out and returned the folder to the table. "Now, let's turn to the object of this private meeting. It is truth or consequences time. You have had the Donovan report for more than a month. Henry has done most of the talking this afternoon, so if you will, Frank, why don't you lead us off on this discussion."

"As you wish, Mister President. Our intelligence professionals in the ONI have serious mixed feelings."

"I'm sure, but they are not the President of the United States, now are they?"

"No sir. Yet, we cannot ignore the reality that those professionals carry the bulk of the intelligence collection and analysis burden."

". . . for the Navy," and he gestured to Stimson, "and the Army."

"Yes sir. They do not see it that way."

"Then, who is going to tell me what Hitler's next move is? Who is going to tell me what the Japanese junta is thinking and why?"

"You are asking questions that have not been asked before," Knox answered, and then added, "to my knowledge."

"These are the times in which we live."

"We accept the need for a strategic, presidential level, intelligence apparatus. You were quite clear about the committee approach. The focus of ONI, at this point, is on the details of how this new entity, whatever it will be, works with ONI, G-2 and State. We have prepared a point paper regarding our perspective and concerns." Knox handed the stapled, half dozen, sheets of paper to Roosevelt.

"Good. I look forward to reading your thoughts. Have you shared your paper?"

"With Henry and the G-2, yes sir."

"With Cordell or Bill Donovan?"

"No sir. We wanted you to pass on it first."

"Thank you for that courtesy, Frank, but please share it with Bill and Cordell, as soon as we are done here."

"As you wish, Mister President."

"So, before I have the benefit of reading your position, please give me the bottom line. Do you see any insurmountable obstacles to a strategic, presidential level, intelligence agency?"

Knox held the President's eyes and took several seconds before he answered. "No sir."

"But, you have reservations?"

"Yes, Mister President, I think that is safe to say."

"OK. Fair enough. Your turn, Henry."

"The G-2 has prepared a point paper as well," Stimson said, as he handed a similar document to the President, "and as you directed, we will share it with the interested parties."

"Excellent. So, I have my reading material and homework to do. Your bottom line, Henry."

"As you will note in our paper, we suggest the new agency operate under the Chief of Staff, since the intelligence apparatus already exists and is fully functional. We can manage it for you and ensure the proper coordination will be accomplished."

"Oh, of that, I am quite certain. So, if I understand your position and that of the G-2, you want to filter and control the information I am given." President Roosevelt stared intently at Stimson. He did not blink or even twitch. He wanted the full effect of his stern gaze to sink into both of them.

"I would not phrase it that way, Mister President."

"Perhaps not, but that is the essential result, is it not?"

"Again, that is not our intent, Mister President. We are only seeking efficiency of operation, sir."

"OK. I am making no accusations. So, for clarity in my mind, do you or the G-2 see any insurmountable obstacles to Donovan's proposal?"

"No sir."

"Very well. Do either of you have anything else you would like me to know?" Both men responded simultaneously in the negative. "Fine. You have had your opportunity to speak. Now, I shall have mine." Roosevelt paused to look into the transfixed eyes of his two military service secretaries. "The status quo, or more specifically any attempt to maintain the status quo will not be looked upon kindly by me. We are going to try Donovan's way, and I need both of you to help him make the new organization work. More importantly, I want it emphatically understood and appreciated by the ONI and G-2 that I shall not tolerate their interference or that of their staffs. The present structure, or anything remotely similar to it, is, simply put, a non-starter. That said, here is what we are going to do. I want the two of you, not your representatives, along with Cordell, to meet with Donovan as soon as the four of you agree upon a time and place . . . and I do mean soon, like this week, or next at the latest. I would prefer just the four of you, initially, and then you can include your intelligence chiefs, as you deem appropriate. This will not be an exercise in wearing Donovan down. Am I clear on this point?"

"Yes sir," they said in unison.

"You are to assume Donovan's proposal will be implemented. For now, since none of the naming suggestions have struck me, we shall call his new organization the Office of Information, and his position will be the Coordinator of Information. The COI and his organization will report directly to me – no filters, no go-betweens, and no layers of bureaucracy. We are going to see how this works. We can adjust once we see how it works. All four of you have two weeks, four at the most, to iron out your working procedures and any other necessary details. I want to implement this information agency yesterday. I have been flying largely blind, and I remain at a distinct disadvantage to Churchill. I have been forced to rely upon his perspective, since he has the information required to substantiate his positions. While I respect Winston, I am not doing my job properly without the strategic information I need." Roosevelt paused. "Do either of you have any objections to the task as I have outlined it?"

"No sir," they again responded together.

"Oh, and one more item . . . the COI will be equivalent to a cabinet secretary, equal to you," he said, pointing to both men, "not at the level of the G-2 or ONI."

"Does Congress need to pass off on that aspect?"

"Good question. Eventually, yes, as it should become a standing agency, funded by general Treasury monies. However, in this probationary period, I will fund the organization out of my discretionary funds. I intend to issue a classified presidential order by June, or July at the latest, to initiate this new agency. Lastly, I will argue that you, each of you in your capacity as Secretary of the Navy and Secretary of War, need this new organization as much as I do, you just may not realize it, yet."

"Henry and I both agree, Mister President. We are not arguing against a strategic intelligence organization, but as they say, the devil is in the details."

"Yes, then, hammer out those details quickly. The British are so far ahead of us in a number of areas . . . this being one . . . and we need to catch up quickly, before we are inexorably drawn into this fight."

"Yes sir," responded Knox. "We will take your mandate and get you a joint solution, hopefully before the end of next month."

"Very well. I am counting on it. Now, I do believe we are done here." Both men stood. "Please send in Harry on your way out. Thank you for your time, gentlemen."

"Thank you, Mister President," Knox said.

They had nearly reached the main office door, when the President said, "I nearly forgot . . . one more topic, if you will." Both men stopped and turned.

"Yes, Mister President," Stimson said.

"What do we know of the situation in Greece and the Balkans?"

"Oh yes, my apologies," Stimson answered, "we have been focused on other things. We received a message this morning from our military attaché in Athens—the latest information we have. I immediately called Cordell. He received a similar message from Ambassador MacVeagh. As you know, the Germans attacked Yugoslavia and Greece on the 6th. Events have moved very quickly. The Yugoslav government surrendered to the Germans on the 17th. Greek Prime Minister Alexandros Koryzis committed suicide on the 18th, as the invasion pressed home. Chief of the General Staff and Commander-in-Chief of the Greek Army General Papagos resigned on the 26th. The remainder of the Greek government capitulated to the Germans on Monday . . . upon signing a dictated armistice, similar to that imposed on the French last year, from what we understand. Cordell might have more by now on the diplomatic side."

"None of that is good for the British situation in the Mediterranean . . . in addition to being on the defensive in the desert on the Egyptian frontier," added Knox.

"Indeed," Stimson continued. "The British began the evacuation of the Commonwealth Expeditionary Force, commonly referred to as 'W' Force, on the 24th, and they will probably close the effort today or tomorrow, as the Germans approach. The destruction of the port facilities at Pireaus at the beginning of the invasion has made the process much more difficult and fragmented."

"They are very proficient at evacuations," observed Roosevelt.

"With all due respect, Mister President," Stimson responded, "the British have been stretched extraordinarily thin from the outset. They have tried mightily to defend their allies against the Germans, even at the risk of their own survival or viability. Churchill was under enormous pressure within the government and the military to not transfer precious divisions from Egypt to Greece, but he pushed it through. We also received word from MI6 yesterday . . . the Germans appear to be at the end of the planning stage to execute an airborne invasion of Crete in the next few weeks."

"Is there anything we can do to assist?" the President asked.

Stimson looked to Knox, and then answered, "Not much more than we are already doing."

"Agreed," added Knox. "We are already pressing the boundaries with Lend-Lease. We are diverting precious supplies to the British that we need for rearmament and mobilization."

"I've not heard from Winston with all this bad news from Greece, but I expect I will any day now. Please keep me informed. I'll be talking to Cordell later today, so I'll get his perspective. Thank you for your assessment."

"Yes sir," the two ministers replied together.

The two service secretaries left the Oval Office and closed the door behind them. Roosevelt wheeled himself back behind the large oak desk. "This should be interesting," he mumbled aloud to himself. Hopkins appeared and closed the door behind him. Franklin recounted the discussion's content, and directed Hopkins to note the direction and deadline he had given Knox and Stimson, so that appropriate follow-up queries could be made to ensure satisfactory progress was made.

—

Chapter 3

A man without a stick will be bitten even by a sheep.

-- Hindu proverb

Monday, 5.May.1941
Standing Oak Farm
Winchester, Hampshire, England
United Kingdom

Brian had not seen Charlotte since his visit on his return from Wichita, five weeks ago. Her abdomen had grown quite a bit since his last visit. Brian could not keep his hands off of her. He thoroughly marveled at the curves and tightness of her belly.

"He kicked you!" Brian exclaimed on feeling the distinct strike.

Charlotte laughed. "He's been doing that for a month or more. He's a very active baby, so I'm told, especially when I am trying to get some sleep. I have to catch as much sleep as I can when he is sleeping. Perhaps, it is nature's way of preparing me for the baby's infancy."

"You are a month away, I think."

"That is what Mrs. Grey claims. I saw her most recently on Thursday last."

Barbara Grey had been retained by Charlotte several months ago to assist her with the birth of their first child. She was a middle-aged, matronly, kindly woman, who was highly recommended and an experienced mid-wife as well as a qualified nurse.

"He kicked, again," Brian said with some surprise.

"He is an active one. I am ready for this pregnancy to be done, and yet, I must admit to some degree of apprehension about caring properly for our infant child."

"I have no worries, Charlotte. You will be a great mother."

Charlotte shot him an icy glare. "Of course you have no worries; you are off flying your damnable machines, while I am here feeling like a bloated cow."

Brian opened his mouth to tell Charlotte about Grace Perkins, Mary Spencer's wet-nurse nanny, but he caught himself, not wanting to open that topic, just yet. "We could hire a nanny to help."

Charlotte laughed at what she undoubtedly thought was a ridiculous suggestion. "We don't have that much income, Brian. We must be careful with expenditures at least until the war is over and we can stabilize things on the farm."

"I can draw on my parent's cash funds until my inheritance has cleared probate. I am far more concerned about you. I want to make sure you are well taken care of and have all the help you need while I am away."

"Thank you, Brian. You are a good man. I am proud to have you as my husband. We shall see. I need to get through the impending birth and see how things go."

"Sure. How is the young boy Jacob working out?"

"Quite well, actually. He's a hard worker and learns quickly. A year of apprenticeship will position him well, to work and eventually manage the farm for us, but in a few more years, he will be of conscription age, and we may lose his employment to the war as well."

"Well, then, we must win this fucking war before then," he nearly spat.

"Brian!"

"Oh, sorry, Charlotte, the vulgarity of my American squadron mates is rubbing off on me, and I forget sometimes to watch my language."

"They really talk like that?"

"Some of them, yes. I try to avoid it, but when you hear profanity all the time, it becomes normal."

"I don't mind profanity myself, but we shall soon have children around us and we must mind what words we choose."

I like that . . . children . . . meaning more than one. "Certainly. Again, my apologies."

"I did like your sentiment, but how realistic is that possibility."

"Well, Fighter Command has started taking the fight to the enemy in France. We are seeing more American military uniforms in London, and they are forming another American volunteer fighter squadron – One Two One Squadron – at Kirton-in-Lindsey. They will be declared operational later this month, they tell me."

"It is nice that more of your countrymen are helping out over here."

"They are your countrymen, now, as well."

Charlotte winked. "There is that, isn't there."

"I'm pleased young Jacob is working out. I'm sure his energy is helping Horace and Lionel, as well."

"Oh my, yes. They can't stop talking about how well he is doing. I just hope he finds satisfaction here and stays with us, at least until the baby does not require so much attention."

"We will try to keep it that way."

"He has asked numerous times about you. Your accomplishments have apparently circulated among the local populace. The local schools have asked if you can speak to their students. You are becoming somewhat of a mythical legend, it seems."

"Nonsense."

"I am just the messenger here. I know he would be thrilled to talk to you . . . some time when it is convenient."

"Sure, but I don't want to give him any ideas."

"You mean like Malcolm Bainbridge gave you?"

Brian laughed hard. "Exactly. We don't need him running off to join this fight with some idealistic notion of nobility and lofty purpose, now do we?" He laughed, again.

They heard the front door open and close. Horace Morgan appeared through the blackout curtains. "We are finished with the afternoon chores, ma'am," he announced.

Charlotte rose slowly and waddled to the narrow, rectangular table beside the fireplace. She retrieved three envelopes and handed them to Mister Morgan. "Thank you, Horace. Here are your pay slips for last week. Please convey my gratitude to Lionel and Jacob."

"I will, ma'am. Thank you. That young Jacob is quite the lad. We've seen nothing like him in quite some time, now." Horace turned to Brian. "He is quite the fan of you, Mister Drummond. He talks about your accomplishments all the time."

"Thank you, Horace," Brian responded.

"Now, if you will excuse me, we shall be on our way. We shall be here for morning chores. Thank you for our pay, ma'am."

"Good evening, Horace."

"Good evening to you, sir."

The blackout curtains settled, and the front door opened and closed.

"How long do you have with me, this time?"

"I have to head back in the morning."

"Would you care for some tea and fresh biscuits?"

"Sure. That would be lovely." Charlotte waddled off to the kitchen.

Charlotte returned, poured a cup of tea for each of them, and handed the plate of oatmeal cookies to Brian before she sat down. They enjoyed their tea and cookies, while they talked about the farm as well as the plans for what he would soon hold in ownership in Wichita. Charlotte asked Brian to describe the city and his childhood on the Great Plains of Kansas. They even discussed the history and the aviation centricity of so many pioneers of aviation in the area. They did not talk about the war, or Brian's participation in the war. Charlotte just never felt comfortable talking about his work. Brian imagined that time in the distant future when they might split their time between Winchester and Wichita, but that was a truly distant future, since there was no end in sight. Germany still occupied and dominated all of Europe from the Atlantic Ocean to the River Bug. Brian saw no signs that situation was going to change any time soon. For Brian, it was good that Charlotte was thinking about Wichita.

Charlotte had apparently reached her threshold, when she stood and reached for the tea service. Brian jumped to his feet and grasped the tray of tea service and remaining cookies, and carried them back to the kitchen. He helped Charlotte wash the cup, clean out the teapot and put away the cookies.

"Now, my darling," she announced, "I may not be able to milk a cow these days, but I can certainly milk you."

"Charlotte!" protested Brian.

"Oh stop the false modesty, my husband. These are perfectly natural processes. Barbara tells me I must help you keep those fluids flowing. I am no longer comfortable taking you as I want to take you, but I can certainly take care of you."

They were off to bed. Brian could not deny his excitement regarding her suggestion. *Who am I to argue with the lady of the manor?*

—

Tuesday, 6.May.1941
Cabinet Conference Room
The White House
Washington, District of Columbia
United States of America
14:15 hours

The President was running late to the schedule in what amounted to a beginning-to-end, action-packed day of meetings and ceremonial events that would run into the evening. Roosevelt had eaten a sandwich with a large glass of orange juice at his desk in the Oval Office for lunch.

The designated attendees for the scheduled meeting had gathered in the Cabinet Room prior to the appointed hour. The subject of this particular meeting, Chief of the Army Air Corps General Arnold, stood in a small group. Others were clumped together and chit-chatting with the essential players: Secretary of War Stimson, Secretary of the Navy Knox, Secretary of State Cordell Hull, Secretary of the Treasury Henry Morgenthau Jr., Army Chief of Staff General Marshall, and Chief of Naval Operations Admiral Stark. Arnold's trip report on his visit to England had been distributed to everyone in the room and presumably had already been read by each of the principals.

The various muffled conversations ceased immediately when the side door opened. Only General Arnold came to a position of attention. Harry Hopkins pushed President Roosevelt in his wheelchair to the center of the long, rectangular, polished mahogany table, and then sat in a periphery chair behind the President. Major General Edwin Martin 'Pa' Watson, USA [USMA 1908], Senior Military Aide to the President since 1933, followed the President and Hopkins, and shut the door behind him. Watson chose a corner, peripheral chair nearest the door.

"Please be seated, gentlemen," the President commanded.

The cabinet secretaries took their usual seats for cabinet meetings, which spread them around the table. The three flag officers took chairs opposite the President, with Arnold in the center, Marshall to the right and Stark to the left of Arnold.

"Welcome home, General Arnold," said the President, once everyone was settled.

"Thank you, Mister President. It is good to be home."

"I think we have all read your excellent report. Let us start with your summary. Then, I have a few questions, and the others may as well." Roosevelt nodded to Arnold, indicating he had the floor.

"Before I begin, I would like to acknowledge the gracious generosity of Prime Minister Churchill, His Majesty's Government, Sir Charles Portal and the whole of the RAF. They made my mission immensely easier to accomplish."

Heads nodded but no words were offered.

"I had several tasks to complete during this visit. We accomplished all of those tasks.

"First, the Royal Air Force is today stronger than it has ever been in history . . . in every respect, I must say. While they may not yet be equal to the German Air Force in quantity and combat experience, they are rapidly growing, gaining the necessary experience, and more importantly, they are implementing technological improvements at an astonishing pace.

"Second, the fighter force has made up all of its losses from last summer and then some. They have added new squadrons, especially night fighter squadrons that are becoming more proficient with each sortie in their prosecution of the night intercept problem. I understand from Secretary Stimson that you have received the latest Rad Lab development results, so I will not duplicate that status here. Let it suffice to say, the British will put the new equipment to good work. I might add here that we have a number of young aviators assigned to a few of their night fighter squadrons to learn firsthand the problems and lessons learned. Air Marshal Douglas and I have agreed to cycle experienced pilots through his night fighter squadrons, so that we can quickly train our pilots in night fighter operations. I doubt we will face the same problem as the Brits face with the terrible night bombing attacks they call The Blitz; however, I am certain we will need that capability, perhaps sooner than we think."

"Good thinking," the President interjected.

"Before I leave the fighter topic, I would like to footnote a conversation Air Marshal Douglas and I had regarding the American volunteer pilots in Fighter Command. He candidly confided in me his displeasure with their performance."

"Really. This is the first I've heard of that. How so? Isn't that young Kansas boy part of that group?"

"Yes sir, I think you are referring to Flying Officer Brian Drummond, quite an accomplished pilot for his age, but he is only one of the surviving volunteer pilots." Arnold looked to both sides of the table. "He is nearly an ace four times over and just turned 20 years old, last month. The King has decorated him twice, so far," he returned his eyes to the President, "but he is not a squadron. My first reaction to Douglas's statement was, send 'em home, if they are not performing satisfactorily. Yet, as we talked, his frustration appears to be largely, if not solely, due to a general lack of basic military training for those volunteer pilots. They are good pilots, but they were not taught to fly as a unit, to fight as a unit. They have come around, more slowly than desired, but Seventy-One Squadron was declared fully operational and moved into Eleven Group just before I arrived. That is their front line fighter group for defense of the Home Islands and taking the fight to the enemy in France. They are forming another squadron of Americans later this month, and they have learned from their experience with the early pilots."

"So, things are settling down in this matter?"

"Yes sir. We have also learned from their Eagle Squadron experience."

"Is that what they call them?"

"Yes, Mister President. Seems appropriate to me."

"Indeed. Continue."

"Third, I also spent time with Bomber Command, Coastal Command and Training Command. We have worked closely with the latter command to ensure the training we are now delivering under the Lend-Lease Act meets their requirements. Of all the sectors of aviation operations, we are the closest in training. Their Coastal Command is more akin to our Navy's anti-submarine air operations, and I know Admiral Stark has good representation with them." Betty Stark nodded his head in agreement. "Bomber Command learned early in this war that daylight operations are very costly. They have focused almost exclusively on night bombing operations on the Continent, including into the heart of Germany. They are using a dead reckoning process to reach their targets. I heard more than a few in government, who are beginning to question the accuracy of their night bombing claims."

"What are we going to do?" asked the President.

"Good question, sir. First, to take full advantage of the accuracy of the Norden bombsight, we need daylight and decent weather – at least broken cloud cover or better. The bombardier must see the target. Second, our heavy bombers, the B-17 and B-24, are heavily armed and can defend themselves against fighter attack. Third, our guys supporting the Rad Lab development work and in direct conjunction with the Brits up there are working on ground-mapping radar techniques to aid the Brits in bombing more accurately at night, and for us to bomb in bad weather."

The President had apparently had enough aircraft talk and moved on. "What of the invasion threat, General?"

"Yes, Mister President. We did discuss that question with the government – the Prime Minister and relevant cabinet ministers – with the operational folks and with the intelligence agencies. They see rapidly mounting evidence the Germans have turned their attention away from England and toward the Soviet Union. They are not yet ready to step back from or soften their invasion defense provisions, but they are certainly looking for that moment. Based on what I saw, I learned and I think of the available information, I believe the invasion threat is approaching zero . . . at least through this summer. Fall and winter weather preclude safe amphibious operations across the English Channel. The Spanish Armada learned that lesson the hard way."

"I'm sure relieved in that. What about The Blitz?"

"The intensity of night bombing of British cities has steadily diminished since the first of the year. The Germans are moving fighter and bomber assets from France, back to Germany for rest and refit, and then into Poland, which is one of the contributing elements of information pointing toward the Soviet Union."

"Should we warn Stalin?"

"That is a political question, well beyond my pay grade, Mister President."

"I am asking your opinion, General?"

"I believe that task, should it be deemed appropriate, should fall to the British, as they are most directly engaged, and thus would have the best credibility. As such, a conversation between you and Prime Minister Churchill on that particular question would be quite appropriate."

President Roosevelt simply nodded his head and gestured for Arnold to continue.

"Lastly, we have engaged with the RAF at all levels from Air Chief Marshal Portal to the cockpit of individual aircraft. Thanks in no small part to their generosity, we are absorbing their experience and learning at a very rapid rate. Most importantly, we have opened intimate lines of communications with all levels. Experienced British combat pilots have begun to arrive in this country to help shape and manage the training programs, and their ambitious acquisitions effort. At the same time, we are learning and improving."

"Excellent," pronounced the President. "I suspect we shall be in need of those very same skills in our intrepid aviators. Learn well!"

"Yes, Mister President."

"Do you have anything else for us, General Arnold?"

"I can go into as much detail as anyone may wish, but that should suffice as a summary. Oh, excuse me, Mister President, yes, there is one more item." Arnold waited for the President's gestured consent. "Prime Minister Churchill insisted that I convey personally the enormous benefit His Majesty's armed forces have gained from your support and the materiel amplification of the Lend-Lease Act."

"Thank you, General. I shall write him this afternoon," he said, looking over his shoulder and gesturing to Hopkins, who in turn nodded his acknowledgement, "and confirm for him, message delivered."

"Mister President," interjected Henry Stimson. The President recognized his secretary of war. "If I may, in a message such as that, it might also be appropriate to illuminate once again our enormous gratitude for Sir Henry Tizard and his team, and the Technical Exchange Mission initiated by the Prime Minister. Our technological advancement has progressed by orders of magnitude and far more rapidly than would have been possible without them. We are both much stronger together, as a consequence."

"Hear, hear," added Frank Knox.

"Very well, then. I shall add a short paragraph to that effect, as well. Anything else?"

"No sir," several said, while others shook their heads in the negative.

"One item, if I may, before we adjourn . . . how are we progressing in the reconciliation of the air forces command structure?" asked Roosevelt.

"We will meet your deadline of June, Mister President," answered Stimson. "Unless you wish to affect this change via an executive or military order, Mister President, I shall issue an Army regulation, to consolidate the Army Air Corps and Headquarters Air Force Combat Command under a new organization called Army Air Forces. Per our earlier agreement, the chief of the Army Air Forces will report to the Chief of Staff and function as a near autonomous service equivalent to the Royal Air Force."

"Excellent. I look forward to seeing the regulation published. I see no need for an executive order. Are we also in agreement as to who shall be the new chief?"

"Yes, we are, Mister President," Stimson responded.

"Then, please allow me to recognize our soon-to-be new chief of the Army Air Forces, General Henry Hap Arnold. Congratulations, General."

"Thank you, Mister President. I am humbled and honored."

"We shall see to your promotion as quickly as possible through the Senate. Won't we, Harry?"

"Yes, Mister President," Hopkins answered. "We most assuredly will."

"Once the Senate has confirmed your promotion, I shall ask for the privilege of pinning your new stars on myself in the Oval Office."

"We'll see to all the arrangements, Mister President," Stimson said.

"Thank you, Henry. Well done on your mission, General, and again, congratulations. We look forward to great things from you and your air force."

"Thank you very much, Mister President."

"Now, if you will excuse me, I must be off to my next appointment."

Everyone stood. "Thank you, Mister President," several men said, not quite in unison.

Each of the men shook hands with and congratulated General Arnold on their way out of the Cabinet Room. General Marshall was the last to congratulate General Arnold.

"We're not going to let that 'autonomous' part get too far afoot, now are we Hap." Marshall said, as a statement rather than a query.

"No sir. I'm just thankful to finally have this Combat Command nonsense resolved."

"You and me both. It's a nice spring afternoon, Hap. How about a walk back to the Munitions Building? I'd like to discuss a couple of innocuous loose ends on our way."

"I'll need to dismiss my driver."

"Both of us. Then, let's be off."

The two military leaders departed the Cabinet Room, leaving the door open to signify not in use. They made their way through the West Wing to the parking lot between the White House and what used to be the State-Army-Navy Building, but now belonged solely to the State Department, having been squeezed out in 1939. Both their drivers departed in their assigned Cadillacs. Off they went at a rather leisurely pace, walking side-by-side, with General Arnold on General Marshall's left. They were followed several paces behind by their aides-de-camp – an Army lieutenant colonel and an Army captain, respectively.

—

Thursday, 8.May.1941
RAF Martlesham Heath
Woodbridge, Suffolk, England
United Kingdom

The day's operations order called for No.71 Squadron to remain at Readiness alert status during a RODEO fighter sweep of Northern France. They were assigned the back-up role for the assigned fighter cover squadron. All of the pilots wore their flying equipment, so that they could respond quickly should the call come. The overcast sky meant they would have formation cloud penetrations to perform, if they were launched and for their recovery. Rain was not forecast, so several pilots relaxed in lounge chairs outside the Dispersal Hut as they waited. After two hours with no change, the grumbling began among the less experienced pilots. Hunter, Whitey and Pete ignored the complaining at least for the time being.

The distinctive telephone ring shot all the pilots bolt upright. All heads turned to Corporal Harris on the operations desk. "Skipper, Group on Blue," he shouted, indicating the commander had a call on the blue colored telephone.

Just as the pilots began to release their wound up tension, Billy appeared at the doorway. "Inside fellas," he said. Hunter and the others joined the rest of the squadron pilots inside their Dispersal Hut. "The RODEO mission has returned without incident, so our coverage mission is not required." Taylor paused. No one reacted. "However, Group just picked up signs of what appears to be a developing daylight raid gathering over Northern France." Billy glanced at the wall clock. "We should expect to be brought to Standby in a few minutes for a potential intercept mission."

"Just to wait some more?" protested Bulldog.

Hunter shot him a disapproving glance.

"Wow," Dusty interjected. "We haven't seen a daylight raid since last October, and that would be September, if we excluded the fighter-bomber attempts."

"That is our . . ." The telephone ring interrupted Billy.

"Squadron to Standby," announced Jimmy Harris.

"There you have it, guys. Let's mount up."

The pilots filed out of the Dispersal Hut. Some of them donned their leather helmets with earphones, and with oxygen mask and goggles attached.

Others carried the last of their flight gear. Each pilot checked in with his crew chief, and strapped into his parachute harness and seat. Once Brian had positioned and checked all his switches were set for start, he looked to his left, and then right to get a thumbs-up ready sign from his wingmen. Once satisfied, Brian held up his left thumb to signal Billy Taylor that Green Section was ready to go.

Brian decided to take advantage of an unknown duration wait time as they sat in their fighters. He leaned his head back against the seat headrest and closed his eyes. Brian had not reached slumber when the scramble bell rang outside the Dispersal Hut door. He immediately shot to alertness, flipped his battery switch on, cracked his throttle and fired off the big Merlin engine.

On signal from their leader, the squadron taxied forward a short distance, turned into the wind and took off across the grass field. They all cleaned up their configuration quickly and tightened up their positions as they quickly approach the cloud deck. Cloud penetrations were never fun and especially as an assembled squadron of twelve aircraft. Blessedly, the overcast layer was only a couple of thousand feet thick, and they soon broke out into nearly clear, sunlit blue skies.

Billy checked in with the group controller. They received their initial intercept vector. Without explanation, they were commanded to stop their climb at 15,000 feet and orbit at their present location and maximum endurance power. Billy put the sections in line to make the process of circling in racetrack orbits over the Kent coastline a little easier for the squadron.

Brian spread out his section and hand-signaled his wingmen to scissor maneuver behind him from the inside to the outside of the turns and vice versa as the turns continued. He did not count the circuits, and he kept his eyes searching for specks on the horizon that might be growing in size. The sky remained clear, although most of England remained covered in a low cloud layer. Brian checked his wingmen as well as the section ahead and behind him periodically, but spent most of his time scanning the sky for enemy aircraft.

Operational communications were by their very nature minimalist—only the essentials. They orbited for what seemed like an inordinate amount of time. They still had plenty of fuel with their Merlin engines turning over at very low power, plus they had numerous divert fields between their current location and home base. They could sustain this loitering orbit for an hour or more, if required. The window of their combat viability would not last that long—perhaps another 30 minutes at their present fuel state. Brian's mind began to wonder how long Billy Taylor would continue to resist asking the sector controller what was going on?

"Eagle Leader, this is Lumba calling."

"Lumba, Eagle Leader, go ahead."

"Return to base."

"Eagle Leader, roger, return to base." Billy continued the orbit to the northbound leg, and then adjusted his heading for RAF Martlesham Heath. "Eagle Flight, Eagle Leader, join up for penetration."

Brian hand-signaled for his wingmen to join up in the usual 'V' formation, as he maneuvered to his assigned position off the right wing of Billy Taylor's Blue Section Right Wing—Pilot Officer Peter Bruce 'Horse' Harrow of Cheyenne, Wyoming. They still had a way to go, as they picked up speed in their cruise descent. Hornchurch Sector Control passed flight control to North Weald Sector Control.

"Cowslip, this is Eagle Leader checking in with you."

"Roger, Eagle Leader, we have you. Vector zero one two for approach to home base. Continue descent."

"Eagle Leader, roger, zero one two, continue descent."

The controller made one more heading adjustment, and Brian also checked to see his wingmen were in their proper positions before they entered the cloud layer. Once in the clouds, each pilot's concentration has to be completely focused on maintaining his position smoothly and precisely, so that his wingmen would have an easier time in their position maintenance task. The controller reported the ceiling in the vicinity of RAF Martlesham Heath was 1,200 feet with visibility of five miles.

When they broke out below the clouds, Brian's altimeter indicated 1,260 feet and the airfield was directly ahead of them. Not bad! Billy continued their descent and landed in the grassy landing area as a squadron in position. They taxied to their assigned positions, parked, and shutdown their aircraft.

Crew Chief Corporal Henry Joyce Jacobs from Birmingham stood on the left wing root just forward of the open canopy, as Brian unstrapped from his seat harness and parachute. "No action, ay Mister Drummond?" Henry asked.

Brian unplugged his radio cords and oxygen mask hose, and then removed his helmet. "Nope. Just more hurry up and wait for the Eagle Squadron."

"That is fine by us," Jacobs said. Brian could hear the result of the crew laughing. "No holes to patch or ammunition belts to load."

Brian nodded his head in agreement. He looked up at Henry. "Quite so. Just petrol today. The aircraft is in perfect shape."

"Just the way we like it."

Brian stood. Jacobs stepped back on the wing toward the leading edge and kept his hands clear of the still hot engine exhaust ports. Brian stepped out of the cockpit and onto the left wing root. The crew rigger, Leading Aircraftman Stephen Hawking from Sheffield, and crew armorer, Aircraftman Stanley George Easton from Blackburn, north of Liverpool, stood behind the left wing.

"Gunport tapes still in place," Easton observed.

"Yep, not a shot fired, or even an opportunity realized," Brian responded.

"We'll fill her up and check her fluids," added Hawking.

Brian shook hands with each man and Corporal Jacobs when he jumped down off the wing. "Excellent. She's a perfect bird, lads."

"As we like it," answered Jacobs.

The crew did not wait for more conversation. The fuel bowser stopped in front of the 'XR-G' Hurricane Mark II fighter. Hawking jumped up on the wing, opened the fuel tank cap, and placed the fuel hose nozzle held up for him by Easton into the tank. Brian knew they would also check the oil and coolant levels as well. Satisfied all was well, Brian returned to the Dispersal Hut. Mission debriefing of the pilots by the intelligence folks was already underway. Brian would have his turn shortly, since the debriefing would be quite simple—we went, we loitered, we came home.

The lesser-experienced pilots had already gathered outside the door to grumble and complain about not seeing action. Brian went inside, hung up his flight equipment on his assigned wall peg, and then went back outside. He found an open lounge chair, reclined and closed his eyes. He ignored the bait cast out by several of the pilots seeking him to engage in their frustration. There was no point to correcting their myopic opinions. He was confident their combat time would come in due course, and they would eventually join the ranks of the veterans who appreciated the times they were not being shot at by the enemy.

—

Sunday, 11.May.1941
Ditchley Park
Enstone, Oxfordshire, England
United Kingdom

As was often the case during the war years, Winston Churchill sat in bed, having eaten his breakfast, and now working on his incessant paperwork with his duty private secretary – today being Jock Colville – along with the relays of stenographers, taking down his orders, messages and notes. A knock at his bedroom door interrupted the process. Colville went to the door, opened it slightly, and whispers were exchanged. Colville turned to face the Prime Minister. "Sir, Director General Petrie of the Security Service has arrived unannounced and asked to see you on a most urgent matter."

"Very well, show him in please. We are done for this morning," Churchill said and offered a shoo gesture. The stenographers returned their chairs to their proper place and departed. Colville followed them, closing the door behind him, to go retrieve the chief of MI5.

A few minutes later, another knock preceded Colville's re-appearance. "Sir, Director General Petrie," Jock announced and ushered him into the Prime Minister's bedroom. "I shall be right outside your door, sir."

"Thank you, Jock." Churchill did not wait for the door to close. "Pull up a chair, David, and let's hear about your urgent matter."

Petrie lifted a chair from the nearest wall to a position next to Churchill's bed and facing him. "Last night, a man parachuted into the hilly terrain of Southern Scotland. He was captured almost immediately. The local constabulary initially held him, and the Security Service took custody of the man early this morning. A field exploitation team from Farnborough is enroute to the crash site, which has been located."

"So far, David, this is nothing new. We have suffered parachute infiltrators since before the war began. Who is this mystery man?"

"He claims to be Walter Richard Rudolf Hess."

"Dear God above, you must be kidding – *der Stellvertreter Führer*, the deputy leader of Nawzee Germany. That is equivalent to Clement Attlee flying solo to Germany."

"So he claims. He has also insisted that he will talk to no one other than you."

"Although my curiosity is piqued, that encounter is not going to happen. I have no intention of giving him the dignity or recognition of such an audience."

"My opinion precisely, Prime Minister."

"What do you intend to do?"

"Just to be clear, I must say at the outset, the aircraft investigation team will have several of our cognizant field agents with them as they examine the wreckage for any clues. One of my agents at the site now believes the aircraft may have been a BF One Ten – the two-seat, twin-engine fighter. The aircraft was pretty much destroyed by the crash impact and consumed by fire, but there were no obvious signs that anyone else was in the aircraft. Until we determine otherwise, we have no choice but to assume he had a collaborator, perhaps the pilot, who may have parachuted as well. We are working with the local constabulary on a rigorous search of the area. The man claims he was alone, but we cannot prove that, just yet."

"Quite right. Good luck with the search. Now, what about the man?"

"He is being moved to one of our secure facilities north of London. He will be kept isolated and incommunicado. Our interrogation branch is assembling a team, as we speak, and an interrogation plan based on the assumption he is telling us the truth. The only other thing he has said so far, other than his full name, is his mission was to deliver a message of peace and reconciliation directly to you from Adolf Hitler himself."

"Unless his message is to accept the unconditional surrender of Germany and relinquishment of all occupied lands, I have no interest in listening to him. Hopefully, you will be able to extract the message in due course. Do you really believe it is Hess? I've actually met the man, you know, before the war, and I do believe he is a qualified pilot."

"I did not know that, Prime Minister. That might prove useful, but we shall hold that option in abeyance and between us, for the time being. We have compared his facial features to numerous photographic images we have of him. He appears to be who he says he is, but we are not yet ready to confirm his identity. That particular aspect of this process should take another few days, to a week perhaps. We will inform you, when we have reached a conclusion on each element."

"Do you think he really is carrying a message of peace or surrender?"

"That is impossible to say at this stage, sir."

"Could he be defecting? Can he be exploited?"

"Again, too early to tell. We shall explore every possibility, with or without his help."

Churchill lapsed into contemplation for a few minutes, and then returned to Petrie. "We must keep this very quiet."

"We have already declared that all information even remotely associated with this event is most secret and under the protection of the Official Secrets Act. We shall keep a close eye on the newspapers and other news agencies, and we will vigorously quash any reference to this event."

"Excellent. Let's keep it that way for now. I shall inform the War Cabinet upon my return to London. I will most likely need to inform Commons, but I can do that in secret session. I should be back in London, at either Number Ten or the Cabinet War Rooms. I would like to meet with you every day at least for the next few days, to ensure I have the latest information."

"As you wish, Prime Minister. I shall coordinate with your schedule through your private secretary."

"That should be sufficient. Now, if I may ask, did you come from London this morning?"

"Yes sir."

"What of last night's enemy bombing raid?"

"I do not know a lot, other than it was a very heavy raid – the worst since the Second Great Fire last December. I do believe they hit Westminster last night and may have hit the Commons chamber."

"I am getting fragmentary reports. I shall leave for London first thing in the morning."

"Yes sir. That is the extent of my report. Is there anything of which I may be of assistance?"

"No, David, I think not. Safe journey back to the city."

"Thank you, sir. Good day." Petrie stood, returned the chair to its place, and departed without another word, closing the door behind him.

Churchill remained motionless and stared at the comforter covering his legs, as his mind churned through the new information and the potential implications of the event. He had been in an unrealized period of deep thought, when a knock on the door brought him back to the present. He needed to get on with this day.

—

Tuesday, 13.May.1941
Cabinet War Rooms
New Public Offices
Whitehall, London, England
United Kingdom

"Gentlemen, please," announced Cabinet Secretary Sir Edward Bridges, "we have a very full agenda this afternoon, and I am certain conflicting appointment diaries as well."

Prime Minister Churchill had been seated for several minutes and was beginning to fidget with impatience. Fortunately, the full War Cabinet and the appropriate Secretariat staff, both military and administrative, quieted and took their seats. Although not a member of the War Cabinet, at present, Home Secretary Herbert Morrison and the three service ministers had been specifically invited to attend this meeting.

"Mister Morrison, you have the floor," Sir Edward stated.

"As everyone knows, we have had a couple of particularly nasty nights. The Germans dropped an untold number of incendiary and high explosive bombs on the city Saturday night and Sunday morning, in what appears to be a clear attempt to replicate the direct and collateral destruction they achieved, to produce the Second Great Fire of London, last December. Fortunately, their incendiary bombs did not touch as much flammable material as they did last December. The ensuing fires were largely extinguished in less than 24 hours. That night, they did hit Westminster Palace and inflicted serious but not catastrophic damage, mostly to the northwest corner and Commons Chamber. There was some minor damage to the Clock Tower, but Big Ben remained fully functional." A few claps punctuated the sentence. "However, they struck again last night in what appears to be the worst attack since the Second Great Fire. This time, the Clock Tower was not so blessed. We expect to have it back in service in another day or so, and the damage mostly repaired . . . or at least the rubble removed.

"It is my sad duty to report . . . the Commons Chamber is now largely destroyed. It is also my grievous duty to report the passing of Lord Suffolk, the 20th Earl of Suffolk, this morning. He perished when an unexploded bomb lodged in the remains of the Commons Chamber he was attempting to disarm detonated, killing him and his 'Holy Trinity' team instantly. I am told by the Bomb Disposal Brigade, he may have been a victim of a new anti-tamper device specifically designed into some of their explosive ordnance for just such post-raid terror effects."

"A word here, if I may," interjected Prime Minister Churchill. The room was stone silent, except for the hushed air circulation system. Winston looked around the room. "Some of you, but not all of you, will recall the exemplary work Charles Howard personally did after the collapse of France, last spring, to evacuate key diplomatic and special operations personnel as well as extricate invaluable diamonds and gold, along with an even more precious substance, from falling into the grotesque hands of the Nawzee invaders. He could have rested on that service to His Majesty's Government alone, for the rest of his life, but no, he chose bomb disposal, of all vital assignments, to further his service to the Crown. Believe it or not, that was a passion of his. Charles Howard was an extraordinary patriot and human being . . . it was my honor to know and be in his presence in this lifetime. May God rest his immortal soul."

The room remained quiet and motionless for nearly a minute, as each man grappled with the information, and for those who knew the Earl of Suffolk personally, with the memories. Clement Attlee was the first to break the silence.

"I took the liberty this morning of talking to Lord Moyne about the tragic events last night. He has agreed to relinquish the House of Lords for Commons use for however long repairs may take. The Peers will meet in Westminster Cathedral, which remains largely undamaged, except for broken windows from the blast concussion."

"Thank you, Clement," Churchill responded. "We are grateful to the Lords for their generosity. We shall need to put their gracious offer to immediate work, since I must address Commons, to inform Members in secret session of recent events. I have asked for a convening of Commons later this evening for just that purpose." The Prime Minister nodded to Sir Edward, to continue.

"Mister Margesson . . . the situation on Crete," Sir Edward said.

"Our situation on the island is less than ideal. General Freyberg has insufficient forces to defend all possible landing areas, and he has minimal support from the Mediterranean Fleet to defend potential amphibious beaches."

Major General Bernard Cyril Freyberg, VC, CB, CMG, DSO** assumed command of Allied Forces Crete and retained his command of 2nd New Zealand Expeditionary Force. He was an accomplished military officer.

"Everyone has insufficient forces, David," Churchill interjected.

"Yes, and I am the messenger here, I'm afraid."

"May I remind you, David, you are the War Minister, not a simple messenger for a general under your authority."

"Yes sir. General Freyberg has been vigorously training his available forces and preparing defensive positions."

"We have had strong indications," Churchill said in deference to many of the attendees, who were not cleared for ULTRA, "for two weeks now of the German intentions to take Crete. Would you and CIGS be so kind to counsel General Freyberg to focus his scant resources at the most likely points of attack? He cannot defend the whole island, granted, but he has sufficient resources to repel an invasion attempt. He must defend Crete, and we believe the Germans are coming. He has had plenty of warning."

"As you command, Prime Minister."

Sir Edward moved on. "The Admiralty is up next, Mister Alexander."

"Admiral Pike reported this morning on field information that suggests the *Bismarck* is making ready for sea. She is down at the waterline, indicating she has recently taken on a full complement of fuel, stores and ammunition. Photoreconnaissance this morning has her still in port at Gotenhafen. The Home Fleet has been notified and returned to action status when we have signs of her sailing, which we expect any day now."

"Keep a close eye on that one," said Churchill. "Those 15-inch main battery guns are to be respected. She must not reach the Atlantic shipping lanes."

"Yes sir. We are ready when she sails. Admiral Pike asked me to pass along to you and the War Cabinet that he misses Agent Diamond and wishes he could have him back."

Churchill had known and watched the exploits of Trevor Andersen, also known as Diamond, for better than three years. He knew how important Andersen's work in Germany and Poland had been before the war, and of the importance of his recent work in France for SOE. There were not enough experienced field agents for any of the intelligence and special operations organizations, however they were all working to recruit and train more men like Andersen. Yet, Churchill recognized more so the significance of Pike's statement. Andersen spoke exceptionally good high German as well as passable Polish and French. He was indeed a rare entity. Churchill sympathized with Jumper Pike, but Andersen's skills were better put to use with SOE at this stage of the war.

"We certainly understand Admiral Pike's sentiment, and please convey to him, the War Cabinet has heard and accepts his statement of need. However, Agent Diamond is doing an exceptional job in his current assignment and doing what His Majesty's Government most needs him to be doing."

"Yes sir. I will ensure he receives the message."

"For now," added Churchill, changing the subject, "I insist that I be notified at any hour when we have evidence of the *Bismarck*'s sailing, and to notify the War Cabinet as soon as possible."

"As you command, Prime Minister."

"Prime Minister," Sir Edward said, "I am mindful of the time. You have your requested secret session with Commons in the House of Lords in less than an hour. I propose we suspend the remainder of the agenda, so that you may make any final preparations for your report, and the Members can make their way to Lords."

"Yes, yes, quite thoughtful, Sir Edward. There is much Commons must hear and I shan't be late. I would also like to go over to Westminster a little early to see the damage to Commons myself."

"No!" exclaimed Morrison. "Sir, the Bomb Disposal Brigade has informed the Home Office just before this meeting . . . there are two additional unexploded bombs in the Commons wreckage like the one that claimed the life of the Earl of Suffolk."

"Is it safe for Commons to meet in Lords?" asked Attlee.

"They indicate, yes," the Home Secretary answered. "Neither of the two bombs in question are near structural elements of the building and they are of average size. Further, the experts are trying to figure out a safe means to disarm those bombs, a process that may take days."

"Then, we shall meet in Lords on the hour," Churchill announced, "and I shall not tour the damage to Commons, just yet. Herbert, I ask you to let me know as soon as it is safe to witness the Westminster damage."

"Yes sir . . . as soon as it is possible."

"With that," Sir Edward said, "we are adjourned."

—

Wednesday, 14.May.1941
RAF Martlesham Heath
Woodbridge, Suffolk, England
United Kingdom

All 12 of the 'XR' Hurricanes landed safely after completing a rather long, cover patrol over the Kent coastline. They quickly dealt with their debriefing smoothly and swiftly. Group released them prior to sunset. They ate dinner together and retired to the Officer's Mess bar for the evening's libation and conversation.

"Gentlemen, if I may have your attention," said Billy Taylor. The room quieted, even the pilots from other squadrons listened. "I just received notification that a new squadron of American volunteer pilots was formed this afternoon – One-Twenty-One Squadron, at Kirton-in-Lindsey."

"It's about time," shouted a British pilot from one of the other squadrons.

"Easy now, mate," Flying Officer Charles 'Rusty' Bateman responded. "We're here, and more Americans are on the way."

"Glad to have the Yanks with us," said another Brit.

Taylor raised his hand. "I am also informed that additional American volunteer pilots are in the training process, and the Air Ministry indicates they expect to form at least one more fighter squadron of American volunteers."

Brian felt good that more Americans were joining the fight. The more that joined, the more he felt vindicated in his decision two years ago. *Malcolm was right. Other Americans would join.*

"Pretty nice, isn't it, Hunter," Dusty Langford said.

"Sure is. Really nice to have more Americans."

"How is it going to change things for us?"

"Your guess is as good as mine, Paul. I suspect, maybe even hope, another American squadron will divert some of the Press and VIP attention from us, so we can focus on doing the job we are here to do. I joined the RAF to fly, not talk."

"I think all of us would agree with that."

"Isn't your baby due right about now?" Bulldog Tolly asked.

"Yes. The due date according to the doctor and Charlotte's midwife is in two weeks, so I guess she could deliver anytime now."

"Are you going to be there?" asked Tolly.

"That depends upon operations and whether the skipper can let me go."

"Where does she live?"

"Hampshire, Winchester actually . . . not far from Southampton."

"Winchester? Like the rifle?"

"Same spelling, but I don't know of any connection."

Squadron Leader Taylor shouted, "I offer a toast to our new brother squadron, here is to One-Twenty-One Squadron."

"To One-Twenty-One Squadron," everyone shouted in response, clinked glasses with those around them, and then took a good, long swallow.

"What are you going to do after the war, Brian?" Paul asked.

"What do you mean? This war is a long way from over."

"It will be over someday."

"I have no idea. Flying is what I know, what I love. I would just as soon continue doing what I'm doing."

Flight Lieutenant Chesley Gordon 'Pete' Peterson must have heard the question and answer. "You know, Brian, the higher you get promoted the less you will fly. I mean, really, how many generals have you seen flying."

"Keith Park flew during the Battle."

"Did he fly combat missions?"

"Well . . . ," Brian thought for a few moments, "actually, no, not that I know of. But, hey, getting shot at is not so much fun."

Peterson laughed. "Yeah, of all of us, you know that best. How many is that for you, so far?"

"How many what?"

"How many times is that now for you . . . having your steed shot out from underneath you?" Chesley Peterson grinned at Brian.

"The number is four . . . sir," Brian said with displeasure in his voice at being reminded about his close calls with death.

"And, how many victories have you been credited with, Hunter?" asked Dusty Langford, probably to counter-balance Pete's dig.

Brian shook his head. *I'm not going to play this game.*

"Oh, yeah, it's 19 bloody fucking Germans," snarled Langford. "Who else around here has 19 victories?" Dusty looked around, as if he was looking for an answer. "Oh yeah, only two other pilots in the whole of the damn Fighter Command."

"I would ease up on the profanity, Dusty," Pete said and turned to the bar.

"Asshole," muffled Langford.

"He is a senior officer, Dusty. Whether we agree or not, we must respect his rank and position."

"Hunter, come on, now. You're one rank above us," he said, motioning to Bulldog, "and, you don't give us crap like that."

"We are not all the same. Let's give him a break. He was correct. I have had four aircraft shot out from under me. I've bailed out twice, crashed once, and somehow miraculously have been thrown clear in an aircraft break-up. Those last times could have and probably should have killed me. I was lucky. He was making the point that our task is to shoot down the other guy and not get shot down ourselves. I failed four times on that task."

"And, succeeded 19 times, for God's sake, Brian."

"OK, let's change the subject. Enough of this!"

The Green Section pilots emptied their glasses and retrieved fresh pints of bitter. Each of them took a healthy swallow. Brian walked to an empty corner table, moved one of the chairs to the corner, facing out, and sat. Bulldog and Dusty joined Hunter at the table. Red Burns came over and filled out the table chairs.

"Hey, Hunter," Red began. "I heard a rumor some group is opening up a new club next month in London for all the Americans now working in London and England."

"I think I heard they dedicated such a club last month," Brian responded.

"Guess there is plenty of Americans showing up over here."

"There were diplomats, journalists, businessmen and such long before I got here."

"You were the first," Bulldog said, "weren't you?"

"So they told me," Brian answered.

"We need to plan to go down there and check it out," Burns added.

"Sounds like a plan," Dusty chimed in.

Brian drained the last of his beer and announced, "I've had enough, mates. I'm calling it a night. We're back on duty in the morning, and who knows what tomorrow holds for us."

"Party pooper," Burns jabbed.

"Yes, and proud of it. Good night, lads."

Brian placed his empty glass on the bar, left the room, went upstairs, undressed and was soon fast asleep.

—

Chapter 4

Thou art slave to fate, chance, kings and desperate men.

-- John Donne

Saturday, 17.May.1941
Chequers Court
Ellesborough, Buckinghamshire, England
United Kingdom

As was so often the case, Prime Minister Churchill, his duty private secretary, this weekend John Martin, and the stenographer relays had made good progress through the morning's paperwork. Working in bed had become normal for Winston and his administrative staff. They were nearly done with the morning's dispatch box of letters, messages, reports and other documents. Their pace came to a sudden stop when the latest bombing report from the Air Ministry reached the top of his stack. Winston quickly read the report.

"OK, John," Churchill announced, "we are done for this morning. This bombing report is just wrong. Would you be so kind to find the Prof and ask him to come see me at his earliest convenience?"

Martin nodded. The two stenographers departed. "Yes sir. Right away."

Churchill re-read the Air Ministry's latest bombing report several more times before Professor Lindemann arrived.

"Good morning, Winston. I trust you are well-rested and fresh."

"Quite so, Fred, and you?"

"Likewise, my liege."

"Thank you for coming so promptly. I've read and re-read the latest bombing report from Bomber Command and the Air Ministry. As we have discussed the issue of night bombing accuracy before, more than a few times as I recall, both of us have been suspicious of their accuracy claims. Yesterday's report has pretty well solidified my suspicion to outright disbelief and distrust."

"May I read the report?"

"Certainly."

Lindemann grasped the proffered document and retreated to a chair against the wall. He carefully read each page, leafing back several times, presumably to confirm connecting statements. Lindemann did not read the report a second time. "I see what you mean and I agree. The photographic reconnaissance, bomb damage report shows the factory the bombers reportedly hit the night before has not a mark on it."

"My read as well. Something is dreadfully wrong with either the aircrew navigation methods, or they are fooling themselves regarding their accuracy, or both. Several months ago, you mentioned a statistical analysis process that you suggested might be applied to a problem such as this."

"Correct. The mathematics will show us the reality regarding accuracy, but it will not and cannot offer solutions . . . perhaps hints, if we are lucky."

"How long will such a study take to complete?"

"Statistically, Bomber Command flies 12 to 15 missions per month, so at first blush, I would say two months to collect the proper data sample, and perhaps another month of analysis."

"We need it today . . . after a report like that," Churchill said, pointing to the papers in Lindemann's hand.

"I understand. I think I have just the man to do this job . . . a brilliant Cambridge mathematician, David Miles Bensusan-Butt."

"Please get him started first thing Monday morning. The War Cabinet charges him, so we will not abide any resistance or interference with his study. I will coordinate with the air minister and chief of the air staff as well as the War Cabinet to ensure Butt receives the support he will surely need. I anxiously await his results as I suspect the results will not be reassuring."

"We have seen evidence over many months that bombing results have been overstated. They cannot claim they hit the target when there is no sign of bomb damage anywhere even near the target."

"Indeed! Work with your man to get the study done as quickly as possible. We need that information."

"We probably should consider what actions the government will take if the study proves what we suspect," added Lindemann.

"I have had some circumspect discussions with Portal, but you are spot on correct. I know the research establishment is working on better, more accurate, electronic navigation means. They need better tactics to use the new technology properly."

"Sounds like you have it in hand, then."

"Not quite. At least they know they need something better than what they are currently using, and I do believe they need a significant improvement over the German Knickebein system." Churchill paused to think. "By the way, are you ready for your investiture next month, Baron Cherwell?"

"I cannot thank you enough, Winston. You have always been most generous with me. To answer your query, yes, I am as ready as I will ever be. Yet, I am also certain my nerves shall flair up as the day approaches."

"You have served His Majesty and me exceptionally well, Lord Cherwell. This recognition is richly deserved. Thank you for your continuing service."

"You are most welcome, Winston."

"Before you go, give me a moment," Churchill said. He looked through the Dispatch Box, finding the folder and a single piece of paper within that folder. "One more item, if you will permit me." He re-read the message, and then held up, but did not hold it out for the Prof to read. "I had a news message that I marked for speaking to you. The message simply stated the Gloster Aircraft Company flew the first flight of their Model E28.39 aircraft two days ago. I believe that is the first turbine engine aircraft on our side."

"Correct. I have not seen a report of the flight, yet, but I believe they flew the flight successfully out of RAF Cranwell in Lincolnshire. I'll check into it on Monday. What is your concern?"

"No concern . . . just curious. I remember the report of the Heinkel turbine engine aircraft first flight last year, so presumably the Germans are ahead of us. We also have intelligence reports they are apparently working on a twin, turbine engine variant . . . well, the Messerschmidt company. I do not know about Heinkel. I was wondering how these various aircraft compare?"

"I will arrange a review of the available information with MI6, the Air Ministry technical intelligence lads and Royal Aircraft Establishment as soon as possible."

"Why is everyone all worked up in a lather about these turbine engine aircraft?"

"One word . . . speed. The fighter pilots have a simple saying they live by—speed is life. The turbine engine propulsion portends significantly greater speed, which in turn will give the pilot more options regarding engagement and perhaps more importantly disengagement."

"I see. So, we should pay attention to these developments."

"Yes, I would say so."

"Then, please do carry out your review of the available data. I eagerly await your assessment." Lindemann nodded his head. Churchill stared at the message and lapsed into contemplation. Apparently satisfied, Winston looked up and said, "Please send in Sawyers upon your departure. I suppose I should get out of bed, have a nice hot bath and dress for lunch. There is much to be done . . . even out here in the countryside."

Frank Sawyers had been his steady and dependable valet for five years.

"You really must make the time to relax, Winston. You work entirely too hard for a man of your years."

"Nonsense. There will be time to rest when this sordid affair is done."

Professor Frederick Lindemann, soon to be the 1ˢᵗ Baron Cherwell and always known as the Prof to Churchill, shook his head in disagreement, as he left Churchill's bedroom to fulfill his friend's request.

—

Sunday, 18.May.1941
Standing Oak Farm
Winchester, Hampshire, England
United Kingdom

Brian arrived in early afternoon on a 24-hour pass. These one-day events were always a bit sporty, since it only took one railway delay or stoppage to consume his available time. On this particular day, the trains ran smoothly and on time, including the station transfers in London.

Charlotte was in the kitchen preparing their evening meal. The three hands were working various odd maintenance chores before the afternoon milking session. Charlotte reminded Brian about talking to young Jacob and indicated his tasks for the afternoon were not particularly important, so today might be a good opportunity to submit to the young man's questions. Brian agreed and went to the barn to find Jacob Holden.

It was a nice spring day—mild temperature, light breeze, and a sky about half full of puffy fair weather cumulous clouds. Brian found Jacob finishing up a repair on the broken gate. He helped Jacob finish the repair in fine form.

"Well done, Jacob," Brian observed, as he stood back admiring the excellent workmanship. "Now, the mistress of the manor has given her consent to your temporary relief from your assigned chores and she has reminded me of your request for a chat. So, what do you say to sitting with me for a bit? I shall do my best to answer whatever questions you might have."

"I am honored, sir. As long as Mrs. Drummond is fine with it, I would enjoy that."

"Very well, then. She suggested it. I am informed that we have a nice, new, sitting area beneath the big oak tree by the lake. Why don't we go out there for our chat."

"Yes sir. I helped her build it."

They stopped at the barn's tool room to replace his tools in their proper position, and then they walked toward the far end of the circular driveway. A new, wide, gravel walkway extended out from the driveway toward the lake and terminated in an oval of flat stones about ten feet by six feet with a large, carved, well-crafted, oak bench on the long edge away from the lake. The entire sitting area sat comfortably under the cover of a large, mature, oak tree.

"Very well done, Jacob," Brian said and motioned for them to sit, which they did. They both sat quietly, looking off to the grassy hill on the far side of the calm lake. "So, what questions may I answer for you?"

"I'm very nervous, Mister Drummond. I have never been able to talk to a famous person like you."

Brian raised his right hand in a stop signal. "Jacob, please, I am not famous and I'm only a few years older than you. Just talk to me as a friend."

Jacob fidgeted a little as he sat an arm's length away to Brian's left. "Why did you come here?"

Brian chuckled softly. "Good question. The short answer is flying. I love flying . . . anything and everything."

"But, the kind of flying you are doing seems very dangerous. Mister Morgan told me you were very seriously wounded last fall and Mrs. Drummond took care of you, and he said Mrs. Drummond saved your life when you landed in her pond last summer after you were shot down."

This time Brian shifted his weight a few times and crossed his legs. "Well, first, Mister Morgan is correct with the facts. Yes, the flying I am doing is dangerous, certainly more dangerous than just flying itself, but it is certainly a necessary risk."

"Why?"

"I do not know how much you are aware of what happened last summer."

"You mean the German invasion?"

"Yes. We flew and fought to deny the Germans command of the air, which was a prerequisite for their invasion of England. We were told that after the Battle of France and Dunkirk there was little of the Army remaining to stop the Germans if they had invaded."

"But, you are an American. Why did you want to take all that risk when your country was not under attack?"

"My mentor and teacher taught me that a loss of freedom anywhere is a loss of freedom to all of us. He also taught me that Hitler had very evil intentions before the war even began."

"Like Mister Churchill?"

Brian looked to Jacob and smiled. "Yes, exactly like Mister Churchill. Mister Bainbridge . . . he was my mentor . . . believed what Mister Churchill was saying. I listened and I decided that I had to do my part in the defense of freedom, just as Mister Bainbridge had done in the Great War."

"What is it like to fly the Spitfire aeroplane?"

Again, Brian smiled, this time quite broadly before he answered. "Like a dream. The Spitfire is the best aircraft I have ever flown and may ever fly. It is fast, powerful and easy to fly."

"Can I fly one?"

"If I can, you can, but I suspect you will have to get a little older."

"Were you the youngest pilot?"

"I don't believe so. Last summer, I think there were a couple of, or perhaps a few pilots who were just a couple of years older than you. My guess is I was probably below the average age of Fighter Command pilots, but I know I was not the youngest. I was the youngest in my squadron."

"Which squadron was that, if I may ask?"

"Six Zero Nine Squadron at RAF Middle Wallop."

"That is just up the road."

"Yes, it is."

"Are you still based at Middle Wallop?"

"No. I'm in Seventy-One Squadron now, an American volunteer squadron, at RAF Martlesham Heath in Suffolk."

"My mates at school want me to ask you to come to talk to all of them. Can you do that?"

"Well, to be honest, I don't know. I will have to ask my commander. It might take a few days or weeks to get an answer, and even if they say yes, it may take a while longer than that to match up your class schedule and my availability."

"We still have a couple of months before our summer holiday break."

"If my commander permits, we should be able to find a day in that amount of time."

"What do I need to do to fly like you are, Mister Drummond?"

"First, you must finish school. Second, I wish I had the ability to teach you to fly like Mister Bainbridge taught me, but these are not my aircraft. Third, I will ask what the current requirements are, and maybe we can . . ." A whistle interrupted Brian's response.

Charlotte was standing the driveway. When she saw both faces turn toward her, she made an exaggerated gesture of milking a cow, and then waved for them to come.

Brian waved back and stood. "We are being summoned to help with the afternoon milking. We can continue after we finish, if you wish." They walked back toward the barn. "I was going to say that we can find a way, if you really want to fly for the air force."

"That would be great, Mister Drummond. Thank you very much for your time. I probably will have more questions, and I hope you don't mind all my questions."

"No, I do not mind in the least."

"Thank you."

"You are most welcome, Jacob, and I truly appreciate your interest."

The two men joined the others for the afternoon milking. Brian kissed Charlotte at the front door before he made it to the barn and took up his share of the task. Brian thought about his answers to Jacob's questions as he squeezed and pulled the cow's teats shooting generous streams of milk into the collection pail. *Maybe I was not supposed to tell anyone where my squadrons were located. Nobody told me not to say it, but I need to ask Billy, if it was OK.* Now he wondered about his other answers. Jacob was certainly no threat to security, but he did not want to get crosswise with anyone. *Someone might think such information is important to protect.* As they worked through the relief of the cows, Brian realized Charlotte had acquired a dozen more cows. *I need to remember to ask her about her acquisitions.* Brian swelled with pride that Charlotte was doing so well, especially being so close to giving birth.

—

Wednesday, 21.May.1941
Cabinet War Rooms
New Public Offices
Whitehall, London, England
United Kingdom

Prime Minister Churchill resigned himself to being a few minutes late for the scheduled War Cabinet, late afternoon meeting. Fortunately, he had already asked to make it a closed session with only the members, and in this instance, the service ministers with the chiefs were invited. All attendees had been cleared and had access to ULTRA.

The Buff Box arrived manacled to an armed guard – an army sergeant – shortly before the hour. 'C' had forewarned him the box was on its way to the War Rooms bunker from Broadway House – Headquarters, Secret Intelligence Service.

Churchill had retrieved the only key to the box before the courier's arrival, so that its location on his person would not be revealed. As soon as he saw the sergeant with the Buff Box in his left hand, Churchill impatiently motioned for him to enter and close the door without the usual arrival announcement by his duty private secretary.

The courier placed the box on the Prime Minister's desk. Churchill unlocked the cover, went immediately to the urgent divider, and withdrew four sheets of paper attached by a paperclip.

MOST SECRET - ULTRA

```
SECRET
DATE: 17 MAY 1941
TO: CHIEF OF FLEET
FROM: HEADQUARTERS NAVY
OPERATION RHINE EXERCISE
BREAK
EXECUTE OPERATION RHINE EXERCISE WITHOUT CHANGE
NO LATER THAN 22 MAY BREAK ALL SUPPORT FORCES
IN PLACE PER PLAN AT YOUR COMMAND BREAK THE
LEADER SENDS PERSONAL GOOD HUNTING BREAK HAIL
VICTORY BREAK HAIL HITLER END
SECRET
```

MOST SECRET - ULTRA

Paper-clipped to the ULTRA decryption message were three additional messages – an unusual arrangement for sensitive decrypted messages. Churchill leafed to the second page. He did not recognize the compartment classification. He guessed it was a highly sensitive source or field agent in Germany – quite understandable.

MOST SECRET - SHINING

```
DATE: 19 MAY 1941
TO: DNI ADMIRALTY
FROM: SWORD
FIELD REPORT 41-23
BEGIN
BISMARCK NOT IN PORT GOTENHAFEN BREAK BELIEVE
BISMARCK SORTIED DURING NIGHT OF 18TH BREAK
```

SEVERAL DESTROYERS BELIEVE THREE ALSO DEPARTED
NIGHT OF 18TH END

MOST SECRET - SHINING

Churchill turned to the second attached message . . . from British Naval Attaché assigned to the British embassy in Stockholm, Sweden.

SECRET

SECRET
DATE: 20 MAY 1941
TO: DNI ADMIRALTY
FROM: NAVAL ATTACHE STOCKHOLM SWEDEN
WARSHIP PASSAGE ORESUND STRAIGHT
BEGIN
SWEDISH NAVAL ASSETS REPORT CAPITAL SHIP
PASSAGE SOUTH TO NORTH THROUGH ORESUND STRAIGHT
DURING EARLY MORNING HOURS PRIOR TO DAWN BREAK
WANING MOONLIGHT INSUFFICIENT TO IDENTIFY THE
SUBJECT SHIP BREAK SWEDISH AIR FORCE RECON
AIRCRAFT MID MORNING DID NOT LOCATE ANY WARSHIP
IN TRANSIT END
SECRET

SECRET

The last message was probably from another source in Norway within the same network. If need be, he could query the First Lord, or the DNI Admiral Jumper Pike, if more detail regarding the sources would be useful. The information from the field sources was complementary but not critical. There was little, if any, doubt remaining that Germany's most powerful battleship was on the move and most likely would soon be hunting in the Atlantic. The Royal Navy Home Fleet had to be successful in stopping this serious threat.

MOST SECRET - SHINING NORTH

MOST SECRET

```
DATE: 21 MAY 1941
TO: DNI ADMIRALTY
FROM: SEAHORSE
FIELD REPORT 41-11
BEGIN
BISMARCK PRINCE EUGEN AND ESCORTS ARRIVED
BERGEN FJORD NOON TODAY BREAK APPEAR TO
BE TAKING ON FUEL AND SUPPLIES BREAK NO
INDICATIONS OF PENDING ACTION END
MOST SECRET
```

MOST SECRET - SHINING NORTH

Before returning the intelligence messages to the Buff Box, Prime Minister Churchill retrieved a blue binder, opened it and leafed through to the page he sought. He wanted to confirm the characters involved. The German Navy's Chief of Fleet was Admiral Johann Günther Lütjens – a name he knew well from his days as First Lord of the Admiralty. The *Bismarck*'s captain was *Kapitän zur See* Otto Ernst Lindemann. "Interesting," Churchill mumbled to himself, noting the name's spelling was exactly the same as that of his principal science advisor. The book had biographies, but he did not have time to read them. Maybe later. He looked farther down the page. The captain of the heavy cruiser *Prinz Eugen* was *Kapitän zur See* Helmuth Brinkmann – a classmate of Lindemann, as he recalled. Churchill left the blue book open on his desk. If he found a few minutes, later in the day, he would read the biographies from the Admiralty and MI6. It was important to understand their adversaries.

Churchill returned the three sheets behind the urgent divider, and as he did so, he noticed another paper marked ULTRA that he extracted and read.

MOST SECRET - ULTRA

```
SECRET
DATE: 20 MAY 1941
TO: OKL OKH AIR FLEET 2
FROM: COMMANDING GENERAL 11TH FLYING CORPS
OPERATION MERCURY
BREAK
INITIAL ASSAULT COMPLETED THIS DAY BREAK
INITIAL OBJECTIVES MET BREAK FIGHTING LIGHT TO
```

```
MODERATE BREAK ENEMY APPEARS TO HAVE LARGER
FORCE THAN ANTICIPATED BREAK CANEA AND MALEME
TAKEN BUT NOT YET SECURE BREAK MALEME AIRFIELD
DAMAGED BUT USEABLE BREAK OPERATION CONTINUES
PER DIRECTIVE NUMBER 28 BREAK REINFORCEMENTS
SHOULD BE PREPARED FOR DEPLOYMENT AS EARLY AS
22 MAY BREAK HAIL VICTORY BREAK HAIL HITLER END
SECRET
```

MOST SECRET - ULTRA

Churchill returned to the blue book. The commanding general of the 11th Flying Corps (*XI. Fliegerkorps*) was listed as *Generalmajor* Kurt Student – another name Churchill recognized.   Student was a celebrated Air Force commander of parachute and glider assault troops.   The newly formed 11th Flying Corps was believed to be comprised of the 1st Parachute Division, the 22nd Glider Assault Division and the 9th Mountain Infantry Division – all of them elite, experienced and accomplished specialty infantry units. The existence of ground combat units in the *Luftwaffe* was a direct artifact of Hermann Göring's ego.

Churchill glanced at his desk clock.   He was already ten minutes late and the War Cabinet was waiting on him.   He returned the message to the Buff Box and locked the cover. "Thank you, sergeant. You may return the dispatch box to Broadway House directly."

"Thank you, sir.   Good day."

The Prime Minister followed the courier toward the entrance to the underground bunker.   While the sergeant began his ascent of the stairway, Churchill entered the War Cabinet Conference Room.

"Please excuse me, gentlemen, 'C' insisted the recent 'Boniface' messages . . . ." He paused long enough to take his seat at the head of the U-shaped table and did not wait for Sir Edward to open the meeting. "The messages considered urgent by 'C' dealt with the breakout of the *Bismarck* flotilla and the enemy's invasion of Crete.

"We have confirmation the *Bismarck* sortied from Gotenhafen in East Prussia, made an uneventful passage through Öresund Straight and joined *Prince Eugen* and their escort destroyers at Bergen." Churchill looked to Sir Archie Sinclair. "We need a photographic reconnaissance flight into Bergen Fjord as soon as light and weather conditions permit." The Air Minister nodded his head in agreement. "According to the information we have, it seems clear to me that the *Bismarck* flotilla intends to make for the Atlantic shipping lanes."

"Admiral Holland has put to sea," offered the First Lord, "aboard *Hood* with his Battlecruiser Squadron. He has deployed the Home Fleet heavy assets to block the GIUK Gap." With the English Channel essentially denied to the enemy by the RAF, the Greenland-Iceland-United Kingdom Gap offered the only access to the Atlantic Ocean from the North Sea. The *Luftwaffe* made any action in Norwegian waters far too risky.

"Let's make sure Coastal Command is on the lookout as well."

"Yes sir," answered Sinclair.

"We must not allow that ship into the Atlantic. Those damn U-boats have a strangle hold on our supply lines. The last thing we need is a major capital ship rampaging those convoys and lightly armed escorts. How many convoys are in the North Atlantic lanes?"

"Six . . . with three more, one westbound and two eastbound, scheduled to sail by the end of the month."

Churchill shook his head. "A.V., please pass to Admiral Tovey, freedom is depending upon his fleet."

"Yes sir."

Prime Minister Churchill displayed uncharacteristic impatience and did not wait for other topics or discussion points. "Now, as if a marauding German battleship is not bad enough, we have a dreadful situation on Crete. That island anchors the Aegean Sea, and in the hands of the Germans, it would present a serious if not fatal threat to the entire Eastern Mediterranean Sea. A 'Boniface' message from General Student yesterday suggests the Germans have successfully landed on Crete just as their operations plan indicated. General Freyberg's Victoria Cross is testament to his courage under fire. We do not question his professional skills, but now, we need him to defend Crete. David, pray tell us what General Freyberg is doing to defeat the German forces that have landed or will land on Crete."

"The 2nd New Zealanders suffered a setback yesterday. They are fighting to regain portions of Maleme."

"They have lost the airfield."

"Yes sir. That information is true, but there is hope."

"Excellent, then by all means, pass along some of that hope."

"Our situation on the island is less than ideal."

"A serious under-statement, it seems to me."

"As we discussed a week ago, General Freyberg has insufficient forces to defend all possible landing areas, and he has minimal support from the Mediterranean Fleet to defend potential amphibious beaches."

"Wait just a minute," Churchill interrupted. He stared sternly at Margesson, and then he looked around the room. "We were informed last weekend, including you, that Bletchley had intercepted and deciphered the entire German operations plan for their invasion of Crete. Why is Freyberg blabbering about defending the island when he knew where and how they were coming?"

"He was suspicious of the intelligence, and we were sworn to not disclose that particular source."

Churchill smacked the table loudly with his right hand. "Damn it all to hell, what more does he want?"

"Two more divisions, four fighter squadrons and a cruiser flotilla."

Churchill actually laughed, and then spoke with a calm, measured voice. "Every general and admiral wants more . . . more of everything . . . more troops, bigger guns . . . I cannot fault him that. Yet, the reality exists there are no divisions and squadrons to spare. Rommel is rampaging across North Africa and threatening Egypt, the Nile and the Suez. Wavell wants more divisions and more squadrons. He cannot have them either, because they do not exist. That said, the really abrasive irritant in this instance . . . Freyberg had the bloody damn German operations plan," Churchill shouted, "and, the worst of it is he apparently chose to ignore the golden egg handed to him on a silver platter two weeks before the German attack. How on God's little bloody green earth can that be acceptable. That is not a query, but a rhetorical statement." Silence dampened the entire room for what seemed like minutes. "There should be no doubt in his mind, now, that the Germans have executed their plan as written. What is he doing to re-deploy his resources?"

"We need to calm this down," interjected Attlee. "This has been a difficult day. You know we are with you, Winston, but crucifying our generals will not improve the situation."

"Wise counsel, as always, Clement. Thank you," Churchill said. The Prime Minister turned back to Margesson. "I did not intend to berate you, David, or General Freyberg. We have had enough for this meeting. Pray give us Freyberg's counter-attack plan as soon as you are able."

"Yes sir."

"With that," said Sir Edward, speaking for the first time, "we are adjourned. Good evening, gentlemen."

The meeting dissolved. Churchill returned to his underground office. The day did not stop.

—

Saturday, 24.May.1941
Chequers Court
Ellesborough, Buckinghamshire, England
United Kingdom
09:10 hours

Jock Colville knocked and entered Prime Minister Churchill's bedroom. He was alone, sitting up in bed and wearing his colorful, floral print robe, and he had just finished his breakfast. Colville did not wait for the usual cordialities. "Sir, I am the bearer of truly tragic news. The Admiralty just called to inform you that HMS *Hood* has been sunk in battle with the *Bismarck* flotilla."

"When?"

"Just over an hour ago . . . in the Denmark Straights."

"Survivors?"

"The Admiralty indicated she was hit and went down in a matter of minutes. Several escort destroyers are searching for survivors, but they do not expect to find many, if any."

"Admiral Holland?"

"Presumed lost."

"Dear God . . . tragic news indeed. May God rest their immortal souls."

"God bless them," repeated Colville.

"Do we know anymore about what happened?" asked the Prime Minister.

"Beyond the notification of the sinking from the Admiralty, the message also states that *Prince of Wales* was seriously damaged in the same engagement and had to withdraw shortly after the *Hood*'s demise. The fleet lost contact with the *Bismarck* and *Prince Eugen*."

Churchill stared at his breakfast tray for several minutes. "I worried about that design when I was first at the Admiralty. The admirals wanted big guns on a fast ship, and in order to achieve fast they seriously lessened the armor for a ship of that class. After the Battle of Jutland, the design was hastily modified to increase the armor, but it was still less than other capital ships of the day. Her keel was laid down in 1916, just after I was canned as First Lord. I truly hope it was not the armor that failed the crew—1,400 precious sailors, dear God above." Churchill lapsed back into contemplation for a minute. He then looked directly to Colville but did not speak.

"It had to be something quite violent. The message said she sank in a matter of minutes . . . that is very quick."

"Indeed! Far too quick for a capital ship! I suspect for an event such as was reported, the Germans may have made a lucky shot and penetrated her main magazine."

"I will inform the Admiralty you wish to know the cause as soon as possible," Colville added.

"No. Do not distract them. The primary, if not sole, objective must be neutralizing or sinking the *Bismarck*. There is nothing we can do for the Hood. There will be time in the future to determine the cause and corrective actions."

"Very well, sir."

"Please inform the First Lord and First Sea Lord . . . I want their briefing on the situation as soon as they have a clearer picture of what happened out there and more importantly what they are doing to stop that bloody Hun battleship."

"Yes sir."

"I also want to know who is now in command of the pursuit of *Bismarck* now that Admiral Holland is presumed lost. They must not allow that warship to enter our vulnerable shipping lanes. They must stop that damn Hun battleship," he repeated himself. "That warship is significantly more powerful than the *Sheer*, and we saw what the latter did to that convoy November last."

"Yes sir."

"Thank you, Jock. Please tell Sawyers I want my bath, now. Today is going to be a busy day and not in a good way."

"Yes sir. Right away."

Colville departed. Churchill did not move as his thought focused on what could have gone so wrong this morning between Greenland and Iceland.

—

Sunday, 25.May.1941
RAF Martlesham Heath
Woodbridge, Suffolk, England
United Kingdom
11:15 hours

The squadron sat at Available status all morning with no missions scheduled for the day. The No.71 Squadron pilots resigned themselves to a no-fly day, as persistent fog kept them on the ground anyway. The Met guys suggested the fog was likely to persist.

Horse Harrow and Red Burns were due back at noon from their 24-hour rest pass, which invariably turned out to be a party pass rather than a rest pass, and thus not particularly restful. Squadron Leader Taylor remained in his office, presumably doing paperwork of some kind. Half the pilots dozed. Brian fought to keep his mind off of Charlotte and the impending birth, and continually tried to focus his conscious thought on his Hurricane fighter and the new bomber formation escort missions they would soon be assigned and had been training to perform.

Red Burns burst into the Dispersal Hut, startling everyone. "Have you guys seen the news?" he shouted, holding up a copy of today's *Sunday Dispatch*. The full page headline in bold print read, "HMS HOOD SUNK – 'BLOWN UP, FEARED FEW SURVIVORS,' STATES ADMIRALTY."

"That's a damn battleship," growled Whitey Whittington.

Billy Taylor appeared from his office. "A battlecruiser, actually, I do believe, Whitey – big, fast and really big guns, but modestly armored compared to a full-up battleship."

Brian remembered that Squadron Leader Taylor had served in the U.S. Navy as well as the Royal Navy.

"What happened?" Whitey asked.

"Doesn't say much other than the Hood took an unlucky hit in a magazine during a naval battle near Greenland. Probably went down quickly, would be my guess," Red offered. "The Germans apparently announced the sinking last night, and I suppose the Admiralty felt compelled to confirm the loss."

"Over a thousand sailors," mumbled Whitey with anguish in his voice.

"Let's see that paper, Red," Billy said.

Burns gave the newspaper to Taylor. After their commander read it, the newspaper was passed among the pilots.

"We don't know what else is going on out there," cautioned Taylor. "We are going to experience wins and losses until this affair is done. Our job is to stay focused on our portion of the task, keep our eye on the target, and kill Germans. Let's not forget that."

The loss of the *Hood* in such dramatic fashion and what was happening in the hunt for the *Bismarck* occupied the conversation of the No.71 Squadron pilots for the rest of their duty day. Harrow returned in time and Sweet Sweeny began his 24-hour pass. Tomorrow would be another day.

—

Tuesday, 27.May.1941
Cabinet War Rooms
New Public Offices
Whitehall, London, England
United Kingdom

Prime Minister Churchill entered the War Cabinet Conference Room with a bounce in his step and flush to his cheeks. As requested, Sir Edward had assembled the entire War Cabinet, the Defense and Chiefs of Staff Committees, and the Secretariat staff. There was no spare room to even stand.

Churchill held the single sheet of paper above his head with both hands, as if everyone could see the small print.

```
DATE: 27 MAY 1941; 12:05 [Z]
TO: ADMIRALTY
FROM: HOME FLEET
BEGIN
DKM BISMARCK SUNK AT 10:40 [N] THIS DAY END
```

Realizing no one could read the message, the Prime Minister hauled it down and read it aloud. "The message simply reads, *Bismarck* sunk at 10:40 this day." Cheers and clapping erupted through the room, and then someone started singing the chorus of "Rule Britannia" that everyone joined in the singing. "That was two hours ago." Again, cheers broke out. Those close to the First Sea Lord shook his hand, patted him on the back and otherwise jostled him in their excitement.

After several minutes of celebration and expressions of relief, Churchill raised his right hand for quiet, and then raised his left hand as well. The room slowly calmed. He motioned for those who could sit to take their seats. The Prime Minister remained standing. "On behalf of the War Cabinet and His Majesty's Government, Sir Dudley, we offer our heartfelt congratulations to the Navy and the heroic accomplishments of our sailors." The First Sea Lord nodded his head in acknowledgement.

"What do we know so far?" asked Churchill.

Alexander nodded to the First Sea Lord.

"Not previously reported," began First Sea Lord Admiral of the Fleet Sir Alfred Dudley Pickman Rogers Pound, GCB, GCVO, "after the *Prince of Wales* had to break off her engagement with *Bismarck* on the morning of the 24th, the ship's damage control parties were able to recover sufficient operational performance to rejoin the chase. Coastal Command appears to get the credit for contact. A PBY Catalina flying boat spotted the *Bismarck* 700 miles west of Brest on the morning of the 26th. Sometime during the prior two days of the chase, *Prince Eugen* split from *Bismarck* with their destroyer escort. We have no idea why. Force 'H' arrived on scene with *Ark Royal* yesterday afternoon, and once in range, launched a squadron of Swordfish torpedo planes. The *Bismarck* was able to evade most of the dozens of torpedoes launched at her. The *Ark Royal* pressed the attack with multiple sorties and managed to gain two hits in the evening hours last night—one amidships and another at the stern. The stern shot apparently jammed one or both rudders as the *Bismarck* began slowly circling rather than making best speed to the safety of air cover and Brest. Admiral Tovey arrived on scene with *Rodney* and *King George V* late yesterday. With failing light and a now-apparent cornered *Bismarck*, Tovey waited until after dawn this morning to press home his attack. Tovey noted in his subsequent dispatch that *Bismarck* put up a valiant fight against impossible odds. She went down fighting with her colors flying."

"An honorable end to a proud warship," Churchill commented. "We are relieved it is done."

"Yes. There is always solemnity to such endings," continued Sir Dudley. "By dawn, she was making headway at a fraction of her speed potential, which suggests she was still dealing with her rudder damage and may have been using differential propulsion to counteract a jammed rudder. Despite her lack of speed or maneuverability, she still had a serious bite. She damaged several of her attackers during the nearly one hour final engagement. Her main battery guns went silent an hour before she disappeared below the waves, and all weapons fire from the *Bismarck* ceased 40 minutes prior to the end. Tovey did not let up and pressed his attack to comparatively short range. They are still working to retrieve survivors; several dozen have been picked up so far; however, there is serious concern that submarines may be in the area or arriving soon. The survivors will be cared for, recorded, catalogued and interrogated to collect more information."

"We have a long way to go before our final report is ready," interjected A.V. Alexander.

"First Lord," Churchill said to Alexander, "we eagerly await the Admiralty's report on the events in the Atlantic this last week."

"You shall have it, Prime Minister, as soon as we can interview the crews and develop a reasonable picture of the protracted action. The full report will most likely take many months, as there were many moving pieces, hundreds upon hundreds of witnesses and participants, and we still have the *Prince Eugen* loose in the Atlantic, U-boats stalking our convoys, and valuable shipping to protect."

"Quite so . . . at least we can rejoice in the moment. To all ministers, please ensure the widest dissemination of this news. I have asked the Speaker to convene Commons this evening, so that I may convey to the full House the success of the Royal Navy this day."

The Prime Minister nodded to Sir Edward, who announced, "We are adjourned, gentlemen."

The First Lord made eye contact with the Prime Minister and nodded his head slightly, gesturing that he would like a private word once the room was cleared. As the attendees left the room, A.V. moved to the chair next to Churchill. Sir Edward was the last person to the doorway. A.V. gestured with his head for the Cabinet Secretary to close the door, which he did. A.V. turned his chair to face Winston, and the Prime Minister did likewise.

Alexander began, "Have you seen the Boniface decrypts from this morning?"

"No, not yet."

"Jumper met with 'C' and GCCS this morning. Yesterday and today have been productive days so far. I knew you would be interested in the last messages from *Bismarck* late last night and this morning."

"Quite so . . . do tell."

Alexander recounted from memory, which reflected upon the importance the First Lord placed on the message. "The first message said, 'Ship unable to maneuver. We'll fight to the last shell. Long live the Leader.' Then, 18 minutes later, near midnight, Lütgens sent directly to Hitler, "We'll fight to the last in trust in you, our Leader, and in rock-hard trust in Germany's victory." The last message was sent just before the final engagement began, 'Send a submarine to save War Diary.' Each of the messages was encrypted. We intercepted no clear text messages throughout the various engagements with *Bismarck*, and there were no transmissions of any kind from *Bismarck* once the final engagement began. Sir Dudley's remarks were shadowed by the Boniface material, which we both were aware of this morning."

"Amazing!"

"We thought so as well. They had to be struggling with torpedo damage all night, and they clearly found at least a partial solution by morning. Tovey maintained radar contact throughout the night," A.V. paused. The Prime Minister asked no questions. "I might add, we have not been able to interrogate the survivors. Sir Dudley immediately issued instructions to not question the survivors, and more specifically to treat them with the respect due them. They will be turned over to the professionals at Trent Park. By the way, as of 30 minutes prior to this meeting, we received from Tovey, they have recovered only two officers, so far; something like 55 at last count."

"Thank you very much, A.V. Please keep me informed as your investigation proceeds. I am truly relieved that this particular threat has been eliminated."

"Yes sir. We certainly will do so."

Both men stood and split outside the conference room.

—

Friday, 30.May.1941
No.10 Downing Street
Whitehall, London, England
United Kingdom

"Clemmie!" Winston exclaimed when his wife appeared in his No.10 office unannounced and unexpectedly. "I didn't know you were coming into town. I was just about to depart and join you at Chequers for the weekend. To what do I owe this pleasure?"

"It was a last minute, spur-of-the-moment decision on my part, Winnie. We are going to Chartwell for the weekend."

"Clemmie, my precious darling, there is no staff there. The house has been closed since the Phony War ended a year ago and the bombing started."

"Yes, I know quite well. These last two weeks have caused me grave concern for your health, Winston. You have not been sleeping well and you have been under enormous stress with this *Bismarck* affair, and the situation in North Africa and the Mediterranean. This is not a discussion, my dear. I have already organized things with Sir Edward, Pug, your administrative staff . . . they are all going to leave us alone the whole weekend – no staff, no paperwork, no messages, no Sawyers, just the two of us, not even Inspector Thompson. You are going to rest, Winston. I shall accept no compromise."

Churchill was somewhat in shock. "I don't know what to say, darling."

"There is only one thing to say . . . yes dear."

"Yes dear."

With that and a last few instructions to his duty private secretary, John Martin, they were off in the Prime Minister's limousine for a weekend of roughing it and relaxation at their closed-up country house. His staff would also enjoy a much needed weekend break from the intensity of their constant running to keep up with the boundless energy of their beloved Prime Minister.

—

Chapter 5

If nature had arranged that husbands and wives
should have children alternately
there would never be more than three in a family.
-- Lawrence Housman

Monday, 2.June.1941
RAF Martlesham Heath
Woodbridge, Suffolk, England
United Kingdom
10:20 hours

Twelve Hurricane fighters landed safely, taxied to their assigned parking spots and shut down their engines to secure the aircraft. The red gunport tapes on all twelve sets of wings were still in place. The mission had been another boring convoy patrol – no action, not even a tickle. The convoy escort missions a year after the intense aerial combat over earlier convoys transiting the English Channel during the summer of '40 were even lower key than training missions.

"No squawks, Henry," Brian reported to his crew chief, Corporal Henry Jacobs, as he removed his flying gloves.

"Excellent, sir. We'll fill 'er up with petrol and give 'er a good look-see in short order."

"Thanks, Henry. I don't know if we have any other assigned missions today."

"No worries. She'll be ready, if something pops up."

Brian entered the Dispersal Hut along with the other pilots.

"A word, Mister Drummond," Corporal Harris said.

Brian removed his helmet with flying goggles and oxygen mask attached, but left the rest of his equipment on. He walked up to the operations clerk's elevated desk. Harris leaned forward, clearly wanting to whisper something to him.

In a very soft, barely audible voice, Harris said, "Sir, the Hampshire Central Hospital called to inform you that your wife has been taken to hospital in labor."

Brian wanted to react but fought against the urge. "How is she?"

"They did not say, sir. They suggested, if you could, your presence might be useful."

"OK," Brian's thoughts raced through all kinds of scenarios. *She was supposed to have the baby at home with Midwife Barbara Grey. Something must have gone wrong.* "I'd better see the Skipper," Brian whispered back.

"That would be advisable, sir."

Brian purposely avoided the eyes of his brethren, who were undoubtedly curious about the unusual, muffled exchange with Corporal Harris. He went directly to Taylor's office. Brian did not have to knock. Squadron Leader Taylor motioned for him to enter and close the door. Brian stood in front of Taylor's desk, not wanting to sit down.

"We do not know anything more than the medical staff believes your presence is important," Taylor said. Brian began to shift back and forth on his feet, and fidgeting with his headgear. "I have placed you on indefinite medical convalescence. One of the spare Hurricanes is fueled and ready. I want you to take it and fly directly to Middle Wallop. I have already arranged for a staff car to be waiting upon your landing, to take you directly to the hospital. Please let us know what is happening and what we can do to help, as soon as you are able. Now, get out of here."

Brian bolted from the office, ran out of the Dispersal Hut, and sprinted to the spare Hurricane fighter. Corporal Jacobs already had the engine running, as if Brian was on a scramble launch. His crew chief helped him strap into the parachute and fasten his seat harness. He plugged in his helmet microphone and earphone jacks, and connected his oxygen mask. Brian shouted over the engine noise, "Thanks, Henry. See you in a few days."

"Good luck, sir." Henry jumped down, ran around the left wing, and then the spinning propeller to pull out and hold up both wheel chocks.

Brian saluted and called for takeoff, which he received promptly. He released the parking brake and advanced the throttle to taxi. He skipped his engine run-up checks but completed his control checks. As soon as he was given clearance to takeoff, he switched his Pipsqueak on, released the brakes and slowly pushed the throttle forward. As the aircraft gained speed and the rudder became effective, Brian pushed the throttle forward to the emergency gate wire – takeoff power. The aircraft literally jumped into the air. Brian retracted the landing gear and kept the nose down to gain speed as quickly as possible. As he approached the airfield boundary, he began a slight climb to clear the tree line and rolled right to a heading of 225 degrees magnetic. He climbed to sufficient altitude for cool air, to allow him to keep the engine at full power. Brian resisted the urge to push the throttle through the wire gate to emergency power, since he felt his situation was an emergency. *No need to punish the engine.* Brian followed the instructions of each sector control station as he flew directly over the center of bomb-damaged London. His mind was too focused on his immediate task at hand and what lay ahead to appreciate the view.

Monday, 2.June.1941
RAF Middle Wallop
Middle Wallop, Hampshire, England
United Kingdom
11:15 hours

The transit and landing were uneventful. He taxied to the base operations building. A staff car was indeed waiting directly behind his parking spot. Brian was surprised to see Wing Commander Herbert Worly, the base commander, standing at the front of the car. He turned his tail, switched off everything, caged his attitude gyro and secured his engine. Brian left his helmet connected, removed it, and placed his headgear next to the gunsight. He unstrapped from the seat and his parachute. Brian stood up, straightened his uniform and stepped out of the cockpit. He closed the canopy.

Worly was standing at the wing root, when Brian jumped off the left wing. Brian saluted and Worly returned the salute, and then the senior officer extended his right hand to Brian. They shook hands. "Great to see you, again, Hunter. We have a car waiting to whisk you off to Winchester. I hope everything is OK."

"Thank you, sir. I don't know and I need to find out. Thank you for the car."

"Get along, now, lad. You have more important things to tend to than blabbering with me."

Brian saluted and jumped in the car. The driver left immediately. He was enormously grateful for the organized and prompt response to help him get to Charlotte.

—

Monday, 2.June.1941
Hampshire Central Hospital
Winchester, Hampshire, England
United Kingdom
12:30 hours

Brian had been to this hospital too many times, but this time it was for Charlotte. He thanked the driver for his expert transportation and released him. He walked into the entrance lobby, not his past point of entry, and had to orient himself. Brian noticed an elderly woman behind a desk, who looked like a receptionist of some sort.

"Excuse me, ma'am. I am looking for my wife," Brian stated.

"May I ask who your wife is?"

"Oh, damn, excuse me . . . Charlotte Palmer, oh wait, no, Charlotte Drummond."

The woman giggled a little. "Which is it?"

"Drummond . . . Drummond, ma'am."

She raised a finger, and then raised the telephone handset from its cradle. She did not dial. "What is she here for?"

"I don't know actually. I was just told a few hours ago that she was brought here. She may be giving birth."

The woman checked a number and spun the dial. She asked for Charlotte, nodded her head a few times, and then looked up to Brian. "She is in labor." The woman gave Brian directions to the maternity ward waiting room.

Brian walked fast, just short of a run, to the waiting room. He asked several people to help him find Charlotte and considered just ignoring the gatekeepers. *It is much easier to seek forgiveness than to gain permission.*

A nurse came to tell him, "Your wife is in labor. We have notified the attending nurses. One of them will come talk to you, when they are able."

"I just want to see her, to support her. She's having our baby."

"I understand, sir. They must keep her mind on the task before her."

"I'll just go find her," Brian said and started to move past the nurse.

She grabbed him by the arm. "I would not do that, sir. This is wartime and we have armed guards, who take their assignment quite seriously."

"My wife needs me."

"The rules are here to protect everyone. Please let the medical staff do their job. Your presence would be a distraction."

Brian nodded his head. "Someone called my squadron this morning and said I needed to get down here to support my wife. Here I am."

"I don't know anything about that. Please be seated. As I said, an attending nurse, or perhaps the doctor, will come out here to talk to you, as soon as they are able."

Brian nodded his head in resignation. The nurse left the waiting room. Brian began to pace back and forth by the door. He hated waiting and this was the worst kind of waiting. Brian did not realize there were two other men in the waiting room, until one of the men stood, approached him and said, "First time?"

"Yes," Brian answered without looking at the man, or altering his nervous pacing.

"A Yank, are you?"

Brian stopped and faced the man. "Yes," he responded with a tinge of defiance in his tone.

"Easy, lad. I meant no offense . . . just curious. You are in the uniform of an RAF officer."

"Yes. I'm a fighter pilot with Seventy-One Squadron."

"Ah, the Eagles . . . according to the newspapers."

"That is what they call us."

"Well, God bless you, son, for your service to the kingdom. Now, if I may be so bold, birthing is usually a long process. You floppin' about like a caged tiger will not help you . . . or her."

"OK. She was supposed to have the baby at home with a midwife."

"The usual way, these days. The midwife probably saw something she did not like and felt your wife should have the assistance of a doctor and proper hospital facility. The same with my wife, I must say."

"Oh, OK. How long have you been here? How many have you had?"

"Since early this morning." He chuckled a little, and then continued, "I've not birthed any, I'm afraid, but this is the fourth for the missus. Now, why don't you take a seat and perhaps try to rest a little. They are not going to allow you back there until everyone is safe and settled, and the mess cleaned up. They don't want you fainting at the sight of her blood."

"Blood?"

He laughed, again. "A normal part of the process, I'm afraid."

Brian nodded his head, and then shook his head, repeating the confusing sequence several times. The man returned to his seat. Brian accepted the suggestion and sat down, but his knees continued to bounce with his nervousness. As time ticked on, Brian began to relax a little. After several hours, a nurse notified the man who had not talked to him that he was now a proud father of a daughter. The notification for the man who did talk to him came an hour or so after that. He was alone in the waiting room. Brian continued to wait. No one had come out to talk to him. He went out to the hallway to the only windows he could find – the sun had set some time ago, as they were past twilight. At least there is no rain. Brian returned to the waiting room. He closed his eyes in an effort to relax.

—

Tuesday, 3.June.1941
Hampshire Central Hospital
Winchester, Hampshire, England
United Kingdom
00:50 hours

Brian bolted upright when someone touched his shoulder. His jerky response startled a new nurse. He rapidly looked around the room to reorient himself, and then looked back to the nurse.

"Mister Drummond, I presume?" she asked.

"Yes."

"I apologize for startling you, sir. I came to inform you that your wife is fine and resting, and your new son is healthy and doing well."

"Oh God, thank you," Brian said, and then sprang to his feet and hugged the nurse tightly. "What time is it?" he asked instinctively, and then looked at his aviator's wristwatch. It was nearly one o'clock in the morning. "When can I see them?"

"Right now, if you wish. Your son was born shortly after midnight – ten fingers, ten toes, a healthy cry, good color, and they have him in the nursery for the moment. I must caution you, Charlotte has been through a difficult and longer than average labor. I urge you to hold down any excitement. She does need to rest."

Brian nodded his head in agreement and understanding. The nurse led him down a corridor, through a larger room with several beds partitioned off by curtains, and into a separate room. Brian saw Charlotte. Her eyes were closed and she was breathing peacefully. He wanted to jump to her and embrace her tightly, but he was mindful of the nurse's admonition. Brian moved silently to her bedside and stood still smiling down at her. *My God, she is a gorgeous woman. How on God's little green earth did I get so lucky to find her? . . . or rather, to save my sorry ass from her pond?* Brian had no idea how long he had been standing there, not touching her and not even swallowing, when her eyes opened slowly, taking a few seconds to focus on him.

"What are you doing here?" she mumbled in a clearly fatigued voice.

"You may not have realized this, but my wife was having a baby, and I told her I would do my best to be there."

"You silly boy," she said. "I did not expect you, Brian, but I am so glad to see you." Her smile vanished in an instant. "Are you mad at me for some reason?"

"No, why?"

"You have not kissed me, yet."

"They told me not to get you excited."

"Oh damn you, don't be so full of yourself. Kiss me, now, Yank!" she commanded.

Brian needed no further prompting. He leaned over and kissed her gently, but she was not satisfied. She wrapped her arms around his neck and held him tightly to her. They kissed passionately. *Damn, I want to touch her. But, what is acceptable to touch?* She must have sensed his thoughts.

"It's OK, Brian. I'm not going to break. I would not touch anywhere near my hips right now, and my breasts are a little sore, but just don't squeeze them." Brian gently touched and caressed her left breast through the sheet and gown. "Barbara tells me I need to start nursing him soon, or they will need to suction my breasts to relieve the pressure."

"I can help you with that."

"Oh you naughty boy. The baby gets first dibs." Brian nodded his head and continued to fondle her firm, engorged breast. "We talked a little about names."

"Yeah."

"Would you be offended if we named him Ian?"

"For your first husband?" She nodded her head. "No, sweetheart, I would not be offended. I've always liked Ian as a name."

Charlotte smiled. "You are a good man, Brian Arthur Drummond. I'd like his full name to be Ian Stanley Drummond . . . for my father as well."

"A noble name, it seems to me."

"Thank you, Brian."

A nurse entered the room, carrying a swaddled, crying newborn. She looked at Brian, "I will have to ask you to leave, sir. Please step out into the hallway."

"No, he stays. He is my husband and the father of this child."

"Very well," she said and handed the baby to Charlotte. The nurse cranked the head of the bed to place Charlotte is a more seated position.

Before the nurse could assist, Charlotte had pulled the sheet and her gown off exposing her left breast. The baby took instinctively to the proffered nipple. Baby Ian seemed to struggle getting what his body wanted. Charlotte's patience and persistence impressed Brian. She squeezed her aerola several times to express some milk for the baby to taste and trigger his instinct. Her commitment eventually yielded results. Brian kissed his son's forehead and smelled that unique baby aroma, as their first-born son remained focused on his singular objective. When Brian eventually shifted his attention, he noticed Charlotte's head was laid back, her eyes were closed and her grimaced expression conveyed pain.

"Are you OK, Charlotte?"

She raised her head and opened her eyes. "Yes. The relief feels so good . . . finally." Charlotte looked down at their son on her breast. "He seems to be doing well," she said to the nurse watching over the process.

"Yes, he is . . . especially for being so new. You might want to see if you can get a burp out of him, and then let him suckle on your right breast to keep them balanced. You don't want one empty and the other one full." The nurse demonstrated the proper technique she espoused.

Charlotte did not have childhood teachers or examples, and she appreciated the assistance. She took to the task with ease. Without a hint of modesty, Charlotte uncovered her right breast and cradled Ian in her right arm. The baby searched for the nipple and found it. Ian had been at her breast for a few minutes and lapsed into slumber. He remained attached for several minutes. Brian kissed the boy's forehead several more times. He could not get enough of his new son. Ian eventually let go of his momma. Both his parents marveled at their creation.

The nurse lifted the baby from his mother's arms. "You really should get some rest, ma'am." She left the room to return their son to the nursery.

"I am very tired, Brian."

"Go ahead, sweetheart . . . sleep. I'll be right here when you wake up."

Charlotte closed her eyes and soon fell into deep sleep. Brian covered her with the sheet. He found a chair and plopped down with an audible thud. He watched Charlotte, to see if he had disturbed her sleep. She was out. It has been a very intense day. Sleep did not take long to claim Brian as well.

—

Thursday, 5.June.1941
Oval Office
The White House
Washington, District of Columbia
United States of America
09:30 hours

Harry Hopkins knocked and entered the Oval Office interrupting the President's reading. Assistant Private Secretary to the President Grace Tully followed him in and left the door open. Both had grave expressions that stopped the President.

Roosevelt spoke first. "What happened last night?" he asked, referring to an official dinner party event.

"Missy collapsed toward the end of the dinner party, last night. Doctor McIntire checked her right then, and believed she was simply exhausted and perhaps dehydrated. He drove Missy himself to her apartment. He insisted she rest at least through the weekend and possibly a week. Doc said he would check on her every few days."

Rear Admiral Ross T. McIntire, USN, MC, had been the Physician to the President since 1933, when Roosevelt took the oath of office the first time.

Grace Tully and Missy had been friends and co-workers for almost as long.

"She told me," Grace added, "that she was feeling light-headed and dizzy right before she collapsed."

"Doc said she regained consciousness rather quickly, once she was prone," said Harry.

"Did she hurt herself?" asked Roosevelt.

"Not that Doc indicated to either of us. I will track him down and see if I can get an update for you."

"Thank you, Harry, and thank you, Grace, for taking over for Missy while she recovers."

"My pleasure, Mister President," Tully responded.

"Now, I would like you to arrange a visit to Missy's apartment this afternoon. I want to check on her myself, to make sure she is OK and getting everything she needs. Move the schedule as you see fit."

"I'll take care of it, Mister President," answered Hopkins. "I'll take a look at the calendar right now, and Grace and I will notify the appropriate people. Would you like me to arrange for Doc McIntire to be present for your visit?"

"That is not necessary. I will be visiting Missy as her friend, not her employer."

"Very well, Mister President," Hopkins said. They both left the President, as he returned to his reading.

—

Monday, 9.June.1941
RAF Martlesham Heath
Woodbridge, Suffolk, England
United Kingdom
10:35 hours

"Welcome back, Hunter," several of the pilots said in unison. The pilots gathered around the youngest of the squadron pilots. They bombarded him with questions.

Brian held up his hands. "Where's the skipper?"

"Gone!" Red Burns exclaimed.

"Gone, what does that mean?"

Whitey explained, "The Air Ministry decided Billy was too old to command a fighter squadron."

"He's been in command since January. They couldn't have figured that out before they gave him command?"

"Apparently not," Whitey responded.

"That sounds like a really bogus reason to remove a commanding officer," Hunter said. "He wasn't a great CO, but he was a good CO." Oddly, the Dispersal Hut remained quiet. *Damn, is there something I'm not being told, that I don't know what's going on here . . . that everyone does not want to mention or discuss?* "Who is the CO, now?"

"For the moment," Whitey said, "I am . . . as the senior officer in the squadron, but acting only, according to Group."

"Are we going to get a new CO . . . not that you won't do, Whitey."

"Well, thanks for that vote of confidence, Hunter," answered Whitey. "That is my expectation, although Group did not tell me anything."

"So, are you a new daddy?" asked Bulldog Tolly, to change the subject.

"Yep, baby boy."

The pilots slapped him on the back or punched him in the shoulder.

"So, that means the drinks are on the new daddy," Pete Peterson added.

Brian nodded his head and smiled.

"And, the missus?" Red Burns inquired.

"She's fine . . . adjusting to having another baby to take care of in the house," Brian quipped.

"You mean, in addition to you?"

"Exactly."

Everyone got a laugh out of that.

"Hey, Hunter," Red Burns began. "Harness came down last week while you were gone. He said something about getting married next month. He wants you to call him, or fly up and see him if you get a chance."

Whitey did not skip a beat. "Do you need a refresher flight, Hunter?"

"I'll never pass up a reason to fly."

"Do you want to take your wingmen with you?"

"Sure, we'll go," volunteered Dusty Langford.

Brian looked to Bulldog, who nodded his head in agreement.

"Yeah, I'll take the section."

"Fine," Whittington said. "Corporal Harris, please notify Group, Green Section will be airborne for refresher training."

"Very well, sir," Harris responded and fulfilled his task.

"Be back before sunset."

"You got it, Skip. Let's go lads." They grabbed their flight equipment and headed to their aircraft.

Green Section took off together in a 'V' flight of three. Once they cleaned up their aircraft, Brian turned them north-northwest. Sector Control gave them a volume of airspace for training purposes. Brian took them through a series of formation aerobatic maneuvers. It felt good to be flying, again.

—

Monday, 9.June.1941
RAF Catterick
Catterick, North Yorkshire, England
United Kingdom
11:50 hours

Green Section landed without incident and taxied their Hurricanes to the Spitfire line of No.609 Squadron. Jonathan was waiting for Brian by the time he reached the ground. They embraced as old friends. Brian introduced everyone.

"Congratulations . . . papa," Jonathan said.

"Thanks mate."

"I guess your lads informed you that I came to visit last week."

"Yep. Was it just a social visit, or was there an objective?"

They followed Jonathan's lead to a set of empty lawn chairs outside the squadron dispersal hut. Jonathan and Brian sat together. Langford and Tolly sat together a couple of chairs distant from Brian.

"I suppose, asked like that, it was both. We don't get to see each other much these days."

"True."

"More importantly, Linda and I decided to tie the knot . . . like you and Charlotte did."

"Outstanding. When?"

"Next month. Linda wanted the ceremony to be at Carlingon, rather than some clandestine affair like you did. We are hoping that you and Charlotte will be able to attend. It will be a simple affair . . . nothing lavish or large."

"Charlotte just gave birth a week ago, so I have no means to predict how she will feel."

"Check with her. Unless you tell me otherwise, we'll plan on both of you joining us."

"Will Rosemary be there?"

Jonathan laughed hard. "I surely hope so. She is my sister, after all."

Brian joined in the laugh. *That did sound rather funny.* Pilots began to file out of the Dispersal Hut. Several who knew Brian punched him in the shoulder. Squadron Leader Jason Billings 'Stack' Long-Roberts, commanding officer of No.609 Squadron, was the last pilot out of the building.

"Great to see you, again, Hunter." Stack said. "We're on 30-minute status, Harness, best get some lunch before we return to Available. Group thinks they might need us for a mission this afternoon."

"OK, Skipper," Jonathan responded, and then said to Brian. "You'd better come along with your lads, so we can get you fed as well."

The walk and conversation brought back memories for Brian. His two wingmen displayed their curiosity and eagerness to learn what they could from Jonathan's experience flying Bf109E-3 and Me110C-1 fighters last year as well as some limited experience flying in the He111 and Do17 medium bombers. Jonathan was a good sport about all the questions. Their lunch was the usual fare – cheese sandwich, potatoes and tea. They returned to the flight line just as No.609 Squadron was posted to Available status, and not ten minutes later, they were raised to Readiness status.

"We'd better get out of your way, looks like you're going flying soon," Brian said.

"So it seems," responded Jonathan. He retrieved his flight equipment from the Hut interior. "Before you go, I was thinking about inviting Air Commodore Spencer and his wife. After all, he was the one who posted me to those exploitation flights. What do you think?"

"They had their baby a month before we did?"

"Ah yes, another of your off-spring."

Brian looked around rather nervously, not expecting Jonathan to be quite so cavalier with his knowledge. *Tolly and Langford heard what he said, but hopefully they did not pick up on what it meant.* "It's your wedding, mate. Invite whomever you wish."

"So, you would have no problem, if we did invite them?"

"No. Why should I?"

"I just wanted to make sure you are on speaking terms with them. I do not want you to feel awkward."

"No worries mate. Now, we'd better launch before you blokes take to the air. I'll talk to you later."

"As you wish, mate. Thank you for making the visit. Great to meet you lads," he said to Tolly and Langford.

The Green Section Hurricanes took flight in short order and headed south. They did not hear Jonathan's squadron on the radio before they switched frequencies. Brian led them through several more formation aerobatic maneuvers on the way back to base.

Once they secured their aircraft and reported their status to their crew chiefs, the three pilots walked back to the Dispersal Hut together.

Dusty Langford was the first to break the silence. "What did Harness mean by 'another of your off-spring'?"

"Nothing," Brian answered, trying to be as calm as he could be. "Just an inside joke. We've known each other and flown with each other since the OTU in the summer of '39. We went through The Battle together in Six Oh Nine Squadron. He likes to joke around a lot . . . more than he should."

The answer seemed to satisfy at least Langford's curiosity, but Brian had no way of knowing for sure, and he sure as hell was not going to raise the topic with anyone. The squadron had remained exactly where they had left them in the morning. *Another slow day. Damn, if Jonathan is going to invite them, I should tell Charlotte before the wedding. Surprising her with that news would not be good. At least, I've got some time to think about it.* Brian settled into a lawn chair slumber in the warmth of the spring afternoon sun.

—

Chapter 6

The enemy of my enemy is my friend.
> -- Proverb
> Derivative of Kautilya, Arthasastra
> (circa 300 BC)

Wednesday, 11.June.1941
No.10 Downing Street
Whitehall, London, England
United Kingdom

"**E**xcuse me, Prime Minister," announced his duty Assistant Private Secretary John Martin, "Colonel Menzies has arrived with the Buff Box."

"Very well, show him in, if you please, John." Martin stood aside to allow 'C' to enter the prime minister's office. 'C' walked directly to the desk. They both waited until the door was closed and latched. Churchill retrieved the only key in the building from inside his shirt, and then he unlocked the case. He retrieved a single sheet of paper from the 'Urgent' folder.

"Station X decoded the message this morning," 'C' said, as Churchill lifted the paper to read.

MOST SECRET - ULTRA

```
SECRET
DATE: 11 JUNE 1941
TO: HGN HGZ HGS
FROM: OKW
BARBAROSSA
BEGIN
EXECUTE DIRECTIVE TWO ONE ON OR ABOUT DAWN 22
JUNE 1941 BREAK REPORT READINESS OF ALL UNITS
NO LATER THAN 14 JUNE BREAK HAIL VICTORY HAIL
HITLER END
SECRET
```

MOST SECRET - ULTRA

"So, it is finally official," Winston said without taking his eyes off the paper.

"The massing of German forces along the entire eastern frontier has only one purpose, and now we know the date they intend to pull the trigger. We have no evidence – zero – that the Germans have deviated from their operations plan. We certainly have no doubt they intend to execute the plan exactly."

"I do so wish we could share 'Boniface' with the Russians. Have we seen any signs the Russians have even heard our warnings of the last few months?"

The code word 'Boniface' had been used since the early days of the British, and to a certain extent their allies, efforts to break coded messages from within the German command structure that had been encrypted by their vaunted Enigma device. Thanks in no small part to the extraordinary work of Agent Diamond (Trevor Andersen) and his compatriot, former and now expatriate Chief of Polish Secret Police Colonel Stanislaus Pordonski, codenamed 'Blocker.' The two men, along with three of Blocker's field operatives, captured an intact Enigma device being transported through Poland during the winter of 1939. After several years of Herculean struggle with large doses of luck, the Government Code and Cipher School at Bletchley Park had begun to read German command message traffic. The yield was still spotty. Once His Majesty's Government began to actually believe they could decode Enigma messages, they collected the work in its entirety under the classified information compartment 'ULTRA.' The very existence of Enigma, ULTRA or 'Boniface' was the most highly protected secret within His Majesty's Government. A tightly controlled list of several dozen intelligence, military and ministerial defense leaders had been read into the program.

"Our sources in the Soviet Union indicate both the NKGB and NKVD have repeatedly communicated their apprehension with the buildup of German forces along the River Bug. So, we know they are aware, but the best we can tell is Stalin remains convinced our warnings are an intentional and deliberate provocation intended to break the alliance between the Soviet Union and Germany. It would seem that Stalin is convinced and has instructed his generals that the Germans would not dare risk opening a second front with an ally no less, as long as Great Britain remains un-subdued."

"My God, his own intelligence people are validating the information we have been providing them."

"Based on what we see, the only thing that might shake Stalin's unitary belief is the artillery barrage soon to come."

"And, perhaps not even then."

"Per your instructions, we gave them the essence of Directive Two One, when we decoded it last December. To my knowledge, Stalin and the Soviet government have steadfastly refused to even acknowledge our warnings. They persist in their demand to know the source or sources of such outlandish claims . . . to use their words."

"Tragically," Churchill said, "we have proof they have been penetrated by the *Abwehr*, so any detailed or specific information shared with the Russians would be tantamount to giving it to the Germans. Unfortunately, the Soviet people are soon to suffer enormously for the failure of their government. It is too bad we could not fly some aerial reconnaissance missions and show them the photographs."

Churchill re-read the message and lapsed into contemplation. He read the message, yet again, before he spoke. "What are 'HGN,' 'HGZ' and 'HGS'?"

"We do not have independent corroborated information to definitively state the meaning. However, our best guess is Army Group North, Center and South . . . in German, *Heeresgruppe Nord*, for example."

"Army groups," Churchill responded with some surprise. "That suggests each group has two or more field armies with each of those with two or more field corps comprised of multiple divisions. Surely the Soviets must have this information."

"At this stage, Winston, I am not convinced they would even believe such physical evidence. Our developed order of battle estimates place in excess of 100 armor and infantry divisions along with three air fleets along the River Bug frontier—perhaps two million men or more. Such massed armed forces are impossible to hide."

"Yes, well, the Germans have apparently done a masterful job in manipulating the psyche of the Russian leadership to convince them they are not seeing what their eyes absorb. Such is the course of fate. I will need to brief the War Cabinet in a few minutes. Would you be so kind to remain and join in that meeting? I shall instruct Sir Edward to make it a closed meeting, so we can discuss this latest information."

"I am at your command, Prime Minister."

"Very well. My inclination is to send a personal, private message from me to Stalin as one last ditch effort to open his eyes."

"I do not need to say that we must be very careful with such a letter. As you say, we know the Germans have penetrated the NKGB and NKVD, and quite likely, the Politburo itself. So, whatever message we send to anyone in the Soviet government will be read at least in summary, if not directly, by the Germans."

"Understood. Now, let us go talk to our colleagues."

Churchill returned the message to the proper folder and re-locked the Buff Box. Once secure, the two men walked the short distance down the hallway to the Cabinet Room. Churchill whispered his instructions to the Cabinet Secretary, who in turn cleared the room. The important meeting began shortly thereafter.

—

Saturday, 14.June.1941
American Eagle Club
No.28 Charing Cross Road
Covent Garden, London, England
United Kingdom

The Green and Yellow Section pilots of No.71 Squadron stood in front of the entrance – a single, black, oak door with the polished brass numerals '28' on the door with lettering on the stone frame above the door that said, The American Eagle Club. The four-story brick building was rather nondescript and looked exactly like so many other buildings in London, at least the buildings not damaged by the German bombing. Various small shops – a haberdashery, a bookstore, and a rather interesting candy store – occupied the ground floor. The Eagle Club entrance filled the space between book and hat stores. Some of the surrounding buildings displayed unrepaired bomb damage from The Blitz. The roadway was surprisingly clear, although the temporary bomb damage repairs made the streets less than ideal.

"This must be it," Pete Peterson announced.

They had all heard about this new club that had opened and was being operated by the American Red Cross since April of 1941. It was said the club was supposed to be a touch of home for the growing number of American citizens who were living, working and fighting in Great Britain . . . well, at least in London. The club apparently occupied all four above ground floors and an unidentifiable expanse of each floor, while the grill featured typically American cuisine.

"I haven't had a real hamburger since I left the States," added Dusty Langford, "when the war started, and Rusty and I went off to France to join *le Armée de l'Air*, or *le Service Aéronautique* as it was known before it became the air force."

"I hope they have enough for all of us," Bulldog Tolly said.

"Only one way to find out," Pete responded. "Shall we get some American grill lunch before the press conference?"

"Sounds like a plan," Sweet Sweeny added. "My Dad said he was going to try to make it to the press conference. We shall see."

Peterson led them into the building and to the only a stairway to climb. On the first floor, a reception / help desk greeted them. Four Red Cross ladies welcomed them. One of the ladies, who happened to be British from Manchester, gave them a quick briefing. They had been expected. The Press Conference was scheduled to begin in 90 minutes in the theater on the fourth floor. Everything was ready for them. The receptionist said they expected several dozen reporters and photographers, primarily from the United States and Great Britain, but some from all over the world, including a "journalist" from Imperial Japan. The grill and dining room occupied the entire second floor, and various "help" rooms, as she referred to them, were located on the third floor. The help rooms had limited shower facilities, sewing and laundry services, travel assistance, a small library, and a reading and writing room, and even a small telephone room. The pilots thanked her for the information and went directly to the grill.

Most of the patrons on this day were enlisted men from each British service as well as a good smattering of U.S. military uniforms. Officers dressed in the uniforms of all three services made up roughly a quarter of the clientele. What was distinctive about these men was the uniquely American language accents – Brooklyn, Chicago, the South. Brian had no idea there were so many Americans in England. It was almost like they were home . . . except for the uniforms.

None of them could believe their eyes. Each of them ordered a hamburger with all the fixin's, French fries, and a Coca-Cola in all its chilly effervescent grandeur. None of them could believe they actually had fresh tomatoes, onions, lettuce and even fresh mayonnaise and catsup. *How did they do it with war rationing all around them?* They savored every bite, as they moaned and groaned through their meal in what had to sound like orgasmic delight. Dusty and Rusty both went back for a second hamburger, while Hunter and Bulldog enjoyed another Coca-Cola.

"We'd better head upstairs, guys," Pete broke the momentary respite. "The afternoon's shindig is supposed to start in 15 minutes or so."

They grumbled, but did as the senior officer suggested. They bussed their plates, utensils, bottles and glasses, and then followed their leader. Surprisingly, there was a functional elevator, or lift as the Brits called it, yet Pete chose to use the stairs and the remainder of the Eagle Squadron pilots followed.

The theater was actually a large room with two rows of support pillars, splitting the room into thirds. A small stage with a movie screen on the wall behind it occupied the center third of the room at the far end from the double doors. A long table with a green cloth draped over it, an American Red Cross banner on the front, and six chairs behind it took up near the entire width of the small stage. A stack of bulky microphones with placards labeled CBS,

NBC, BBC, ITV and other not easily discernible markings stood at the center of the table. Folding chairs provided the seating for the audience.

A young, attractive, brunette woman in a blue dress covered by a large white apron with a large red cross on it approached them. "Y'all must be from the Eagle Squadron and our guests today," she said.

"Yes ma'am," answered Peterson.

"I'm Julie Browning," she said and shook hands with each pilot. "I'm the duty manager today. I will introduce you for the press conference. We are expecting a couple of dozen, maybe more, journalists and photographers for this event." Each pilot introduced himself. She ticked off the names on her clipboard. "Excellent. You are all present."

"Any relation to Browning Arms?" Rusty asked.

"No, not that I'm aware. I do get the question quite a lot. I wonder why?"

"Because of our machine guns," Rusty answered. Everyone laughed except Bateman, who was the only one who did not get Browning's faux question.

"I'm sorry Mister Bateman. I was only kidding. I know quite well who makes the guns in your wings."

Rusty was clearly embarrassed but fortunately did not react to being the unaware jester.

Miss Browning escorted the pilots to the table. Peterson wanted them seated by rank seniority inside to outside. Brian ignored the suggestion and took an end chair before anyone else did. He wanted to be as far from the microphone stack as possible. Brian was quite content to let Pete Peterson and Rusty Bateman do all the talking. They seemed to enjoy it. Pete started to insist but quickly backed off. Brian was clearly the most accomplished of the Eagle pilots and also the least interested in publicity. In that sense, Malcolm had been a very good teacher.

Surprisingly, the room began to fill up. Film and still photographers lined both sides. Reporters nearly filled the first three or four rows of chairs. By the time Julie began her introductions, the theater was full, with many standing at the back. Brian recognized Edward R. 'Ed' Murrow in the third row toward the end closest to him. He made eye contact and nodded his head in recognition. Murrow responded with a smile and a nod of his head.

The press conference progressed as all the press events did for the Eagle Squadron. The questions were the same. The answers were the same. Several attempts were made by various reporters to engage Brian, but he would only respond with simple, non-specific answers that were often supplemented by Pete or Rusty. Flash bulbs burst bright light into the room throughout the session, and every so often when a moment of quiet occurred, the whirr of the film cameras could be heard. The whole affair lasted over an hour, and then mercifully, Julie Browning interjected to end the press conference, so they could get ready for a movie showing scheduled in 30 minutes.

The other pilots rose and mingled with reporters and members of the audience.

Brian remained seated for several minutes to let the crowd dissipate. Before his threshold of departure arrived, Murrow approached, clearly intent upon a word. Brian stood and stepped down off the stage. The two men shook hands.

"I guess you were not feeling particularly talkative today, Brian."

"No sir. I would not have been here, if I had not been ordered to attend. Just not my cup of tea."

"I am sorry to hear that. You are a great story."

A medium build man, older than Brian, approached them, dressed in the uniform of a Royal Navy Volunteer Reserve lieutenant with the peculiar double zigzag, sleeve stripes and wearing standard, black-rimmed, military issue eyeglasses.

Murrow saw Brian's eyes focused on the man and looked over this shoulder. "Draper, great to see you, again."

"Nice to see you, Ed."

"Brian, may I introduce Lieutenant Draper Kauffman, a countryman of ours serving with the Royal Navy. Draper, this is Flying Officer Brian Drummond."

The two men shook hands.

"Well, this is a most impressive gathering," Ed announced. "Brian, you may not be aware of Draper's accomplishments, please allow me to sing his praises. He volunteered for the American Volunteer Ambulance Corps in France in February of last year. He served valiantly during the invasion, was captured in mid-June, held by the Germans as a POW, was released in August, made his way to England, and has been a leading explosive ordnance disposal specialist through The Blitz – a very busy man."

"Wow! That is impressive, Mister Kauffman."

"Draper, please."

"Brian is . . . ," Ed began but stopped when Kauffman help up his hand.

"No need for the résumé, Ed. I have been following his accomplishments since your first broadcast report about him last year. It is a genuine honor to meet you, Brian."

"The honor is mine, sir."

"Again, please, Brian, let's keep this on the familiar. I understand you are the youngest American pilot over here right now. How on earth did you manage that?"

Brian told him the abbreviated version of Malcolm teaching him to fly, leaving home after high school, and joining the Royal Air Force before the war started in Europe.

"So, you flew through the entire, what they are now calling, the Battle of Britain."

"That's true."

"I wanted to fly, but my eyesight was just not good enough. In fact, my damn eyesight kept me from gaining a commission when I graduated from the Naval Academy in '33."

"I did not know that little fact," Ed interjected.

"Not something I'm particularly proud of, actually, but it is a fact nonetheless. Anyway, I cannot imagine how you guys flew so much, against those odds, day in and day out."

"It was not easy, believe me. I don't think any of us have ever been that tired and spent. I am ashamed to say we all breathed a huge sigh of relief when the Germans turned their attention from our airfields and fighters to London."

"A dreadful exchange," Kauffman said.

"Yeah, which is why we all feel shame in that trade-off."

"You shouldn't, Brian. As General Sherman so aptly said, 'War is hell.' And, indeed, it is. I heard tell among the senior officers that Fighter Command was within days of collapse when the Germans turned on London."

"We were too tired and too busy to know, but I have heard that said. The Blitz must have kept you very busy as well."

"Perhaps an understatement, I'm afraid."

"Did you know the Earl of Suffolk?" asked Murrow.

"Oh my, yes, helluva man; the Holy Trinity we called them. Our bomb disposal group is a fairly tight-knit bunch of chaps, as you can understand."

"What happened?" asked Ed, as the reporter in him took over.

"The Germans mixed special bombs that we had never seen before. They had a delay fuse with a rather ingenious anti-tamper device incorporated. Charles was working on the fuse when the bomb exploded."

"Tragic," Murrow said, with genuine sadness in his voice. "Have you bomb jockeys figured out how to deal with this new device?"

Kauffman stared at Murrow, undoubtedly thinking about what he should say. "I have probably said too much already and our counter-measures are definitely beyond the public domain. If you intend to print any of this, I must ask you to clear it with the Admiralty."

"Certainly, Draper. I have no desire to cross you."

"Thank you, Ed."

"I can't imagine having to face a bomb that was dropped in anger and failed to detonate," Brian said.

"Well, I can't imagine sitting in a cold, confining cockpit with bad guys out there trying to drill holes in you," responded Draper.

The three men laughed.

"I really must be on my way," Kauffman said.

"It was great to meet you, Draper. Thank you for saying hi."

"It was a true honor, Brian. I will continue to watch your accomplishments in the air. Great to see you, as always, Ed. Good day, gentlemen." Kauffman departed smartly.

Brian and Ed worked their way down to the ground floor reception / lobby area. The other pilots appeared to be waiting rather impatiently.

"'Bout time, ace," said Rusty Bateman.

"Sorry about that, guys. Allow me to introduce CBS News London Bureau Chief Edward R. Murrow." Ed shook hands with and addressed each pilot by his correct name.

"Great to meet you all," Murrow said. "I hope you remain safe. I shall bid you adieu."

Brian wanted to head back to Martlesham Heath, but he was odd man out, so he decided to stay with his comrades. They began walking toward Shepherd's Pub for more to drink than a Coca-Cola. A light drizzle dampened the street and their uniforms. A generous cabbie offered them a ride. It was quite crowded with six men inside the cab, but they made it. They paid the man, even though he refused compensation. Brian knew this was going to be a long night.

—

Friday, 20.June.1941
Office of the Secretary of War
Munitions Building
Constitution Avenue & 20ᵗʰ Street Northwest
Washington, District of Columbia
United States of America

"Thank you for taking the time to see me," said Secretary of War Henry Stimson, as he shook hands with Major General Hap Arnold, USAAF and motioned for the general to sit on the long couch in his office. Stimson took the overstuffed chair next to the couch. "It is official, Hap," he added, as he held up a document, and then handed it to Arnold. "That is Army Regulation 95 dash 5. The Army Air Corps is now the Army Air Forces, and you are officially the Chief of Army Air Forces, Chief of the Air Staff in British parlance. Congratulations."

"Thank you, sir."

"I know it has been a long hard struggle to get through the War Department and administrative bureaucracy, but your perseverance has paid off. As the President indicated last month, your name has received presidential endorsement for promotion to lieutenant general, as soon as we can get the Senate to approve the promotion list, and since we are asking for an out of sequence action, the process may take a little longer than usual. We expect to promote you to full general as soon after that as possible."

"Again, thank you very much, sir."

"I wanted to take this opportunity for a private chat between us . . . not without General Marshall's knowledge and concurrence, I must add. We expect you to function as the direct equivalent to Air Chief Marshal Sir Charles Portal, and that means managing the aviation assets of the United States."

". . . including the Navy and Marine Corps?"

Stimson laughed. "I misspoke . . . the aviation assets of the Army – the Army Air Forces."

"I just thought we needed to be clear."

"Exactly. It may come to that point eventually, but that is a fight not worth fighting at this stage of the war. It is my opinion, and the President's I might add, that the Air Force deserves to be an equal, independent service, but again, that is a *mañana* matter to be dealt with in the future, when the time is right. The hardest part of your task is going to be the rapid modernization of our Air Force and finding the necessary, capable leaders required to run the Air Force properly. We do not have much time left to prepare. It is only a matter of time before the Germans turn their guns on us. They will eventually find our support for and collaboration with the British as intolerable. Further, the situation with Japan is deteriorating rapidly as well. The Tripartite Pact between Germany, Japan and Italy does not leave any of us with a comfortable feeling. When the war comes to us, we will need a powerful and effective Air Force, and that my dear general is the Herculean challenge before you. Use your time wisely; there is not much left."

"Yes sir."

"Do you have any questions for me?"

"No sir. Well, I guess one . . . have all parties been informed of this change and direction?"

"The regulation in your hand covers all elements of the Army – military and civilian. So, yes, all parties have been informed. The Navy Department was consulted and informed of this change. If you should have any problems, please see me . . . or General Marshall, directly. We shall not countenance resistance to the march of progress. This is a change long overdue."

"Yes sir."

"Well, if you have no questions, I do have one. It is simply a point of curiosity at this juncture." Arnold nodded his head. "National Airport opened three days ago. American Airlines landed a commercial flight there, shortly after the dedication and opening. Have you seen any impact on operations at Bolling Field and Anacostia?"

"The only impact, so far, has been positive. The traffic control procedures worked out ahead of time seem to be working perfectly. And, more importantly, the pilots like having another airfield option in the DC area."

"Excellent." Stimson paused. "Please let me know as soon as possible if any conflicts arise."

"Yes sir."

"Oh, maybe one more thing," Stimson chuckled. "The appropriations order was signed once the Air Force was approved, to establish the Army Air Force headquarters at Bolling Field. I hope you will enjoy your new office when it is done."

"Thank you very much, Mister Secretary."

"Again, congratulations. You have a very big task ahead of you. Good luck, Hap."

Arnold stood, saluted and left the secretary's office with a new, broader charge in hand.

—

Sunday, 22.June.1941
Ditchley Park
Enstone, Oxfordshire, England
United Kingdom
04:10 hours

"Excuse me, sir," said duty Assistant Private Secretary Jock Colville into the darkened room. Churchill grumbled and struggled to sufficient alertness after only two hours sleep.

"What is so damn urgent, Jock?" mumbled Prime Minister Churchill.

Colville turned on a small table light to see what he was doing, and just in case Churchill wanted to read the message, he could get to a proper reading light next to the bed. The young assistant was still in his pajamas and covered by a gray robe. "We just received an urgent message from MI6."

"Yes, yes, what does it say, for God's sake?" Churchill said with impatience and irritation in his voice.

"It says, from multiple reliable sources, the Germans have executed a broad, general, combined arms invasion of the Soviet Union. Large armor forces crossed the River Bug an hour ago behind a heavy artillery barrage and aerial bombardment in advance of their armor units."

"What about the Russians?"

"According to the message, they were apparently caught completely by surprise. Would you like to see the message?"

"Not at the moment, but don't let it go. I will want to read it carefully, once I am fully awake."

"Yes sir."

"So, Operation BARBAROSSA is finally underway, and despite months of warning Stalin and his general staff of what we saw coming, they ignored our warnings. We must now pray the Russians gather the strength to stop the Huns before they reach Moscow and the Caspian oil fields."

"Yes sir. What would you like for me to do?"

"We need to collect up the Defense and Chiefs of Staff Committees along with the Joint Intelligence Committee as soon as it can be arranged. I do not care about location . . . just get me to where I need to be. The War Cabinet should be called shortly, after we have a handle on what is happening and what options we have in this situation."

"Yes sir. I will jump on this immediately."

Churchill chuckled slightly. "I don't think it is that urgent, Jock. You can get dressed and have a bite of breakfast, if you wish." Both men laughed.

"Well, I certainly do not want to embarrass any of the staff." They laughed, again.

Churchill lapsed into contemplation. Colville recognized the Prime Minister's expression and that it was not time to leave him alone. Colville stood patiently for the Prime Minister to return. Several minutes passed, and then Churchill's eyes re-connected with Colville.

"At least there is some good in this disaster."

"What is that, if I may ask, sir?"

"We are no longer alone in this fight with the bloody Huns. You know Jock the thought has come to me, if Hitler invaded Hell, I would make at least a favorable reference to the Devil in the House of Commons." Churchill scowled, glanced at Colville, and then grinned slightly.

"Then, we must be grateful for small blessings from rather bizarre sources."

"Indeed. Now . . . off with you. I'm going to try mightily to get a little more sleep."

"Yes sir. I'll make sure Sawyers knows that I disturbed your sleep."

"Thank you, Jock," he said and lowered himself to a recumbent position under the sheets and duvet. "You're a good man."

"Thank you, sir." Colville switched off the table light and closed the bedroom door.

—

Sunday, 22.June.1941
Residence
The White House
Washington, District of Columbia
United States of America
08:00 hours

Harry Hopkins entered the President's bedroom along with the chief steward pushing the breakfast cart. President Roosevelt had already been helped to his wheelchair although still dressed in his pajamas and plush robe. Harry was fully dressed in a light grey business suit with a solid, medium blue necktie.

"Good morning, Mister President," Harry said cheerfully.

"Is it?" Roosevelt grumbled.

Hopkins allowed the steward to arrange two place settings on the table serving breakfast to the President and his aide. When he was complete, the steward departed without a word and closed the door behind him. "As in all such situations, whether I am the bearer of good or bad news depends upon one's perspective."

"Thank you for the philosophy lesson, Harry." Roosevelt took a bite of his scrambled eggs and hashed brown potatoes. "Now, get on with your news."

Hopkins had not touched his breakfast, yet. "At 03:15 this morning, a massive German army crossed the River Bug and invaded the Soviet Union. We received essentially the same information in various forms from HMG, MI6, COI, G-2, and State."

"So, Winston was correct, once again."

"It would appear so, Franklin." Harry began eating in eager fashion.

"I know he did not want to be correct, and he tried monumentally to convince Stalin without compromising the intelligence he held. So, I suppose in that sense, he failed . . . failed to convince Stalin of the impending attack. Now, there is no doubt."

"What time was it here?"

"I am told the attack began at 20:15 our time, last night. The first messages did not arrive until after midnight."

"I presume you have arranged for a defense review," Roosevelt said.

"Yes, I know you would want the latest information. State, Army, Navy and COI will gather in the Cabinet Room at 10:00 this morning."

"That should work."

Hopkins pushed his half-eaten breakfast away. "Now, for the bad news."

"Bad news . . . you mean worse than the German invasion of the Soviet Union?"

"Perhaps not worse, but certainly far more personal."

This time Roosevelt finished eating his breakfast. "Oh dear, what on earth could that be?"

Hopkins looked away, swallowed hard, and then looked back to Roosevelt's waiting and anticipatory eyes. "Also last night, Missy had a very serious stroke. She was taken by ambulance to Doctors Hospital at 9:39 last night. She has not regained consciousness as of an hour ago and is listed in critical condition. Doctor McIntire is with her and assisting in her treatment."

Roosevelt blankly stared at Harry, lost in his thoughts. "I need to go see her. Please arrange it for this afternoon."

"She may not regain consciousness by this afternoon."

"It does not matter. I need to see her."

"Very well. I will arrange it. Have you decided on Independence Day?"

"Let's take an overnight train on Thursday. Do you want to go to Hyde Park, or would you prefer to spend the holiday with your family?"

"Given last night's events, I think I should stay with you."

"Thank you for that, Harry. I do appreciate it."

"Next question," Hopkins said. "When do you want to return?"

"How about a late night train, after the fireworks, to arrive back in DC on the morning of the 5th. I suspect we are going to have a lot to deal with."

"I'll take care of arranging things today."

"Thank you, Harry. Now, time marches on and I must get dressed, if I am going to make the 10 o'clock meeting."

Harry Hopkins departed the residence and presumably made his way to the West Wing. Franklin tended to his attire, while his mind churned over this morning's news. The Soviet Union was a long way off and in the other direction from England; however, Missy was far closer and more personal.

—

Monday, 23.June.1941
RAF North Weald
Epping, Essex, England
United Kingdom

No.71 Squadron moved to yet another air base. North Weald was not closer to France than Martlesham Heath, but it was closer to London. Epping was the northeast terminus of the Underground Red Line, so travel into and across London would be much less complicated. That observation meant it would take Brian less time to reach Standing Oak Farm.

Whitey decided they would land as sections in trail, which gave each section leader a clear view of the airfield area. The regeneration of spring had eliminated the last vestiges of bomb damage from last summer's mortal battle. North Weald had an unusually broad, grass landing area. They could takeoff or land in virtually any direction, which meant no crosswind landings. The solid, permanent hangars, operations building and tower, and other support buildings made their new air base feel more substantial and durable. Whitey must have known their new location. He taxied without hesitation or a guide truck toward the end to the right branch V-shaped arrangement of typical flight line buildings. Once he swung his tail pivoting into a parking spot, the rest of them knew where to place their fighters.

The ground crews had not yet arrived. Brian shutdown his 'XR-G' Hurricane fighter, unstrapped, stepped out on the left wing, closed the cockpit access panel and canopy. He found a set of chocks behind his aircraft and placed them on either side of his left main wheel. Brian joined the rest of the squadron outside what had to be their new Dispersal Hut.

An RAF flight lieutenant stood next to Whitey. He was introduced, but Brian was not really paying attention. He had been through so many of these base-transfer orientation briefings. The briefer took a slightly different tack when he asked the pilots join him inside. A detailed map of the base and associated adjacent buildings had been hung on the wall next to the operations clerk's desk. The flight lieutenant pointed out key buildings on the base of interest to the pilots—the officer's mess, flight equipment storage, operations building, medical facilities, and designated hangar. Several questions were asked and answered.

As the officer approached his closing, he looked to Whitey. "With the commanding officer's consent," he said and waited for an approving head nod, "you may not have heard the latest news." No one spoke or made a sound. "Yesterday morning, the Germans invaded Soviet territory across a very broad front. The information from the Air Ministry indicates it is far too early to tell the extent of the invasion. So far, they have only entered territory occupied by the Soviets two years ago, when they joined Germany, parsing Poland and occupying the Baltic countries."

"What does that mean to us?" asked Pilot Officer Arnold Samuel 'Salt' Morton—Left Wing, Red Section.

Whitey stood to take the question. "It is too early to tell, Salt. Their operation might be a limited action to simply take the rest of Poland and perhaps the Baltic states, but it also could be something much bigger. The Soviet Union is a very big country. If their mission is the latter—to take most or all of the Soviet Union—it will mean fewer Germans in France and the Low Countries. I think you, all of us, can see the consequences to us." Whitey looked to the other flight lieutenant.

"Very well said, Mister Whittington. I have nothing to add. Are there any other questions?"

"Where are the best pubs?" Rusty asked. Everyone laughed and jabbed at their squadron mate. The war may have taken a monumental shift to the east and away from them, but Bateman was more concerned about beer.

"Actually, a good question, it seems to me. The King's Head is an old, traditional pub to the east of the south gate. Garnon Bushes is just about the same distance to the west, out the south gate. Those are the closest pubs, and they are both quite good. There is a wider selection in Epping itself. Several are close to the Underground station in Epping."

"There are still Germans shooting at us," interjected Peterson, apparently not comfortable with the diversion. "Let's keep focused on the job we are here to do."

"Pete is correct," Whitey added. "We have a job to do. I intend to take the generous offer of Group and Fighter Command to make an orientation flight this afternoon. I expect to be at Available status tomorrow morning, although we have not been given any missions, as yet. We need to get a couple of operating days behind before we resume issuing rest passes." The grumbling was not serious and probably expected. "I would also appreciate everyone remaining on base until the weekend, or until we get two missions successfully completed, not counting our orientation flight." More nondescript grumbling.

"Are we restricted to base?" Red asked with a rather challenging tone.

"Did I use the word restricted?"

"No."

"There is no reason to restrict anyone to base," Whitey continued. "I am simply concerned about Fighter Command's perception of this particular squadron. You cannot go carousing around in each new place we show up and expect to be taken seriously by those who decide our assignments." Whitey paused to give anyone an opportunity to speak up. "Very well, then. Let us go check in at the Mess and avail ourselves of the early sitting for lunch."

Not a word was spoken as the pilots hung up their flight equipment. The rack of wall pegs was the same in every Dispersal Hut, and the pilots were creatures of routine. They did not need labels to know which peg was theirs. They walked as a group past the various buildings of an air force base and found their way to the Officer's Mess easily enough. The check-in process had become routine as well.

—

Friday, 27.June.1941
Springwood Estate
4097 Albany Post Road
Hyde Park, Dutchess County, New York
United States of America

Federal Bureau of Investigation Director J. Edgar Hoover and another distinguished man, both dressed conservatively in suits and ties, arrived and were shown into the President's study – a moderate sized room, with richly polished woods, walls of books, and leather upholstered furniture. The President positioned his wheel chair between two opulent, leather chairs, and was waiting for his guests when they entered the room.

"Welcome back to Hyde Park, Edgar."

"Thank you, Mister President. I trust you are well."

"Quite well, thank you for asking, and I hope your journey to New York was comfortable."

"It was, yes sir. Mister President, may I introduce Supervisory Agent in Charge Edward Connelly. He prefers 'E.J.'"

"Great to meet you, E.J. Welcome to Hyde Park, as well."

"Thank you, Mister President. It is a genuine honor to meet you."

Roosevelt waved his hand dismissively. "Please be seated. Would you care for coffee . . . or tea, perhaps? Did you get lunch?"

"No, thank you, sir," both men responded in unison. "We did get lunch before coming up here from the station," Hoover added.

"I presume you have a matter of urgency that induced you to make the journey up here," Roosevelt said and motioned for his guests to be seated, which they did.

"Yes sir. I briefed the Attorney General before I left Washington, so he is up to speed. I picked up E.J. in Manhattan, when I transferred trains at Grand Central Station. I thought it important that you have direct access to E.J. in this instance.

"As you may recall, we discussed the large scale German espionage campaign in this country last summer." The President nodded his head in agreement. "We have aggressively investigated what we are now calling the Sebold case. A week ago today, we had evidence at least two of the subjects were moving swiftly to flee the country, which gave us the impression at least some of the spy ring may have become suspicious with respect to our surveillance. We have seen no evidence they were tipped off. We arrested both men at the dock as they were about to board two different ships. At that point, the risk to our investigation exceeded our threshold of tolerance. I have asked E.J. to supervise our nation-wide round-up operation we intend to execute Saturday night and Sunday morning. Over 250 agents as well as local police in some instances will be deployed. I asked E.J. to summarize the operation for you." Hoover gestured to Connelly.

"As the Director indicated, Mister President, we will execute arrest warrants on 33 suspects in 11 major cities and locales with most, not quite a majority, of those suspects being located in the greater New York City area. The entire operation will be coordinated from the Bureau's Foley Square office. We have coordinated with the supervisory agents in each location. Our objective remains simultaneous execution. We have deployed one quarter of our entire field agent force to this task. Yet, there are practical logistics obstacles that will likely interfere with our simultaneous execution objective."

"What about Sebold?"

"Once the arrests were made last week, we carried out a planned sequester of Bill for his safety. When we execute the plan tomorrow night, he will be with me in the Foley office. We will continue to do our utmost to protect him. His identity will become known publicly as we prosecute these spies."

"Will we be able to determine what has been compromised by this spy ring?" Roosevelt asked.

"The evidence teams will deploy with the arrest teams," E.J. responded. "The evidence teams have detailed instructions and training to look for tradecraft evidence or any possible signs of inappropriate material in their possession."

"I will also add, Mister President," interjected Hoover, "I have personally coordinated with David Petrie, the new director general of MI5, first to make sure they are aware of what we are doing, and secondly and more importantly, several of our suspects implicated potential alien agents in Great Britain – two in England, one in Scotland and one in Ireland."

"Ireland proper, not Northern Ireland?"

"The Republic of Ireland . . . near Dublin, by our knowledge," Hoover added. "David knows our execution time. I believe it is his intention to also execute arrest warrants at dawn on Sunday."

"Excellent, well done, Edgar, and good luck tomorrow night, E.J. I shall anxiously await your report of results," the President said.

"Thank you, sir. You will have our arrest phase results as soon as we have concluded the operation, which should be Sunday morning, at least by noon, we hope."

"Is there anything else I need to be aware of in this exercise?"

"No sir."

"Fine, then please allow me to ask a couple of relevant questions?" Both men remained silent and motionless. "Attorney General Jackson and I had quite the debate last year regarding your request for warrantless wiretaps. What part did your wiretaps play in this investigation?"

Attorney General Robert Houghwout Jackson of New York had been serving in his position since January 1940. He was a Democrat, the former Solicitor General of the United States, and an essential legal advisor, as the government began to supply Great Britain during the previous summer.

Connelly looked to the Director, since he probably had no desire to answer the President's query. Hoover smiled. "Simply put, those wiretaps were vital. We would not have the knowledge we have of this network without the flexibility and agility to move against these individuals."

"Do you have sufficient evidence not derived from those warrantless wiretaps to prosecute these spies?"

"We believe so. The Attorney General and relevant assistant attorneys general have been involved with this investigation from the outset. I think it safe to say they are satisfied with the evidence they have to date, so whatever we acquire in the following days will be icing on the cake."

"I hope and trust you are correct, Edgar. Our British friends have been critical of our slowness to respond to this threat. Now, last question, do you think these 33 are the full extent of German agents operating on our soil?"

"I would never be so bold to make such a claim. We believe we have the full extent of the Sebold network, but we have no way to know about and no indication of other networks operating in this country. I can assure you we shall remain vigilant and will aggressively run to ground any clue, tip or sign we might acquire."

"We can expect nothing less, Edgar. Well, then, thank you, again, for this briefing. Good luck tomorrow night."

Both men stood. Roosevelt extended his hand to each of his guests and shook hands firmly.

"Good day, gentlemen."

"Thank you, Mister President. Good day."

They departed. Harry Hopkins was waiting outside the President's study. He said good-bye to both visitors, escorted them to the front door, and then returned to the President for any follow-up that might be necessary.

—

Chapter 7

The pride of the peacock is the glory of God.
The lust of the goat is the bounty of God.
The wrath of the lion is the wisdom of God.
The nakedness of woman is the work of God.

-- William Blake

Wednesday, 2.July.1941
RAF North Weald
Epping, Essex, England
United Kingdom
04:40 hours

The early morning mission came as quite an adjustment for the No.71 Squadron pilots, just a week and a half after the squadron moved farther south and closer to the action. They were now on the outskirts of London. Each of the pilots had taken the Skipper's caution to not party or drink heavily last night. They had been awakened by the mess stewards prior to twilight, walked silently to their Dispersal Hut, completed their final mission briefing, gathered up their flight equipment including pistols and ammunition, and then mounted their trusty steeds. Morning twilight offered sufficient visibility to see they had scattered, fair weather cumulous clouds. Fortunately, the weather guessers had been correct.

Whitey Whittington remained their acting commanding officer with no clues from No.11 Group or Fighter Command regarding a permanent CO. Whitey was doing a good job, despite the uncertainty of command. The pilots respected him. Half the pilots including Brian thought they should just make Whitey the permanent commander, but no one asked them.

They took off as sections in order. As their mission brief dictated, they kept their throttle settings comparatively modest in a slow cruise climb to the south. This was going to be a long mission and they needed to conserve fuel.

Ten minutes after takeoff, over the Kent countryside at 5,000 feet, they spotted the object of today's mission – No.21 Squadron Bristol Blenheim light bombers. The sun poked its first rays over the eastern horizon.

Their target was a railroad, switching yard just south of Calais city center. The weather over the target area was also good – scattered fair weather cumulous clouds, some haze and light winds.

Whitey led 'B' Flight as the acting squadron leader. Hunter Drummond led 'A' Flight with Horse Harrow and Pilot Officer Andrew Adam 'Rocket' Downing flying as his second section. The squadron took its position 1,000 feet higher and half a mile behind the bombers. A mile separated 'B' Flight and 'A' Flight on each side of the bomber formation, to give them greater freedom of action. The weather suggested they would quite likely face German fighters. For the less experienced of the squadron pilots, this would be their first combat fighter engagement. Brian knew they had to be excited by the prospect and apprehensive regarding how they would perform in aerial combat with enemy fighters . . . and, over enemy territory.

They were well past halfway across the Channel and approaching the coastline of German-occupied France. "Tally-ho," Hunter broadcast when he spotted a squadron of fighters. "Bogeys two o'clock high." Per their mission plan, the bombers descended to wave top height, while the fighters maintained their position to protect the bombers. The Blenheim pilots would have their hands full flying at such low altitude. They would transition to treetop height at landfall to minimize their exposure to anti-aircraft fire. The fighters would face different risks as they flew evasive maneuvers to avoid anti-aircraft fire, remain in position with as much energy as they could maintain, and remain vigilant for German fighters.

"OK, lads. Here we go. Go get 'em, Hunter. We'll shift over. Let's keep the bombers clean."

Brian signaled for his flight to deploy for aerial combat, pushed his throttle up to the emergency gate, and climbed to meet their attackers. Brian looked over both shoulders. Everyone was in position.

The German formation split with half their squadron rolling hard into a steep dive toward the bombers. The remainder faced Brian's 'B' Flight in what would be a head-on initial engagement. Both groups opened fire at essentially the same time, seconds before they passed through the other's formation with a closure rate in excess of 500 miles per hour. Brian pulled up hard and pushed his throttle through the emergency power gate wire. The Merlin engine responded with robust energy, but the Hurricane just did not have the excess power margin of the Spitfire. Brian's head and eyes moved rapidly under the strain of 'g' forces, as he sought to keep track of his brethren and find a target. Airplanes were everywhere. Radio calls became a jumble of noise, as the less experienced pilots conveyed their fears. Brian was nearly on his back, with the nose still coming through the horizon, when he saw a green spinner 109 passing just in front of him. He fired a quick stream of bullets, saw the flashes and sparks of several hits, but he had to roll hard in the opposite direction to avoid a collision. Brian felt the aircraft shudder, dangerously close to stall. He released most of his back stick pressure and pushed hard against the throttle mechanical stop. The engine was giving him all it had to give. As his aircraft shuddered and approached the upright position, Brian lowered the nose well below the horizon and continued his roll at a slower rate. He quickly gained speed. Another 109 was chasing 'XR-E', Horse Harrow. Brian rolled hard and pulled his nose down to near vertical. He dove to intercept Horse's attacker. Before he could get within gun range, the 'XR-E' gyrated a few times, and then exploded. The German must have seen Brian closing. He rolled sharply into Brian's pursuit angle to increase Brian's attack angle. As his airspeed increased rapidly, Brian pulled his throttle back to near idle. He rolled hard to press his attack on the German, and pulled as hard as he could against the heavy 'g' forces of high-speed flight. He could not get his nose up fast enough and overshot his pursuit line on the German. By the time Brian got his nose up, the German was gone.

"Eagle, break it off," came the command to end the engagement.

The rejoin process took half the Channel to accomplish. The bombers were all still flying several miles ahead of Whitey and 'B' Flight, but there was another Hurricane missing – 'XR-D' Rusty Bateman. They would have to wait until after landing to find out what happened. They were all too short of fuel to make it back to RAF North Weald. Whitey diverted the squadron to RAF Lympne, just beyond the coastline, while the bombers continued to their base.

While the ground crews scurried to refuel and rearm their aircraft, the pilots stood behind the line of fighters to compare notes. Rusty Bateman had lost his aircraft and successfully bailed out. A German patrol was waiting for him, as he descended under his parachute. Whitey and others of 'B' Flight made several passes in an effort to disperse the welcoming party. Rusty was taken prisoner as soon as he hit the ground. Brian reported the explosion and death of Horse Harrow. The bombers had successfully hit their target and caused several secondary explosions – probably ammunition railcars. While the mission had been successful, it had been a costly endeavor.

The short, return flight to RAF North Weald was silent and uneventful. The intelligence debriefings took several hours to complete and were interrupted only by a lunch break. They would remain at Available status until sunset, before they were released and could make their way to the Officer's Mess bar to celebrate and mourn the loss of their comrades. Missions into France made the specter of capture and imprisonment a significant added risk the veterans had not faced during the Battle of Britain the previous summer.

—

Wednesday, 2.July.1941
Royal Society
6-9 Carlton House Terrace
St. James's, Westminster, London, England
United Kingdom
14:30 hours

The entire committee of British physicists, chartered by His Majesty's Government last year, assembled at the headquarters building of the Royal Society. Their designated mission focused completely upon their expert critical assessment of the analysis and conclusions presented by German refugee physicists Otto Frisch and Rudi Peierls in their memorandum of April 1940. The committee formed less than a month after Sir Henry Tizard received his secret copy of the Frisch-Peierls Memorandum.

The composition of the committee of prominent nuclear physicists included:

-- George Paget Thomson, FRS – committee chairman, 1937 Nobel Laureate (Physics) for his demonstration of the wave-like properties of electrons;

-- Australian Marcus Laurence Elwin 'Mark' Oliphant, FRS;

-- Patrick Maynard Stuart Blackett, FRS;

-- James Chadwick, CH, FRS – 1935 Nobel Laureate (Physics) for his discovery of the neutron;

-- Philip Burton Moon, FRS; and,

-- John Douglas Cockcroft, FRS.

In an odd twist of history, as the memo circulated at high-levels of the government's scientific apparatus, the Germans had invaded and subdued Denmark in less than a day. Esteemed Danish physicist Niels Henrik David Bohr sent a telegram to his British colleagues with reference to John Cockcroft and Maud Ray Kent. Maud was Bohr's English housekeeper from Kent. At Cockcroft's suggestion, the committee chose the name Maud, to refer to their group and work, in partial recognition of Bohr's telegram of warning and encouragement. As is so often the case in government circles, people began speculating what Maud stood for in the context of the committee's charter. A more popular acronym quickly bloomed – the Military Application of Uranium Detonation (MAUD) Committee.

"Gentlemen, may I have your attention," announced Thomson. He waited for the chitchat to die out and each of his colleagues to take their seats around the rectangular table in one of the Society's modest conference rooms. "Each of you has a copy of our draft report before you, and I trust you have carefully read its contents." Heads nodded, but George did not poll his comrades. "You will also note, I have taken the liberty of adding a cover letter, which you have not seen, yet, as a means of succinct executive summary, to which I seek your concurrence."

MOST SECRET – TUBE ALLOYS

2 July 1941

Home Secretary the Honourable Mister Morrison

Dear Sir,

Attached, please find the final assessment report Sir John Anderson chartered last year on behalf of His Majesty's Government. The Committee's work under the charter is hereby complete.

The Committee wishes to emphasise that the Germans were first to scientifically demonstrate fission, i.e., the splitting of the uranium atom and consequent release of energy, in late 1938. We know they have placed high national priority on development of fission as an energy source and have pursued atomic energy with urgency. We also know a large part of the Kaiser Wilhelm Institute in Berlin has been set aside for uranium research. They have also sought acquisition of uranium ore supplies, not least of which are available in German-occupied Czechoslovakia, as well as the monopolisation of heavy water supplies in German-occupied Norway. The progress of the Germans in nuclear weapons development must not be underestimated.

As indicated in the attached report, the Committee offers this conclusion and recommendations:

(1.) The committee considers that the scheme for a uranium bomb is practicable and likely to lead to decisive results in the war.

(2.) It recommends that this work be continued on the highest priority and on the increasing

scale necessary to obtain the weapon in the
shortest possible time.

(3.) The present collaboration with America
should be continued and extended especially
in the region of experimental work. The
collaboration initiated by Sir Henry Tizard's
mission to the United States should be
solidified and expanded.

The Members of the Committee stand ready
to assist, as the Government deems appropriate.

On behalf of the MAUD Committee,

Respectfully submitted,

George P. Thomson

George P. Thomson

Chairman

MOST SECRET – TUBE ALLOYS

"I know we have a few related items to discuss before we close our charter. However, I do believe we can approve the draft report to be submitted to the government, as fulfillment of our task. As such, if you would be so kind to raise your hand . . . ."

"Before we vote to approve," interrupted Blackett, "what is the significance of tube alloys?'

"I was informed that it is a classified compartment for segregating all materials associated with this subject and project."

"Yes, I deduced that much, but why that designation. Is there some connection?"

"Not that I am aware of, only that they wanted something metallurgical and sophisticated sounding," answered Thomson. The scientists chuckled at the notion.

"I suppose they achieved that."

"Any other questions?" Thomson scanned the distinguished group. No one spoke up. "Very well, then, as I was saying, if you concur with the final report and cover letter, please raise your hand." Every member raised his right hand, and Thomson did as well. "Very well, we shall record a unanimous concurrence with the final report." George shuffled a small stack of papers before him to find the specific paper he wanted. "Now, onto the related items,

John and Mark, I do believe you wanted to discuss the American position a little more."

Cockcroft nodded to Mark Oliphant.

"John and I have spent the most time with the Briggs Committee, and I do believe we are in agreement," Mark looked to John, who nodded his head in agreement, "the Americans, at least as controlled by Lyman Briggs, do not feel the same urgency that we do. I think we both agree as well that many of the working scientists do feel and appreciate our concern for the German advantage. John and I discussed this matter numerous times in America, with our counterparts, as well as during our discussions in this matter. We also agree with the Committee's position regarding the wording of our final report. We have gone as far as we should go in an official report that will certainly be shared with the Americans. We raise this matter to urge each member of the committee, when we are queried regarding our deliberations, to raise the matter of a perceived paucity of urgency across the Atlantic."

John Cockcroft interjected, "I suspect some of this slowness could be attributable to the public isolationist mindset of many of their countrymen. They may feel they might be able to sit this one out."

Mutters and grumbles punctuated John's statement.

"I do not know who will heed our recommendations, or more specifically, who will listen, if we are to gain an audience. However, if I judge Mister Churchill correctly, I suspect I shall be asked to brief him and Professor Lindemann, at a minimum, and probably the War Cabinet."

"Then," interjected Oliphant, "we must ensure they appreciate our dilemma."

"To be clear," Thomson responded, "that dilemma is?"

"We need the industrial capacity of the United States, but the Americans do not see or feel the threat as we do."

"Understood. If we should find the opportunity, I shall do my best to impress upon them the dilemma you have illuminated and to seek their political influence on our American cousins to improve their commitment to this project." All of the scientists nodded their heads. "I do believe the next item belongs to you, John."

"Yes, well, thank you, George," Cockcroft began. "Our report clearly states our position, but I think Mark and I would agree it understates the position of the Americans and the difficulty we may have with the core material decision and more importantly the production process to be employed."

"In what sense?" Patrick Blackett asked.

"The matter of the core material is going to present some serious challenges for us . . . and by us, I mean the British people. I believe we are in general agreement as a group that plutonium two three eight will likely have comparable neutron flux as enriched uranium two three five, and thus comparable explosive potential. However, our collective ability to produce the necessary amounts of each substance is not equal. Ironically, this decision is also going to boil down to time . . . the time necessary to develop the production process and generate sufficient material for the required experimentation and of course at least a few functional units. While our calculations establish the critical mass for a functional explosive potential at roughly ten kilograms, we will need three to five times that amount for experimentation . . . to prove the theoretical and conceptual principles. In this, I can only speak for myself; I cannot see the reactor production of plutonium two three eight being sufficient for these purposes, or the thermal diffusion, electromagnetic, or the centrifuge methods producing sufficient uranium two three five, being sufficient for that matter. Somehow, we must convince the Americans to focus their industrial energy on the gaseous diffusion method on a massive scale in order to have a functional unit in time to positively affect the outcome of the war."

"What do you propose?" asked Thomson.

"Well," Cockcroft paused, "Mark and I have tried to crack through their interest in plutonium without success. It is like some superior force has calcified their minds. Their working level scientists cannot articulate that commitment, for some odd reason. I am quite reticent to ask for political intervention, but that may be our only hope."

"Why is this such an issue?" asked Blackett.

"We do not have the industrial capacity to carry out the necessary rate production, and even if we did, we could not keep the facility safe from German aerial attack. We certainly have the technical capability, but it would take us too long . . . and I suspect would take a prohibitive amount of our scant resources to produce the necessary amount of core material. The question centers upon inspiring the Americans to make the requisite commitment of intellectual and industrial energy to get this project done soon enough to be of value to ending this dreadful war. To put a sharp point on this, we have but to look out the window to see the terrible destruction being inflicted on our beautiful city by the Germans. It does not take much imagination to see the consequences of the Germans achieving an aircraft deliverable fission device before us. We must beat the Germans, and unfortunately, we need the Americans to feel the same threat and commitment to being successful."

The room remained stone silent as each of the physicists considered Cockcroft's words. Eyes remained distant for several minutes, and then slowly those distant eyes began to re-connect.

"John is spot on," Mark Oliphant added. "To complement John's words, I would like to add that we are farther along in our technical assessment of the fission potential of high neutron flux matter; however, we have not solved many of the scientific and engineering challenges that block our path to what we know is possible. We need their intellectual contributions as well."

"Hear, hear," Cockcroft expressed his concurrence. "So, to answer your query, George, I would offer for our collective consideration a multi-level approach to this particular problem. We must be deeply integrated in all aspects of whatever joint program evolves from the culmination of our effort. We have clearly established and documented the science behind the massive explosive potential of fissile material. Now, we must convince the Americans . . . not so much the physicists, I think they are convinced, but engineering, manufacturing, financial, and most importantly the political force to make it all happen. The Briggs Committee and I believe other circles of the U.S. government are waiting for our report. We will likely have a rather narrow window to impress upon them what is not in the report."

"Well said, John," commented Oliphant. Heads nodded in agreement.

"Very well, then. I shall do my part as I am able or given the opportunity. We are unanimous in our endorsement of our final report. I would propose we add another paragraph to the cover letter asking for an audience with Minister Morrison, and if he deems appropriate, with the War Cabinet, to present the report and answer any questions. It would be proper to raise these supplemental opinions at that time. We should and must be unanimous in our articulation of these supplemental items as well. Are we all in agreement?" Thomson looked each scientist in the eyes and received a confirmation sign or words. "Excellent. Is there any other business to discuss?" Everyone shook their heads in the negative. "Our work is done. I thank each of you for your extraordinary efforts and contributions to this task. May God bless us on our journey."

"Well done, George," said James Chadwick, as the men stood, gathered their papers and prepared to depart.

Several stood in social intercourse before leaving the room. A few would visit colleagues in the building. The historic meeting closed. Others would now have to pick up the gauntlet and drive the ultimate project to its conclusion.

Wednesday, 2.July.1941
Cabinet War Rooms
New Public Offices
Whitehall, London, England
United Kingdom
17:30 hours

"**S**o," began Attlee before Sir Edward could convene the combined War Cabinet and Defense Committee meeting, "we have the shopping list from the Russians."

When ministers were like this, worked up in a lather, Sir Edward knew he should just remain clear and ensure the minutes were recorded properly.

"Yes, we do, Clement," responded Churchill. "It just arrived an hour ago, which is why I called this unscheduled meeting."

"I imagine they were not bashful," Arthur Greenwood added.

"No, they were not, especially when viewed in the light of their date of issue—the 30th of June. The invasion was only a week old."

"Before we jump into this supply matter, perhaps an update of the ground situation on the eastern front would be warranted," offered Eden.

"Quite appropriate, Anthony," Churchill answered. "Proceed."

"The Germans have made shockingly swift progress and appear to have caught the entire Red Army napping. They have overrun and captured entire divisions, and their armor forces are literally vastly superior in the region. We do not have independent, corroborating information to confirm the enemy's objectives, however, the evidence points strongly toward Leningrad for Army Group North, Crimea and the Caucasus for Army Group South, and Moscow for Army Group Center."

"All of the Soviet Union," mumbled Greenwood.

"Yes, it would appear so . . . or at least as far as the Urals and the Caspian. We have been suspicious since the Tripartite Pact last September that the three Axis powers may have an, as yet, unknown or even unwritten plan to divide the world among them—Italy with Africa, Japan with Asia and the Pacific, and Germany with Europe. We continue to look for physical or corroborating evidence of their master plan. We are convinced one exists, but we have no proof. The three of them would likely further divide North and South America. The Japanese sector would likely extend to the Eastern Urals. We have confirmation that Finland has joined Germany for unknown reasons beyond their anger at and resentment of the Soviets. The Baltic States and probably all of Eastern Europe will join Germany against the Soviets, which is a measure of the hatred they hold for the intimidation and oppression of the Soviets and the Russians before them."

Lord President of the Council Sir John Anderson, GCB, GCSI, GCIE, PC, Member of Parliament for the Combined Scottish Universities gasped, "Good God, man, you can't be serious."

"John, we have glimpses of evidence that suggest those objectives, thus the assessment of the enemy's ultimate objective. We have been suspicious since we learned of the Tripartite Pact; however, it was BARBAROSSA that provided the biggest chunk of the puzzle, so far. We have no proof. This is only an educated guess by experts, who study such things. However, I think we can all agree . . . the facts we have demonstrated before us tend to substantiate their assessment."

"OK," interjected Churchill, "enough gloom and doom for this afternoon. We have the Russian supply request on the table before us." Churchill nodded to Sir Edward, who passed the single sheet of paper to his right. "You will notice they have not left much off the list." The Prime Minister waited for each minister to read the message. Several of the men reacted with simple physical gestures of amazement, incredulity or disapproval, and no words. "Now, before we react too negatively . . ." he was cut short.

"Just a week ago, the Russians were aligned with Germany," Margesson interjected, "and, a year ago, Germany and the Soviet Union divided up Poland, while the Russians occupied the Baltic States."

"And now, they want us to supply them with precious aircraft we need for our home defense," added Sir Achie.

"As well as, tanks, rifles, ammunition, uniforms . . ." David continued.

"And, food while our citizens are enduring serious rationing," Greenwood chimed in.

"Winston, they are correct," said Clement Attlee. "How can we spare any of these items? We have no surplus . . . of anything . . . not one item on their list."

Churchill raised both hands for the onslaught to stop. "First, I will send this list," he said, holding up the circulated message, "to the President and seek the additional assistance of the United States. However, I must caution, we are not likely to gain a favourable response in the short term, as our supply needs are not being met, as yet."

"We can hardly afford any of this," commented Chancellor of the Exchequer Sir Howard Kingsley Wood, Kt, PC, Member of Parliament for Woolwich West.

"We must convince the Americans to expand Lend-Lease to include the Soviet Union. Until then, it is my opinion we have no choice. They are ill-prepared to deal with the German invasion."

"So, we must suffer because Stalin ignored the warnings we have given them," Sir Kingsley responded.

Churchill considered the words. The ministers gave him the time. "I understand and appreciate the sentiments expressed, but as I said, I do not see that we have any choice."

"Prime Minister," the First Lord said, "I am compelled to remind you that while we have made significant gains in the Battle of the Atlantic, we are still a net negative in off-loaded tonnage in the Kingdom." Alexander paused. No one spoke up. "We have been net negative for nearly two years. We may achieve our first positive month in another few months of brutal combat. Diverting any of our precious supplies to the Soviet Union will set us further back in achieving a positive flow."

"Here is my reasoning," Churchill said. "Let us assume the Germans are successful in taking the Soviet Union, with its vast agricultural, industrial and petroleum resources. Where are they likely to turn their attention next? If Anthony's assessment is even remotely accurate, how long do you think the Germans will tolerate a free England? The long and short of it, my esteemed colleagues, our defense depends upon sustaining the Soviets and occupying the preponderance of the German armed forces."

Silence dominated the conference room.

Clement Attlee broke the quiet. "As regrettably as I must say, I'm afraid Winston has driven the nail squarely . . . in one stroke. Given the German action to the east, our defense lays in sustaining the Soviet Union."

"So you are throwing your lot in with this notion," Greenwood said.

"Yes, Arthur, I am. His reasoning is precisely correct. We must find a way to support the Russians."

Eden raised his hand and spoke. "It has been just over a week since the invasion began. There is no sign the Germans are slowing down. There is also no sign that Stalin is taking this invasion seriously and mobilizing his nation to defend the Motherland."

"I would add to Anthony's rather dire prognostication," Margesson contributed, "at the rate of current advance, the Germans will reach and likely take Moscow by the end of October."

"And, if Moscow collapses, the Russians will probably sue for peace or surrender," added Eden.

"Are there any objections?" Attlee called the motion.   "Hearing none, the War Cabinet and Defense Committee have endorsed the Prime Minister's policy proposal."

"Then, what do you propose we do to enable this policy?" Sir John asked.

"We should entertain all options to support the Soviet effort. My suggest is we examine the next several eastbound convoys and divert cargo on their requirement list to Iceland, where the ships can be resupplied for an Arctic transit to Murmansk or Archangel."

"That will take a lot of effort to organize," the First Sea Lord responded. "Adding Arctic route convoys will significantly alter our escort cycle time and will reduce the available transport capacity for the North Atlantic route until new ships can be delivered from the United States."

"Yes, yes, Sir Dudley," Churchill answered, "but we shall have to do what we have to do to get them supplies as quickly as possible."

"What timeframe are you thinking?" asked Admiral Pound.

"Within a few weeks . . . by the end of the month"

"Prime Minister, that is impossible without seriously disrupting the North Atlantic convoys."

"Admiral, your task is to provide a plan that accomplishes the mission and informs us of the costs and risks."

"Yes sir. The Admiralty will provide the War Cabinet a skeleton of a plan by the weekend, and our best effort plan in a fortnight."

"You must do better than that, Sir Dudley, but we will go with that for a start. We can take more time to adjust for the second and third convoys, but the first one will be symbolic. The Russians must know we are taking their needs seriously. I shall do my part and engage President Roosevelt for American assistance."

"We understand the need, Prime Minister. I also know you will not take kindly to an increase in our loss rate on the North Atlantic route because we stretched our escort warship coverage too thin."

Churchill smiled, and then scowled. "Your assessment is correct, Sir Dudley, but that does not alter the task before us. Stalin needs demonstrative evidence that we are with him, now. I eagerly await the Admiralty's plan."

The moment of silence and the apparent conclusion stimulated Sir Edward to jump in. "Are there anymore questions, opinions or statements on this issue?"

"Excuse me," Churchill said. "Yes, one more statement for the record," he looked to the First Lord, "please ensure the appropriate ministers are fully involved—Shipping, Supply and Transportation at a minimum."

"Certainly, Prime Minister," responded First Lord A.V. Alexander.

"Very well. You may proceed, Sir Edward."

"We are adjourned, gentlemen."

—

Friday, 4. July. 1941
American Eagle Club
No.28 Charing Cross Road
Covent Garden, London, England
United Kingdom
17:30 hours

Charlotte Drummond had met her husband at Waterloo Station. They had taken the Underground to Leicester Square Station and walked the short distance to the Eagle Club. Baby Ian remained at the farm with Midwife Barbara Grey and a wet nurse she employed from time to time. This was also the first weekend she felt some semblance of normalcy since Ian's birth. Brian's invitation came as a welcome first excursion from their newborn child.

It was a fine, near cloudless, summer day in all of Southern England. Brian stopped several yards from the front entrance. A half dozen servicemen in mostly American uniforms passed them to enter the Eagle Club. Brian turned to face Charlotte.

"Are you sure you want to do this?" he asked.

"Why wouldn't I?"

"My countrymen can get a bit rowdy, and this is our Independence Day celebration from Mother England."

"Nonsense. It is part of history we cannot undo. This is also part of who you are, and thus now, is part of who I am. We will celebrate together."

"I just don't want any of these guys grabbing you."

"Not to worry, my darling husband. Your British brethren have been known to be a bit grabby from time to time as well. I am not offended. If anything untoward happens, I shall let you know immediately. Then, again, if they grab the wrong thing, it might get a little messy."

Brian stared at her in puzzlement. She glanced down, but he still did not know what she was referring to by her comment.

"You silly ninny. It has been too long since you have been home. This is also the longest I have been away from Ian. I could use . . . nay, I must have your help in expressing my milk. I shall be full to the gunwales and nearly bursting by the time we reach the hotel."

"Oh, damn, I should have thought of that. We could have tended to nature before coming here."

"No need to be too eager, my young buck." She laughed rather heartily. "Now, let's get this going and done before I spring a leak."

They entered the multi-story building. It was far more crowded than the last time Brian had been to the club. He deposited her bag and his peaked, service cap with the hatcheck lady, and took the proffered check stub.

"Brian," came the distantly familiar voice.

Brian turned toward the source of the call. He quickly picked out of the crowd the chiseled features and closed cropped blond hair of Colonel Roy Geiger in his distinctive, forest green uniform, resplendent with his stack of ribbons and gold naval aviator wings above his left breast pocket.

"What, are you shy, young man?"

"Great to see you again, Colonel."

Geiger turned to Charlotte. "And, you must be the gorgeous Mrs. Drummond."

"Indeed, I am, sir."

"Charlotte, may I introduce Assistant Naval Attaché Colonel Roy Geiger of the U.S. Marine Corps."

"Pleased to meet you, Colonel."

"An honor to meet you, my dear. It is not quite dinner time, but may I treat the two of you to a hamburger and a soda?"

"That would be delightful, Colonel," responded Charlotte without consulting Brian. Geiger motioned for them to lead. "I'm afraid I do not know where we should be going."

"To the second floor, Charlotte," answered Geiger from behind the Drummonds.

Charlotte saw the stairway and made her way toward the stairs.

Brian leaned forward and whispered in Charlotte's right ear, "The first floor to you, my sweet." He chuckled, and she laughed. The Europeans called the first floor the ground floor, and thus shifted subsequent floors by one. They ascended the stairway.

The dining room was only about half full. They made their way to the counter and ordered their hamburgers with the works and fries, along with an ice-cold Coca-Cola. It was Charlotte's first hamburger, so Brian and Roy talked her through the finer points of a good hamburger.

"I was hoping to see you at today's celebration," Roy said to Brian, as they waited for their food.

"Is there something I can do for you, sir?"

"Well, actually . . . ," Roy began, but was interrupted by the clerk surprisingly calling out their order number. "That did not take long."

They retrieved their meals, made their way to an open table near the far wall, and took their first bites.

"Oh my," Charlotte spoke first, "this is delicious."

"These are not the best," Roy said, "but, they are good . . . quite passable."

"You were saying, sir," Brian sought to return to the open topic.

Geiger finished his bite. He placed his elbows on the table, either side of his plate, and interlaced his hands. "Before we get to business . . . please excuse us, Charlotte," he paused for her consent, "I attended the ceremony at St. Paul's Cathedral to commemorate the new plaque in honor of 'Billy' Fiske. Air Minister Sinclair did a fine job with the dedication. I did not see you there."

"I could not attend, sir. I heard about Fiske, but I never met him."

"I hadn't met him either, but he did have a reputation and was well thought of in the RAF."

"Yes, that he was." Brian did not want to press Geiger.

The loudspeaker system began to play "Yankee Doodle Dandy." Three young women, dressed in rather scant, matching, "Stars & Stripes" outfits entered the dining room carrying a large sparkler in each hand and pranced around the room. Rowdy cheers from the Americans present added to the festive tone of the little celebration. When the song finished, the women virtually ran from the room, and an announcer boldly said, "Happy Independence Day everyone."

"I guess that is the most fireworks we will get in war-torn London," observed Colonel Geiger.

"We don't even get fireworks on Guy Fawkes Night these days," Charlotte added with a little laugh.

Roy looked to Charlotte. "Again, please excuse me, Charlotte." She nodded her head in consent. Geiger looked back to Brian. "The last time we had a chat, I offered to help you transfer to the Marine Corps."

"Yes sir, I remember and I appreciate your offer of assistance."

"But . . . ?"

"Sir, I have no objection to transferring to and flying for the Marines; however, the United States is not in this war."

"Yet!"

"Perhaps tomorrow will change. Malcolm told me years ago that we were headed toward another world war, and the United States would not be able to avoid the fight."

"After you mentioned Mister Bainbridge . . . and Air Commodore Spencer, during our last talk, I took it upon myself to learn more about both men. You are a very fortunate young man to have such mentors. I can certainly understand your affinity for and loyalty to the Royal Air Force, from Malcolm and John's service alone. I happen to agree with Malcolm's prediction. The United States will not be able to avoid this war. I suspect President Roosevelt feels exactly the same. Somehow, Japan is going to strike out one too many times. The Marine Corps will be an important player. To be blunt, the Marine Corps needs pilots like you. What can I do to convince you to join us and fight the Japanese?"

"Colonel, I truly appreciate your confidence in me, but I must answer by saying, my wife is sitting right here listening to this."

"And, his wife is none too pleased about the prospect of having her husband disappear for even longer periods of time into the very distant Pacific Ocean."

"My apologies, Charlotte. The stark reality is, your husband possesses unique skills few in this world have." Roy looked back to Brian. "I do not want to leave you with the image that we are desperate, but we are building a substantial fighter force for the upcoming war in the Pacific, and we cannot train pilots fast enough."

"As I said, I appreciate your confidence in me, but these are my brothers-in-arms. While I am not flying Spits anymore . . . at least for the moment at hand . . . I am happy doing what I am doing." Brian noticed the American volunteer, ordnance disposal man he met at the Eagle Club the month prior walking toward them. He stood to greet him. Roy and Charlotte both looked to see the approaching Royal Navy Volunteer Reserve officer. "Great to see you again, Mister Kauffman." Both Roy and Charlotte stood as well. "May I introduce Assistant Naval Attaché Colonel Roy Geiger and my wife Charlotte. Colonel, Charlotte, this is Lieutenant Draper Kauffman."

Greetings were exchanged.

"Would you care to join us?" asked Geiger.

"No, thank you, sir. I'm afraid I must go. I saw Brian as I was leaving and just wanted to say hello. It seems the Emergency Services folks discovered a little present the Germans left us during The Blitz . . . well, apparently, not so little."

"Then, perhaps you should get to it," Roy said.

Kauffman said his good-byes and departed.

Brian looked at Charlotte. *Should I really explain what he does?*

Roy preempted Brian's concern. "I take it he is a bomb disposal man."

Brian saw the concern in Charlotte's eyes before he turned to answer Colonel Geiger. "Yes sir, and apparently a fairly good one, from what I am told."

"I don't know how they do it."

"Me either."

"Well," interjected Charlotte, "I do not know how either of you men fly those machines of yours, either."

Roy and Brian laughed, and Charlotte eventually joined them.

"Would you two care to enjoy a few beers at Shepherds Pub?"

Brian looked to Charlotte, who responded for them. "Sure. That would be lovely. Thank you, Colonel."

"Before I forget," Geiger said and handed a folded slip of paper to Brian, "that is my contact information. That will work no matter where I am in the world and what assignment I have. I am quite serious, Brian, you belong in the Marine Corps, and I will do whatever I can to make it happen, whenever you tell me you're ready to transfer."

"Thank you, sir. I am most grateful and honored."

Geiger nodded his acknowledgment and stood. They bussed their dishes, utensils and glasses, and then departed the Eagle Club, while the Independence Day celebration continued for the Americans serving in London and surrounding facilities.

—

Friday, 4. July. 1941
Shepherds Tavern
No. 50 Hertford Street
Mayfair, London, England
United Kingdom
20:45 hours

Squadron Leader Mud Morrison and Flying Officer Jonathan Harness Kensington were in the mix of a couple of dozen, rather raucous, air force officers in the bar area and closest to the entryway. They noticed immediately as Roy Geiger, Charlotte and Brian entered the pub.

"Colonel Geiger," shouted Mud.

Greetings were exchanged, as if they knew each other well. Brian had not realized that even Jonathan had met Colonel Geiger on a previous visit.

"Where is Linda?" Charlotte asked.

Jonathan smiled. "She's in the back, catching a bite with a few of the other ladies."

"If you gentlemen will excuse me, I will leave you to your boy-bravado foolishness, and join Linda and the other ladies."

Morrison did not wait for Charlotte's scent to fade. "Lads," Mud shouted, again, and held up his arms, "may I introduce Colonel Roy Geiger of the U.S. Marines. Colonel, these," he said, swinging his right arm as a gesture to the RAF fighter pilots before them, "are the chappies who won the Battle of Britain."

"Hear, hear." "Amen." "Spot on, mate." The shouted replies filled the bar. Pints were thrust into their hands.

"Gentlemen," Mud continued, "may I remind you, this is a glorious day of celebration for our colonial brothers in arms, as they remind themselves of their rejection of King George the Third's magnanimous generosity all those years ago."

"And his tyranny," Geiger joined in, "let us not forget that little tidbit."

"Hear, hear," came the unison shouts, with a mixture of jeers . . . all good hearted and spirited.

The pilots drank their beers, laughed, told stories, and regaled in the splendor of flight.

Brian found a quiet moment. He pulled Jonathan aside. "Only two weeks to go, my brother."

"Spot on!"

"Are you ready for this?"

"As ready as you were last Christmas." They laughed.

"Point taken, my friend."

"Well, I suppose the more important question, is Linda ready for you and this life?"

"She says she is, but I guess we never know until we are in it. How has Charlotte handled it? You have been married for more than six months now."

"We should really ask her. There is no doubt that she resents the war and my service, or rather my absence. This is her first time away from Ian, and regrettably, I have not had that much time with my baby boy."

"Does she seem happy?"

Brian laughed hard. "Do we ever know if they are happy?"

"Good point."

"The bottom line is, I love her. I believe you love Linda. That is all that should matter. The rest will sort itself out in time."

"Both of you will make it to our wedding?" Jonathan asked.

"Yeah. This weekend is a trial run . . . to see how Charlotte does without Ian, and vice versa."

"Well, Linda and I both hope your trial goes well. We really would like you both with us . . . even though you did not give us the same courtesy," he said and laughed hard.

"You don't have to be mean about it. It was very spur-of-the-moment. She was ready, and I could not let the opportunity pass."

"Oh, I'm only kidding you, mate. Of course. We both understand. I know a little of how hard it was for her to get to that point, so . . . perfectly understood."

Charlotte appeared with Linda. The ladies smiled and traded barbs with their pilots. Charlotte grasped Brian's right arm and pulled him toward her, so she could whisper in his ear. "My breasts are leaking. We need to go, now."

Brian grinned from ear to ear. "Excuse me, gentlemen. We must go."

Jonathan chimed in, "We need to go as well."

They paid their respects and stepped out of the pub, onto the still darkened streets of London. At least the streets were peaceful. The last major German air raid had occurred on the night of the 10th of May, with all bombing ending a couple of weeks later. The Germans had turned their attention eastward to the Soviet Union and began their invasion on the 22nd of June. Nightlife was returning to London, after nine horrific months of The Blitz, although darkness prevailed outside.

"We need to let Charlotte go," Linda announced, probably aware of the reason for their early departure.

"We hope to see you at Carlingon in a couple of weeks," added Jonathan.

"Thank you, Linda, and yes, Jonathan, we expect to attend your special day."

They hugged and separated, going in opposite directions as the Drummonds headed to their hotel, and Jonathan escorted Linda to her apartment for the night. They planned to meet in the morning for breakfast before Charlotte headed back to her farm and the pilots took different trains to their squadrons.

—

Chapter 8

Other sins only speak;
murder shrieks out.
-- John Webster

Monday, 7.July.1941
Oval Office
The White House
Washington, District of Columbia
United States of America
12:30 hours

Secretary of the Navy Knox, Secretary of War Stimson and Colonel Bill Donovan were ushered into the Oval Office by Harry Hopkins. The President focused on one last piece of the endless stream of paper that crossed his desk each day. Donovan and Knox stood in front of the right couch, while Stimson stood alone before the left couch. Hopkins pulled up a straight back wooden chair to a position to the left and behind where the President usually positioned himself. Everyone remained quiet while the President continued his reading and notation.

"Good afternoon, gentlemen. Please be seated," Roosevelt said, as he wheeled himself from behind his desk to the spot at the head between the two facing couches. The others offered their greetings. By prior arrangement between the two men, Roosevelt looked over his left shoulder to Hopkins. "Harry, where are we on the notification of Congress?"

"The written notice to Congress was delivered this morning, Mister President. The United Kingdom has formally transferred defense responsibility for Iceland to the United States."

"Thank you, Harry." The President looked to Knox. "Iceland's defense now rests with the Navy and your Marines, Frank."

"Yes, Mister President. We have the task in hand."

"Very well." Without looking this time, he said, "Harry, if you would be so kind . . ." Hopkins stood and handed a single seat of paper to each man. They each scanned the document as the President continued. "This is my draft order for the new strategic intelligence office. For the time being, we will call this position and office the coordinator of information, since that is how I envision this organization. We have discussed this matter in various forms since Bill's first trip to the UK a year ago. I believe we have exhausted the concerns. Nonetheless, I offer each of you one last chance to make your case for any changes you, your chiefs or your respective intelligence departments feel are appropriate. If you have any alterations, please let us know as soon as possible. Harry," Roosevelt said and looked over this shoulder to get a confirmatory nod from Hopkins, "has scheduled us our final meeting of this phase on Friday. I intend to sign this order at that time, and as noted, it will be effective immediately upon my signing the document."

"The three of us have met a half dozen times since our last meeting, Mister President," Knox began. "Speaking for the three of us, I believe we are agreed that we have resolved most of the operating issues. Furthermore, we recognize there will be issues arising as we work together. However, we are convinced we will resolve any working problems without your intervention."

"Excellent."

"I fully endorse Frank's assessment," added Donovan. "I can add that I have also met with Cordell twice, and I think we are good there. I also met with Attorney General Jackson once and Hoover once separately. Based on that latter meeting, the COI's relationship with the FBI may become a little more problematic."

"Do you think you can maintain a sufficient working relationship with Hoover?" asked the President.

"I would like to say, yes, absolutely; however, I am not that naïve. Fortunately, based on the current mission sketch, our spheres do not intersect to a significant degree, so I feel we should be able to handle any rare conflicts. Further, Bob Jackson has offered this assistance, when necessary."

"Are you OK with that?"

"Yes, Mister President."

"You've been unusually quiet, Henry. Do you want to add anything to what has been said?"

"No sir. I concur. I might add one comment in deference to the task ahead of Bill. I am grateful I do have to deal with Edgar, especially on matters in this arena." They all chuckled.

"Very well, then. We will meet again on Friday. Good day, gentlemen."

Knox, Stimson and Donovan stood and departed. When the door closed, Roosevelt turned his wheelchair toward Hopkins. "Do you think this is going to work?"

"Yes, Franklin, I do. I hold a great deal of respect for Bill Donovan, and I believe Knox and Stimson are sincere in their support."

"Time shall tell the tale. Let's have some lunch, shall we?"

Hopkins left the Oval Office to coordinate with Grace and the chief steward.

—

Tuesday, 8.July.1941
RAF North Weald
Epping, Essex, England
United Kingdom
19:00 hours

Standing at Available all morning and half the afternoon, and then being scrambled to intercept a raid that never materialized did not leave the pilots in a good mood when they landed. Brian was perhaps one of the exceptions; he always appreciated the opportunity to fly, especially when not being shot at in any form. Blessedly, Group released them shortly after their quick mission debriefing. They made the early sitting for supper, and then gathered in the Officer's Mess bar. They would make it an early evening, since Whitey informed them to expect an early morning mission tomorrow.

"Where have all the Germans gone," Red Burns quasi-sang out.

"Gone to fight Russians," Pete answered.

Brian felt the urge to join in such chatter, but he always tried to listen and pay attention to the mood of the other pilots.

"Are they going to come back for us?" asked Bulldog.

"Who knows?" responded Pete.

"My guess," added Whitey, "is, yes. We are gaining strength and they know it. I was informed at the commander's meeting last week that Bomber Command has been flying nearly nightly raids into Germany . . . taking the fight to the enemy."

"I wonder when we will go to Germany?" Salt said.

"We don't have the range," Whitey said. "If we did, we would be escorting the bombers, like we do on CIRCUS missions into France. Nasty business from what I hear. They take a real pounding every time"

Beer was flowing freely. It was a nice break from the tension of combat operations.

"You're always so quiet," Dusty more to Brian than the rest of the pilots.

"I don't have anything to say," Brian replied.

"Do you think we face fewer Germans today?" Dusty pressed.

Oddly, the bar nearly hushed upon hearing the question and perhaps wanting to hear a rare answer.

"Compared to what we faced last summer," began Brian, "absolutely . . . much less. I think every pilot who survived The Battle last summer would agree."

"Yep," interjected Red.

"Yes, without question," Whitey added.

"The evidence of what we face today is a pretty clear statement, it seems to me," Brian continued. "Not only do we face fewer numbers of fighters, their pilots are noticeably not as experienced as those we faced last summer. Heck, they have bombed—day or night—since last May. Just from what we see, I'd say the Germans are focused very heavily on Russia at the moment."

"If the Germans finish off the Russians," Whitey added, "they will have no choice but to come back for us. They will not know peace as long as we are flying on their western territory."

"We are not out of this, yet," Pete contributed.

Two pints of bitter were enough for Brian. He drained the last couple of fingers of beer in his glass and turned it over on the bar to signal he was done for the night.

"Leavin' us so early, Hunter?" asked Pete.

"Yep. I've got a couple of letters to write before sleepy-bye time."

"See'ya in the morning," offered Dusty.

Brian was not entirely sure he would write a letter to Gerty Bainbridge and another to Bobby Joe Sales, but he felt the urge, and that was enough for him to call it an early night. He could hear the boisterous pilots joking and laughing as he ascended the stairway to his small room. *Another day, another dollar!*

—

Thursday, 10.July.1941
No.10 Annexe
New Public Offices
Whitehall, London, England
United Kingdom
17:30 hours

SIS Director-General Menzies arrived at the Prime Minister's temporary residence with the Buff Box manacled to his left wrist.

"Good afternoon, Stewart," Churchill said.

"Good afternoon to you, Prime Minister. Thank you for the private meeting."

Churchill waved his hand dismissively. The duty assistant private secretary John Martin shut the door to the reception room. The Prime Minister gestured toward the two facing, over-stuffed chairs. He retrieved his key and unlocked the case. Only one message occupied the candy-striped folder.

MOST SECRET - ULTRA

```
SECRET
DATE: 7 JULY 1941
TO: FM FO
FROM: CHARGE USA
SUBJ: ARRESTS
BREAK
RECEIVED OFFICIAL NOTIFICATION THIS MORNING
BREAK RECENT ARRESTS OF OVER 100 GERMAN
NATIONALS ACROSS USA ON ESPIONAGE CHARGES
BREAK INDIVIDUALS INVOLVED ILLUMINATE GRAVE
DEFICIENCY IN ACTIVE INFORMATION ACQUISITION
ACTIVITIES BREAK TO A MAN THESE INDIVIDUALS
WERE TOTALLY UNQUALIFIED AND INCOMPETENT BREAK
THIS EXPOSURE THREATENS OUR POSITION WITH THE
AMERICANS BREAK THE INFORMATION CAMPAIGN IS TOO
IMPORTANT TO BE ENTRUSTED TO SUCH INDIVIDUALS
BREAK I URGE YOU TO INFLUENCE FOREIGN OFFICE
AND MILITARY INTELLIGENCE TO DO MUCH BETTER FOR
ASSIGNMENTS TO THE USA BREAK HAIL HITLER
END
SECRET
```

MOST SECRET - ULTRA

"This is Thomsen, again."

"Yes sir."

"He is none too pleased with his countrymen we rounded up at the end of last month here and in the United States."

"Based on preliminary interrogation results from the Security Service and shared with us from the FBI, I think it safe to say that Thomsen is correct. They were all rather inept and deficient in their tradecraft."

"Have we learned anything meaningful?"

"Not as yet. MI5's interrogators have implemented a multi-level program similar to what has been so productive with captured senior military officers at Trent Park. I must say, on the interrogation timescale, we are in the very early stages of the process. If there is anything meaningful in these German agents, it may take weeks and more likely months to yield positive results."

"Are you getting reasonable cooperation from the FBI?"

"Yes, from my perspective. David Petrie may have a different view, and we should ask him the same question. The embassy legal attaché is an invited witness to most of the interrogations . . . at least those he chooses to attend. We have not placed special constraints on the derived information."

"Excellent," Churchill said, and then changed the subject. "Do we have any more information from Thomsen's disclosure in his April 17th message?"

"Without being able to confer with the Americans on the topic, we do not believe so. We did confirm the German transmission via their diplomatic code to the Japanese government, so we know the Japanese were notified. That was the last message of any type related to this topic, not even a thank you. It would appear the Japanese were not impressed with Thomsen's source and may have ignored the information."

Churchill smiled, "We can only hope. If true, we and our compatriots may have dodged a potentially lethal bullet."

"We continue to remain vigilant for the slightest clues. I might also say, Thomsen's April 17th message is high on our priority list for discussion when we are finally able to share ULTRA with them."

"Understandable and appropriate. That was the only message in the ULTRA folder. Do you have any other topics?"

"We have a large volume of traffic that I did not think worthy of your time. In summary, German Army Group South is making rapid progress. They may be in Kiev within a month or two. Red Army resistance is just not substantive. At its current rate of progress, Army Group Center will reach Moscow by October. Army Group North appears to be gaining significant assistance from former military and police personnel in the Baltic States. They harbor considerable animosity toward the Red Army, the Communists and the Russians in general. The local populace is acting like the Germans are their liberators."

"How little they know what is coming for them," Churchill observed.

"We did decode a rather cryptic message within Army Group North about units they refer to as special. We are not sure if they are armed forces derivatives or attached from the SS, and we do not yet know what the purpose or mission of these special units might be."

"Curious! Please keep an eye on that matter," he said and smiled, "along with everything else you are watching. I do have one matter to inform you. A week ago, we received a rather long materiel request list from the Soviet Government. The War Cabinet directed we attempt to fulfil their request. The Admiralty presented a skeletal plan two days ago that was barely adequate to the task. I am trying to push the Admiralty and Ministry of Supply to begin convoys into the Arctic for delivery to Murmansk and Archangel to sustain the Soviets. I tell you this; one, to make you aware of what His Majesty's Government is doing, and two, to inform your field agents so that they may be attentive to such materiel and proper application or usage."

"I had not heard. I will certainly inform our field agents. Based on what we have seen so far, I am not sure there is any amount of aid we can provide to sustain the Russians. The Germans are very effective with their rapid armor tactics. The Red Army has yet to generate any meaningful resistance, and they are a long way from a productive counter-attack. The numbers of captured Russian soldiers are so large they are difficult to count or corroborate. Even the Germans have been surprised and perhaps even overwhelmed with the numbers of prisoners of war they are dealing with in their campaign. I must say, there are scattered indications a significant number of Russians may be joining the Germans, apparently unhappy with their life under Soviet domination."

"Somehow, we must sustain the Soviets at least until winter. The famous Russian winters should slow the Germans down, if not stop them entirely, and give the Red Army some breathing room to regroup and find their strength."

The two men talked for another ten minutes on predominately administrative and routine matters. Both of them exhausted their immediate topics, and Menzies departed the Annex for Broadway House.

—

Friday, 11. July. 1941
Oval Office
The White House
Washington, District of Columbia
United States of America
14:50 hours

"Are you ready for your journey, Harry?" asked President Roosevelt of his trusted assistant Harry Hopkins.

"As ready as I will ever be, I suppose."

"When do you leave?"

"I will take the train to Boston, tonight. The plan calls for me to meet a B-17 heavy bomber at Westover Airfield tomorrow. The aircraft is en route from Seattle to England for delivery to the Royal Air Force. They modified their planned route a little south to pick me up."

"That should be quite the adventure. You will get a first-hand, up close look at what will be the mainstay of our heavy bomber force."

"One perspective, I am certain. It will most assuredly not be the luxury of the flying boat I enjoyed last winter."

Roosevelt laughed softly. "It can't be peaches and cream every day, Harry."

They both laughed hard. "As you say, Franklin."

"I know you will convey our respects to Winston. The Riviera conference is set for early next month. I am most grateful that you can make this journey and quietly convey our agenda. I recognize Winston will not be happy about our objective of their divestment of empire, but we must stand for freedom of choice. So, what do we have next?"

"According to the calendar, it is the intelligence business. The proposed and negotiated order is on your desk."

"Excellent. Yes, I have read the order. I think it does what I need in this matter, but I really need to make sure Hull, Stimson and Knox are going to support this action."

"Stimson and Knox should be here shortly, along with Donovan. Cordell has already signed off on the change and he is working on the Soviet situation."

A knock at the door interrupted their conversation. Personal Private Secretary to the President Grace Tully stepped into the Oval Office and closed the door behind her. She had replaced his long-term secretary Missy LeHand just last month, after Missy suffered a serious stroke at just 42 years of age, leaving her unable to speak. Tully had served as the assistant to LeHand since Roosevelt attained the presidency. "Excuse me, Mister President, Secretaries Knox and Stimson, and Colonel Donovan are here for their scheduled appointment."

"Thank you, Grace. Please show the gentlemen in." Hopkins pulled the President's wheelchair back, and then pushed him to his usual position between the long couches.

Stimson led the group into the Oval Office. "Good afternoon, Mister President."

The men traded appropriate salutations and settled into their places on the couches. Hopkins pulled up a straight back, wooden chair behind and to the left of the President.

"Before we get to the principal topic of this meeting," began Roosevelt, "I would like the latest intelligence assessment of the situation in Russia."

"All of the information we have is not good," began Secretary of War Stimson. "The Germans continue to roll back the Red Army, destroying whole infantry and armor divisions in their path. They have captured more Russians than anyone has yet counted. At their current rate of progress, they may well take Leningrad and Moscow before winter sets in. The Red Army has shown no signs of mustering an adequate defense to even slow down the Germans."

"What if the Soviet Union falls?"

"That is a scenario a little too frightening to contemplate."

"Well, from your description, it sounds entirely possible."

"And," interjected Donovan, "a natural progression for Hitler." Everyone looked to Donovan, as if he had just committed some serious faux pas. "Hitler has been an avowed anti-Communist since they created the National Socialist German Workers Party. This is an ideological fight for him. He had effectively subdued the Soviet Union with their mutual Non-Aggression Pact, nearly two years ago."

"They stand to gain vast resources they desperately need for their war machine," Stimson added. "Not least of which may well be the Caspian oil fields at Baku."

"I received a note from Churchill two days ago. He is diverting some of his precious supply ships to the Russian Arctic ports in a desperate effort to sustain Stalin. He has asked for our contribution . . . to help the Russians, he says."

"What is he asking for?" Secretary of the Navy Knox asked.

"Ships, armaments, food stuffs, everything he is asking us to supply for Great Britain. He is asking us to extend Lend-Lease to the Soviet Union. He fears a rejuvenated and resupplied Germany, if the Soviet Union is subdued."

"An understandable fear, I'm afraid," Stimson said.

"What is Stalin going to do?"

"We have no way to know," responded Stimson.

Roosevelt smiled. "Therein lies the perfect segue. I need the answer to that question, among many other strategic questions, or at least the best guess based on the facts we have." No one moved, blinked or spoke. They waited for the President to continue. "It is my understanding that the two principal service secretaries and their intelligence bureaus have successfully negotiated the charter for the office of the coordinator of information." Roosevelt extended his left hand without looking for Harry Hopkins. The President's assistant placed the order in his hand. "Are you ready for me to sign this order?" he asked, holding up the order, and then handed the paper to Frank Knox.

The White House

Washington, DC

July 11, 1941

ORDER

By virtue of the authority vested in me as President of the United States and as Commander

in Chief of the Army and Navy of the United
States, it is ordered as follows:

1. There is hereby established the
position of Coordinator of Information,
with authority to collect and analyze
all information and data, which may bear
upon national security; to correlate such
information and data, and to make such
information and data available to the President
and to such departments and officials of the
Government as the President may determine; and
to carry out, when requested by the President,
such supplementary activities as may facilitate
the securing of information important for
national security not now available to the
Government.

2. The several departments and agencies
of the Government shall make available to the
Coordinator of Information all and any such
information and data relating to national
security as the Coordinator, with the approval
of the President, may from time to time
request.

3. The Coordinator of Information may
appoint such committees, consisting of
appropriate representatives of the various
departments and agencies of the Government,
as he may deem necessary to assist him in the
performance of his functions.

4. Nothing in the duties and
responsibilities of the Coordinator of
Information shall in anyway interfere with or
impair the duties and responsibilities of the
regular military and naval advisers of the
President as Commander in Chief of the Army and
Navy.

5. Within the limits of such funds as may
be allocated to the Coordinator of Information
by the President, the Coordinator may employ

necessary personnel and make provision for the
necessary supplies, facilities, and services.
        6. William J. Donovan is hereby designated
as Coordinator of Information.
                        S/

All three men completed their reading of the order.

"So, you are agreed," said the President. He looked to each man and received a confirmatory nod. "Very well, then. I will sign and we will issue this order when we are done here. I have no doubt with respect to your sincere support, and I trust you both appreciate the difficulty you will have in shepherding your intelligence services. Paragraph four clearly states the COI will not interfere with the functioning of the service chiefs or their intelligence directors. Paragraph two directs all information to be shared with the COI for strategic analysis. I would like to hear how you intend to ensure compliance of your intelligence bureaus with paragraph two."

Knox answered immediately. "We cannot ensure there will be no stumbles from time to time, Mister President. We acknowledge, recognize and agree there are gaps in the intelligence analysis we can and do provide you, and further we agree Donovan's proposal makes sense, especially from your perspective. The Chief of Naval Operations supports this change as long as the Office of Naval Intelligence is not restricted or interfered with in developing intelligence for the Navy."

"I do not see these operations as exclusive in any form, but rather complementary to the whole." Roosevelt looked to Donovan, who nodded his head in agreement. "The world situation is complicated enough. We do not need parochial turf battles to blind us." Everyone nodded. "Henry, your turn."

"What Frank said . . . Mister President. The Army and specifically the Chief of Staff see potential positives from additional eyes looking at things, and possibly from a different perspective. Perhaps, Bill can develop an organization to fill in the gaps you noted, and we all agree. We must fill those gaps."

"How do you propose to resolve conflicts or disagreements in the assessment of common facts?"

Knox responded, "Bill has suggested an expansion of the Joint Intelligence Committee to include the Office of the COI, and we both," he said, gesturing to Stimson, "endorse his proposal."

"Yes, yes, but that is the obvious, Frank. I want to know how the three of you are going to make this work? There are going to be conflicts, disagreements, offended sensibilities, and I do not want you running back to me to referee your spats."

"Quite understandable, Mister President. Speaking for myself alone, I can only assure you of my belief in and commitment to your strategic intelligence apparatus . . . ."

"Our."

Knox nodded his acquiescence. "Yes, certainly, our strategic apparatus. We need Bill to be successful. I cannot and thus will not offer any guarantees the process will be devoid of difficulties, but I know the Chief and I will do our part to enable his success."

Roosevelt simply nodded his head, and then looked to Stimson. "We have our reservations," Henry began. "Any new start-up by its nature should garner apprehension. However, as Frank stated, we are committed to success."

The President nodded his head, again. He looked out the open window to the green of a fine, sunny, summer day. "Neither of you answered my question, but I also acknowledge the difficulty of asking a clairvoyance question. I believe all of you appreciate my earnestness in this matter. I trust you will do your part to make this happen. I shall sign this order shortly, and it will be distributed to cognizant executives this afternoon . . . effective this day. Thank you, gentlemen." The three men stood. Hopkins remained seated and the President did not move. "Bill, if you would be so kind, I would like a word."

"Good day, Mister President," Knox and Stimson said in unison, and then departed closing the flush door behind them.

The President gestured for Donovan to be seated. "I needed to make an implied statement."

"Yes sir."

"You report directly to me. The other intelligence chiefs do not. I doubt this direct reporting link will last; but, until we are in this war, which I believe is inevitable, now, it will remain. I want you to use this time to get as far down the road as possible. You will be funded from my discretionary monies, but that cannot last either. I have already approved your initial budget request with only a few slight pluses and minuses. Please sit down with Harold Smith as soon as you are able to solidify your funds and familiarize yourself with the annual budgeting process."

"I'll tend to it immediately, Mister President."

"You have met Harold Smith?"

Harold Dewey Smith had served as the Director of the Bureau of the Budget in the Office of the President of the United States since 15.April.1939.

He managed the extraordinary funding requirements of the Lend-Lease Program and of the unprecedented military services expansion begun the previous year.

"Yes sir. He has been most helpful in preparing the current and extended budget plan."

"Excellent . . . good man Harold. I will take this opportunity to once again thank you for your selfless efforts to resolve this matter. I anticipate great things from you, Bill, and the organization you are now chartered to form and operate. My earlier question is a good place for you to start. As I said, Churchill has asked us to extend Lend-Lease to the Soviets, but we must know if they can survive the current German onslaught. We cannot tolerate our supplies falling into the hands of the Germans, if the Soviets collapse."

"The signs at present are not good."

"An understatement of gargantuan proportion I'm afraid. I know you have your hands full just creating a functional organization. Yet, I need that answer today. By the end of next week, I would like to have some idea how long it will take for you to develop the answer to that question."

"Yes sir. Thank you for your confidence, Mister President. This is going to be an adventure. We all need a world-class strategic intelligence agency, and we do not have much time. I trust you recognize, we will be dependent upon the British and MI6 in the short term."

"Yes, I do, but that dependence cannot endure. While I generally agree with Winston, he remains focused upon preserving the British Empire, and that alone portends conflict with the British in the future. We must be prepared for that friendly struggle, and the confrontation may come sooner than we expect. Harry," he said, gesturing to Hopkins, still quietly seated behind him, "leaves tonight on his latest journey to England. He will prepare the British for our agenda for the RIVIERA conference at Placentia Bay early next month. We are going to insist the British divest themselves of their empire in favor of free choice of governance for the indigenous peoples. That alone may generate palpable conflict. We will not know until we get there, but we must be prepared."

"Do you need COI to prepare a strategic assessment for that matter as well?"

Roosevelt chuckled softly. "I appreciate your energy, Bill. I can probably handle Winston for the time being. After all, they are becoming quite dependent upon us for supply and finance. It is the Stalin and Soviet matter I am the most concerned about and have the least information to help us form a proper policy position. You need to build upon your relationships with MI6 and the security services as quickly as possible, but focus your analysis effort on the

Soviets. We must find a way to sustain them, not unlike we did last year when the British were hanging on by a thread."

"Understood, Mister President. There has been no mention of attending RIVIERA with you."

"No need, Bill. Winston will not have his intelligence officials with him, either."

"Very well . . . just thought I would ask."

"Quite appropriate and thank you. Now, I believe I have kept you from your important new duties. Good luck, Bill. You are going to need it in the daunting task before you. As we in the naval services say, Godspeed and following winds."

Donovan stood and smartly saluted the President. "Good day, sir." Roosevelt nodded his acknowledgment, and then Donovan departed.

Roosevelt waited for the door to close, and then turned to look over his shoulder to Hopkins. "Let's get this order signed to make it official what I suspect will be an historic step on our journey."

"Certainly," Harry answered, as he rose to wheel the President back to his desk.

—

Saturday, 19.July.1941
Carlingon Castle
Newcastle-upon-Tyne, Tyne & Wear, England
United Kingdom
15:30 hours

Charlotte and Brian had managed to meet at Kings Cross Station before they boarded the afternoon train to Newcastle. Amazingly, the entire rail network had continued to operate normally throughout the German onslaught, adapting promptly when bomb damage disrupted a particular line. The journey north had been comparatively quiet and uneventful. They had their compartment to themselves for the majority of the trip, allowing them to freely catch up on the affairs of their lives. The trial run a fortnight earlier had gone exceptionally well, and had given Charlotte the necessary confidence to make this multiple day excursion. The Drummonds had considered including Ian, but Charlotte instinctively knew an infant would have made the long weekend far more complicated. They spent a passionate and welcome night in a Newcastle hotel close to the main railway station.

Brian wore a new RAF flying officer's uniform with his full medals arranged properly below his bright gold embroidered pilot's wings and his peaked cap.

Charlotte wore an elegant, apricot colored dress that accentuated her exquisite body, a nice, matching color, summer hat. She had arranged her hair drawn up in a delightful interlaced fashion off her neck and shoulders. The bright sun gave her hair a light blond appearance. Charlotte was a strikingly gorgeous woman.

Jonathan had offered to send a car for their transportation, but Brian had respectfully declined. The Kensingtons had more than enough to deal with in all the arrangements for Linda and Jonathan's wedding. The taxi ride from the hotel to Carlingon Castle had been a first for Charlotte to the area, and she had marveled at the scenery outside the city, just as Brian had done on his first trip. The summer scenery was decidedly more dramatic than the winter version. The clear, nearly cloudless sky made the landscape all the more brilliant. The full foliage of the trees and bushes made the opening upon the view of the Kensington home truly awesome. Charlotte had gurgled joyously at the magnificence of the stone building and the well-kept grounds.

The Kensington's butler Mortimer stood at the main entry door to receive the Drummonds. "Welcome back to Carlingon, Mister Drummond," he greeted Brian, "and this lovely lady must be Mrs. Drummond."

"Thank you, Mortimer. Nice to be back and especially for such a noble occasion. Yes, may I introduce my wife Charlotte."

"Welcome to Carlingon, Mrs. Drummond."

"Thank you, Mortimer."

"Our guests are gathering in the study before the ceremony, which will be outside in the garden."

"I believe you know the way, Mister Drummond, or would you like me to show you?"

"Yes, I remember the way to the study. Thank you."

"Excellent. Please proceed," Mortimer said, as he opened the massive oak door easily.

Brian took Charlotte's hand and led her into the Kensington home, through the foyer, down the long right hallway, and into the study. Jonathan saw them immediately, excused himself from his conversation, and walked briskly to the Drummonds. He was in a fresh RAF uniform as well. Jonathan touched both cheeks with Charlotte, kissed her right cheek and gave her a warm hug. He embraced Brian and whispered, "Thank you for coming, brother."

"My honor, Jonathan. We would not miss this grand occasion. Plus, I wanted Charlotte to see Carlingon Castle, after I have had such good things say about my visits here."

"Hopefully," Jonathan responded to Charlotte, "I can give you the cook's tour before you leave."

"That would be sweet, Jonathan, but that is not the purpose of this visit. You are about to marry a lovely lady. That is your only focus on this day."

"Thank you for yo . . . ," Jonathan spoke and did not finish, when Rosemary appeared, as if from the ether in a modest but nicely tailored, floral print dress.

Rosemary hugged Charlotte, and then said, "So great to see you both." She hugged Brian and quickly kissed him on the lips. "Welcome to our humble home."

"Modest as always," mumbled Jonathan.

"Stop being a drip, Jonathan. I love you both."

Jonathan's parents – Theona and George Kensington – walked over to their children and the Drummonds. "This must be the renowned Charlotte we have heard so much about," Theona boldly said.

"Yes ma'am," responded Charlotte.

Theona Kensington embraced Charlotte, touched cheeks, and stood back surveying her guest. "You are far more lovely than your Brian's description. We are so glad you were able to make the journey north for our son's special day."

"Thank you, ma'am. It is an honor to be invited. You have such a delightfully gorgeous home," she looked to Jonathan, "and I am certain just the setting alone will make your wedding most memorable."

Jonathan smiled and nodded his concurrence.

George Kensington shook hands with Brian and extended his hand to Charlotte. She placed her hand in his. He bent over and gently kissed the back of her right hand. "As you see, we are a household of strong women," he said.

Charlotte chuckled softly. "Yes, Mister Kensington. I think I have known that from my first meeting with Rosemary."

"Well," Theona said, "you are family, now."

Please don't ask her about her family. I've told you the essence of her story. She doesn't need to be reminded of her losses. They must have felt Brian's thoughts.

"Come," continued Mrs. Kensington, "we need to introduce you to Linda's parents and family." She hooked Charlotte's arm and led her into the room. Brian glanced to George and Jonathan, shrugged his shoulders and followed the two women.

Charlotte and Brian were introduced to Linda's parents, siblings, cousins and extended family. They also met more of Jonathan's extended family, as well as invited friends. Several, both women and men, wore service uniforms of all three British military services. One of Jonathan's university classmates wore the forest green uniform of a Royal Marine lieutenant – a first for Brian. In all, Brian estimated there were 50 invited guests for the wedding.

A well-dressed, matronly woman appeared at the study exterior door and announced the need for everyone to take their seats outside, so the ceremony could begin. Charlotte and Brian waited until they were nearly the last to step outside.

"Oh my goodness," exclaimed Charlotte, as she surveyed the scene.

The arrangements were almost angelic. A few, modest, fair weather cumulous clouds dotted the otherwise flawless azure sky. Blessedly, the temperature remained comfortably moderate despite the brilliant sun. A large, open-sided tent covered most of the stone patio. On the grass just beyond the patio, two traditional sections of chairs were arranged in bride and groom halves under a not quite taut cover that appeared to be lace with hanging, multi-colored flower arrangements masking the support poles. By the time they reached the remaining open seats at the back, Jonathan was standing with two of his cousins dressed in matching tuxedos at the altar – a floral arch that framed the distant mountains. The Church of England minister dressed in full, purple regalia stood under the arch, behind Brian.

"This is unimaginably gorgeous," Charlotte whispered.

"It is rather impressive, isn't it," Brian responded in comparable tone. "It makes me feel rather embarrassed that I did not do better by you, my darling."

"Stop that instantly!" Charlotte protested. "I am happy. I think you are happy. That is all that counts."

"But, still . . ."

"No, buts, Brian. It is not fair to either of us if you regret our simple ceremony. We are just as married as Linda and Jonathan soon will be. I urge you to expunge any hint of inadequacy from your thoughts. This is Linda's first marriage. If anything, I should apologize to you, since your first marriage was so simple."

Brian smiled and kissed Charlotte on the cheek.

A cellist played a soft, exquisite song that Brian did not recognize. After a short pause, the cellist began the distinctive chords of Felix Mendelssohn's famous Opus 61, the incidental music for an 1842 rendition of Shakespeare's *Midsummer Night's Dream* and known more popularly as *The Wedding March*. The gathering stood and turned to an open doorway adjacent to the now closed study double doors. A young girl, perhaps six or seven years of age walked cautiously from the doorway toward the altar via the center walkway, sprinkling rose petals as she walked. A boy about the same age, carrying the couple's rings on a pillow the ring bearer held with both hands, followed the flower girl. Two young ladies dressed in matching light blue dresses began their broken, paced walk to the tempo of the music. The Bride Linda Mason appeared in a brilliant white, lace dress. Her father stood beside Linda. A large veil completely covered her head nearly obscuring her face and head. A floral crown held the veil in place on her head.

The ceremony appeared to transpire flawlessly, although neither Brian nor Charlotte could hear the words being spoken. Yet, they both understood the essence of what was happening, even without the words. The whole process took a little over 15 minutes to complete. Linda and Jonathan kissed, turned to their witnesses, held hands and raised their free hands like triumphant athletes to the applause and cheers of their guests. The newlyweds led the wedding party down the aisle, followed by family and then friends. Most of the adults gathered in the shade of the tent. A light breeze and moderate temperatures made it almost cool. The children ran through the gardens and grounds, as children will always do.

The household and supporting staff served *hors d'œuvres* – rather delicious – and drinks. They both took flutes of amazing champagne—rare for ration-starved England. Brian was surprised, perhaps even shocked, how many people he did not know had more than casual knowledge of what he did in the Royal Air Force. Several recognized the distinctive shapes and colors of his three awarded medals under his wings. Brian began to feel uncomfortable with the recognition and attention being devoted to him. He was a guest at the wedding of his best friend . . . other than Charlotte. Brian did not want the attention. The more awareness that came, the worse he felt.

Charlotte must have sensed her husband's unease. "Are you alright?" she whispered.

"No, I'm not. This is embarrassing. Perhaps . . . ," Brian broke off his sentence, as yet another man approached and thanked him for his service, and asked nearly the same question: how can a man so young be so accomplished? He offered the same answer; he was just doing his job. When the man moved on, Brian continued, "Perhaps we should leave."

"Brian, my dear, we are expected for supper. It would be even more awkward, if we suddenly excused ourselves before their celebratory dinner. We haven't even congratulated Linda and Jonathan, yet."

Brian looked into Charlotte's eyes. "I just love your eyes."

"Oh, Brian . . ."

"I know, stop gushing. And, of course, you are correct. Maybe I should take off my medals . . . that seems to be what is attracting all this attention."

"That is your choice, my darling, but you earned them. You should be proud of what you have done for this country and for freedom itself. I am proud of you."

Brian nodded his head in acknowledgment. Fortunately, several people chose to ask Charlotte about her farm, which made Brian feel a little better.

Mortimer appeared and rang a simple hand bell once. "If you would be so kind, please make your way to the dining room. Dinner is about to be served."

The Drummonds followed the throng of guests. Brian was rather surprised. He did not remember the dining table being so large, but then again, the room had always been large enough for such a table, just not the oak dining table as he recalled it. The table setting and floral centerpiece arrangements were exquisite. Place cards marked the sitting assignments. Brian and Charlotte were seated across the table and four places apart. For wartime England, the meal was extraordinarily sumptuous – mixed green salad with an olive oil and unusual vinegar, followed by roast duck, green beans, new potatoes and creamed corn, with a glazed custard dish for dessert, all served with copious amounts of an exceptional Chardonnay.

To Brian's surprise and relief, the dinner conversation, between the toasts to the bride and groom, and their parents, proved entertaining, informative and enjoyable. Brian repeatedly looked to Charlotte, who appeared to be enjoying her dinner conversation as well. The more he learned and became aware of Jonathan's family, the more he loved them. His discomfort of earlier was probably just a passing moment.

After the delightful dinner, the evening celebration transitioned to a party with Victrola music, dancing, laughing and general frivolity. Linda and Jonathan were clearly enjoying themselves. They did not want to leave and had to be reminded by Theona Kensington that no one else would leave until they were sent along their way. It was after midnight when the newlyweds finally left the party. The guests began to dissipate rapidly thereafter. Many of the two families would stay at Carlingon Castle, as the newlyweds would. Jonathan had to be back on duty Monday by noon, just as Brian did. Fortunately, the Kensingtons had graciously arranged for hired cars to transport the remainder, like Charlotte and Brian, who could not stay at the Kensington home, to their hotels in Newcastle.

The Drummonds literally collapsed into bed at 01:30. They had reservations and tickets on the mid-morning train from Newcastle to London. They planned to spend one last night together in London before they parted, again, Charlotte returning to the farm and Brian returning to his squadron.

—

Sunday, 20.July.1941
Chequers Court
Ellesborough, Buckinghamshire, England
United Kingdom
20:30 hours

"Welcome back to Chequers, Harry," Churchill greeted Harry Hopkins, as the American exited the limousine, and extended his right hand to his guest.

Hopkins grasped Winston's proffered hand. "Great to be back. Thank you for having me."

"They are holding supper for us, so if you are agreeable we shall skip the cocktail hour and sit down to dinner."

"I am famished. It seems like I have not eaten in days."

Churchill laughed. "I take it your journey was rather arduous."

"A rather large understatement," Hopkins laughed.

"Shall we," Winston said, gesturing to the open main entryway door.

They walked down the central hallway to the left and entered the main dining room. The massive table was set for only two at the far end.

"It will be just us tonight?" asked Hopkins.

"Yes, if that is acceptable. I believe we have plenty of topics to discuss."

"Certainly."

They sat at adjacent seats on the corner of the table. Their first course, a rich, vegetable soup with barley, was served.

"I was informed," Churchill began, "you came across the pond in one of our new B-17C Flying Fortress Mark One heavy bombers."

"Yes sir. I was told it is the last of your original purchase of 20 aircraft. It is a most impressive aircraft, but an airliner it is not."

"Do tell, Harry."

"I was advised to wear a thermal flight suit and I was issued several blankets. My bones are too old for that kind of cold. The young aircraft crew handled the flight quite well, and they worked hard to make me as comfortable as they possibly could . . . to no avail, unfortunately. We landed at RAF Polebrook earlier this morning.

"Northamptonshire?"

"Yes, near the village of Oundle, so the signs said. The base commander graciously had me driven to the nearest railroad station, and the embassy staff retrieved me at Kings Cross Station. I checked in with the ambassador, picked up the latest messages from Washington, took an all-too-brief nap, a quick shower, change of clothes, and here I am. While the B-17 is a great machine for its intended purpose, it is just not for my tired, old body."

"I am sorry you had such a tortured journey, my friend."

"Thank you for the sentiment, Winston. I have finally vanquished the chill and I shall endeavor to remain awake for our discussions this evening."

"I know the fatigue of travel, Harry, so when you have had enough, just wave your hand and we will get you to a nice, warm, comfortable bed for the night."

"Thank you very much for your understanding."

"By the way, I believe the squadron your aircraft was delivered to was Ninety Squadron, the first of our B-17 heavy bomber squadrons. They flew their first combat mission on the 8th of this month, a daylight raid on the graving docks at Wilhelmshaven."

"Was the mission successful?"

"No, unfortunately. The crews indicated they found excessive cloud cover for accurate bombing."

"Too bad."

"Yes, quite. As the Air Ministry has informed me, they have another pending daylight mission to Brest. The *Scharnhorst* is in port for repairs."

They finished their meal and the staff cleared the dishes. "Let us retire to the library, far more comfortable there – a good fire, some cognac to warm your cockles, and comfortable chairs for conversation between friends."

They rose from the table. Harry followed Winston. Churchill's valet Frank Sawyers offered the humidor. Both men took cigars. Sawyers assisted the preparation and lighting of the cigars, and then poured a snifter of Hine cognac for both men.

"Will there be anything else, sir?" Sawyers asked.

"No, Sawyers. Thank you. That will be all."

Churchill waited for his valet to leave and close the door behind him. He began, "Pray tell me, how is our friend Franklin?"

"He is in good health and doing exceptionally well, I must say."

"Excellent. So, what news have you?"

"First, Franklin wanted me to confirm for you personally, not in message traffic, his agreement to the RIVIERA conference next month. We have very tightly controlled access to the whole picture for his travel arrangements. Franklin authorized me to inform you of his plans. Other than the Secretary of the Navy, the Chief of Naval Operations, the Secretary of State, the President and myself, you are the only other person on earth who will know the whole picture."

"Thank you, Harry."

"Next week, Franklin will head north by train for what the press will know is a routine summer vacation—a coastal cruise of New England. He will board the presidential yacht *Potomac* at New London, Connecticut, since the city pier is very close to the rail station. We wanted it to be a very public boarding. The public statement will be that the president will take an extended cruise of the New England and the Mid-Atlantic coast before returning to Washington. A look-alike will represent the President onboard the *Potomac*, just in case photographers happen to take pictures when the ship is near enough to the coastline. What only a handful of people will know, the heavy cruiser *Augusta* will rendezvous with the *Potomac* 20 miles out to sea in a closed military training area. The President will transfer to the *Augusta* and proceed directly to Newfoundland. The *Augusta* should be anchored in Placentia Bay when we arrive on the 9th. He will transfer back to the *Potomac* before the presidential yacht enters the Chesapeake Bay and docks at Anacostia."

"Excellent. We will have similar operational security on our end. We will board *Prince of Wales* on the 4th. Our cover is a naval tour of the Mediterranean."

"Thank you for allowing me to travel with you, Winston. I am certain I shall be warmer and more comfortable than I was on the trip here."

Churchill chuckled. "Indeed. Then, the agenda is agreed?"

"Yes."

"Well done, Harry. Thank you."

"You are most welcome, Winston. The President also wanted me to convey his grave concern about the situation in Russia. He would appreciate an off-the-record, frank, private discussion with you on that topic."

"He is not alone with his concern. The information we have from Russia is not good. The Germans are making steady, relentless progress toward Leningrad, Moscow and the Caucuses. We have diverted as much of our supply as we dare to their Arctic ports to sustain them. We have provided Marshal Stalin with the best intelligence we are able to offer. In my personal opinion, if they can make it to the winter, they might stand a chance. If the Germans take Leningrad and Moscow before winter, I fear the worst. If the Soviet Union is subdued, the Germans will inevitably return to us. We must keep the Russians in this fight."

"The President shares your assessment. To be candid, his concern, as it was last summer with you, remains supplies that could fall into German hands with a Russian surrender."

"It is a risk we must take, Harry. We have asked for extraordinary sacrifices and hardships of the British people as we divert precious supplies to Russia. To be blunt, Harry, we need Lend-Lease extended to the Russians."

"The President understands your request. That is the essence of your discussion with him at Placentia Bay."

"I am prepared for that discussion. I will ensure we have the latest intelligence available to help us."

"We also understand from Ambassador Leahy, the Vichy French have acquiesced to pressure from the Japanese and Germans, ceding control of French Indochina to Imperial Japan. As a consequence, a U.S. oil embargo against Japan will become effective at the end of the month."

"We shall join you in that embargo."

"The President doubts the Japanese will respond positively to the threat or the embargo. The President's preeminent concern in the Far East is for Malaya and Singapore, given Japanese occupation of Indochina, and to a lesser degree the Philippines."

"I am assured by the Army and the Navy that Singapore is secure."

Hopkins contemplated Churchill's statement. "The oil embargo will be made public next week. We must anticipate the consequences."

"We are ready."

"Likewise, we have alerted the Pacific and Asiatic Fleets and the Philippines command, as well."

"We must remain vigilant. I have two items I intend to raise off-the-record as well." Hopkins nodded his head. "Do you recall our MAUD Committee?"

"Yes."

"They completed their assessment earlier this month. I will deliver a copy of their final report. You are welcome to read it, if you wish. The committee concluded that an atomic explosive is technically feasible. We do not have the industrial capacity to perform the enrichment process in a timely manner. I intend to propose to the President a joint development program for building an atomic explosive weapon. The committee's concern and mine is the Germans may well be ahead of us, and given the potential of this new weapon, whomever produces an operational unit first is likely to win the war. We must not allow the Germans to be first."

"The President is keenly aware of the threat. As you may know, Professor Einstein wrote a letter two years ago warning the President of the German development efforts."

"Yes, and we have been working with your Briggs Committee since that time. However, to speak frankly, candidly and privately," Churchill paused, to receive a confirmatory head nod from Hopkins, "we are concerned about the degree of urgency in this endeavor."

"Understood. That is an appropriate topic for your private discussions."

Churchill took another sip of his cognac along with several puffs on his cigar, as he studied Hopkins' face. "We also have a rather disturbing item of intelligence that I need to discuss with President Roosevelt."

"May I ask . . . ?"

"We have indications from several very reliable sources that the Germans have initiated planned operations by what they call special operations groups – *einsatzgruppen* is the German word. We are still trying to develop more first-hand evidence, however, it appears these are SS killing squads that are targeting Jews, communists, Soviet leaders and others of what they call undesirables in the Baltic States and Ukraine. Worse, there is direct evidence these groups are employing indigenous personnel to do most of the blood work under SS supervision. The Germans have apparently stimulated local vengeance against the Soviet occupation and those who supported them."

"That is rather foreboding information. To my knowledge, we are not aware of such conduct. That sounds like an appropriate assignment for the Coordinator of Information. I understand your government was notified a week ago that the President signed an order last week, to create the Office of the Coordinator of Information under Bill Donovan. The purpose of the COI is to develop strategic intelligence for the President."

"Yes, we were informed . . . good man, Donovan. He will do well, and we look forward to working with him and his organization. I shall have 'C' contact 'Big Bill,' as soon as possible to bring your COI up to date with where we are on this."

"Thank you. I know the President will be most interested in your discussion."

"We must consider what options we have available to halt or slow down these killing squads. We must assess what we might be able to ask of the Soviets."

Hopkins drifted off into thought. He was showing signs of his fatigue. His eyelids were noticeably heavier. "Why have the Germans turned to such obscene behavior?" Hopkins asked.

"One word . . . Hitler . . . well, and his henchmen. He told us all what he was going to do in 1925. We did not listen. Now, we reap the consequences."

"As I recall, Winston, you tried valiantly to sound the alarm. Unfortunately, few listened."

". . . and the rest is history." Churchill studied President Roosevelt's special assistant. "Now, my friend, I laud your noble effort to remain awake, but I am afraid I need to be the courteous host and shuffle you off to bed for a good night's sleep. We can renew our discussions tomorrow morning. I shall stay here at Chequers a couple of extra days to allow you to recover properly from your tortuous journey."

"You are most gracious, Winston."

They adjourned for the night. Churchill escorted his guest upstairs, and deposited Hopkins in the spacious and opulent guest suite. Having been in that state a few times, Winston knew it would not be long for Hopkins to be sound asleep. There were sufficient hours remaining for the Prime Minister to knock out some more paperwork. Churchill asked Sawyers to find his duty private secretary – this weekend it was John Martin – along with the duty stenographers. He could still manage another few hours of work before he too would find slumber.

—

Chapter 9

Anybody can make history.
Only a great man can write it.
                -- Oscar Wilde

Friday, 1.August.1941
RAF North Weald
Epping, Essex, England
United Kingdom
14:45 hours

Green Section, No.71 Squadron descended effortlessly through the comparatively thin overcast layer covering the Essex countryside. The brilliant greens of the English summer landscape with a few browns of recently harvested fields added dimension to the scene.

"Lumba, this is Eagle Green leader calling," Flying Officer Brian Drummond broadcast.

"Eagle Green Leader, this is Lumba. Go ahead."

"Eagle Green is clear of the clouds, aerodrome in sight."

"Roger, Eagle Green Leader, switch to Heavy. Thank you for your work today."

"Eagle Green switching." Brian checked both wingmen, looked down to his right console, grasped the appropriate knobs, and switched his radio to the North Weald control tower frequency. He looked back to each wingman and received a thumbs up gesture of affirmation from each of them. "Heavy, this is Eagle Green Leader, three fighters for landing."

"Roger, Eagle Green Leader, we have you in sight. No traffic. You are cleared to land. Runway Three Zero in use. Winds light and variable. 'Q' 'N' 'H' Two Nine Seven Three."

Brian acknowledged the primary runway and altimeter setting. He checked and set the barometric pressure setting in his altimeter. They were not quite lined up for runway 30. Brian adjusted their flight path and descent rate to be in a proper position for a straight in approach to landing. He lowered his landing gear, checked his gear down confirmation pins were extended above the wings, lowered his flaps to the landing position, and unlatched and slid his canopy back. The fresh air felt good, even though he still had his oxygen mask in place and his goggles down. One last time, he checked his wingmen were in tight position slightly behind and below his wings, and their wheels were down and canopies open. The runway had been finished a few months earlier, after the worst of The Blitz had passed, and was comparatively wide, but still not boundless in width.

Dusty and Bulldog touched down safely and smoothly, and Brian followed shortly after them, having flared his Hurricane fighter to arrest his remaining descent rate, pulled his throttle the remaining way back to idle, and touched all three wheels simultaneously and ever so gently. He let most of his speed dissipate naturally and applied just a little brake toward the end of the runway to achieve a good taxi speed. Dusty and Bulldog took up position in trail, since the taxiway was not wide enough for all three or even two of them. Brian led them back to the squadron area, parked, switched off his electronics, and shut down his engine.

"No joy, ay sir," pronounced Corporal Jacobs, as he glanced at all eight squares of red tape covering the Hurricane's leading edge gun ports.

"Nope," Brian responded. "The bastards turned and ran as soon as they saw us roll to engage them."

"No shots . . . but no holes either."

Brian laughed hard. "Well, Henry, you do have a point there, don't you."

Flight Lieutenant Whitey Whittington stood outside the door to the Dispersal Hut. Several of the squadron pilots reclined in lawn chairs, although the sky remained overcast but relatively warm. "Couldn't find them, Hunter?" poked Whitey.

"Oh, we found 'em easy enough, Skipper . . . flight of four one tens. As soon as they saw us roll to engage, they turned and ran for home. We gave chase, but they were apparently light. We weren't gaining on them. I began to feel an ambush with the weather around. We broke it off and here we are."

"Wise move, I should think. Make sure you add that for Royster," referring to the squadron's assigned intelligence officer Flying Officer James Royster.

"Sure thing, Skipper."

The Green Section pilots completed their mission debriefing in short order. Brian ensured he conveyed his apprehension chasing and not gaining on the flight of Me110s. They had not closed sufficiently to tell which model the intruders were, but they were distinctly Messerschmitt Destroyers.

After returning their flight kit to their assigned pegs, they grabbed chairs and went outside to join the others.

"Hey, Hunter," Red Burns said, "you missed the announcements while you guys were out gallivanting around the skies."

"What happened?"

"Most important, the Skipper informed us that we start receiving new Spitfire Mark Two's next week."

"Excellent. It'll be nice to get back in Spits."

"We also received word the Air Ministry formed the third Eagle squadron – One Thirty-Three Squadron at RAF Coltishall in Norfolk. They have Mark Two 'B' Hurricanes, with tail designator letters 'MD' and callsign 'Falcon.'"

No.121 Squadron was the second Eagle squadron of volunteer American pilots, and had been formed the previous May. They were currently flying out of RAF Kirton-in-Lindsey with tail letters 'AV' and callsign 'Hawk.'

"Well, how about that . . . not bad."

"Now, we have three American volunteer squadrons," added Salt Morton.

"How many more are we going to get?"

"As many as they have pilots to fill," answered Whitey, "of that I am certain."

Sweet Sweeny added, "My father and his uncle say there are many more coming."

"What does that mean?" Pete Peterson asked.

"Well, they never said precisely, but I understood them to mean like hundreds."

"If that is true," said Whitey, "you bloody Americans may form your own wing or group." Everyone laughed.

"At least I am not the only one," Brian added.

"No, you are most definitely not," responded Pete.

"Speaking of bloody Americans," the acting squadron commander said, "I neglected to inform you chappies earlier, the Air Ministry declared One-Twenty-One Squadron operational 11 days ago. They have now officially joined the fight for freedom."

Several of the pilots clapped. All of them cheered the news.

"Skipper, Green Section was actually out on the line defending freedom," Brian said, only to be interrupted by jeers from his brethren. "Before I was so rudely interrupted," he continued, only to stimulate more jeers, "you announced we would be transitioned back to Spits. I wanted to ask, what is the plan and expectations for the switch?"

"Well, I did not say, actually, but since you asked, I am informed by Group that we should expect half the aircraft next week, and the other half plus spares the following week. 'A' Flight will transition first, and then 'B' Flight. Since most of us have flown Spits, Group expects us to be fully operational by the end of the month. We'll focus on the newer pilots who have not flown Spits. The tactics and procedures will remain the same, but the Spitfire is a little more sophisticated, shall we say."

"Well said, Skipper," Brian added.

They sat at Available status for the remainder of the day. The talk centered upon the new Spitfire fighters due in next week. Brian contained his enthusiasm as best he could, but there was no question in his mind, he preferred Spitfires. He would not let others see his eagerness, but it was prominent in his thoughts.

—

Sunday, 3.August.1941
Wendover Railway Station
Wendover, Buckinghamshire, England
United Kingdom
13:00 hours

The limousine ride from Chequers to the railway station for Clementine and Winston had been quiet but picturesque. They both admired the English countryside untouched by the violence of the war they lived with in London. They both noted the welcome respite and scenic beauty they were passing through.

The Prime Minister's special train waited on a siding at the station. Most of his entourage for this journey had boarded the train in London. In fact, the Prime Minister would be the last to board for the long overnight train journey to Thurso, Caithness, Scotland. His duty private secretary for the journey, along with his valet and personal physician, followed in a second automobile.

"Before we arrive, Winnie," Clementine finally broke their silence, "I want to wish you a most successful journey, and above all, please come home safely to me and the children.

Churchill smiled and gave his wife a gentle kiss on the cheek. "You know I shall do my level best to comply, Clemmie, my dearest. You know how important this first face-to-face meeting is between the President and me."

"Yes, I most certainly do, but you are still my husband."

"And, the King's prime minister."

"No, Winston, you are my husband first, and the father of our children. I would be speaking an untruth if I said I was not concerned. I know how much you have been vexed by the German submarine threat. I cannot and will not claim understanding, but I have been through both your stints as First Lord of the Admiralty with you. I am neither blind nor unfeeling. While this may be a powerful warship, it is still a ship, and the North Atlantic is still a very cold body of water."

Churchill nodded his head. "I shall make no attempt to assuage the sense of foreboding you feel and the risk of which you speak. It is altogether real. However, the *Prince of Wales* is the newest, fastest and most powerful battleship on the high seas. We intend to make a rather fast transit, sea states permitting, which will mitigate that risk to the greatest extent possible. It is a very big ocean. Only a handful of senior people know our destination, and even fewer, our purpose. No submarine can even come close to matching the speed of the *Prince of Wales*. He would have to be extraordinarily lucky."

"And, luck still exists, does it not?"

"Yes, my darling, and the risk will never be zero."

"Thank you for the reassurance, Winston. I do love you so . . . and I am far too young to lose you." Clementine leaned toward her husband and kissed him on the lips. "I must kiss you, now, as it will soon be inappropriate."

"Thank you, my darling Clemmie. We shall make quick work of this. I shall return to you in a fortnight."

"I shall hold you to that," Clementine Churchill said and kissed her husband, again. "Here we are. Good luck, Winnie."

"Thank you, my dearest."

The Prime Minister's black 1939 Rolls-Royce Phantom III drew to a stop on the far side of the tracks from the main station house. Chief of the Imperial General Staff General Sir John Dill and First Sea Lord Admiral Sir Dudley Pound meandered slowly on the siding platform, until they noticed the Prime Minister's arrival. He would have only the Chief of the Imperial General Staff and First Sea Lord with him for the summit conference code-named RIVIERA.

The second car disgorged its passengers, who moved quickly to board the train, as Winston and Clementine exited the elegant limousine. Dill and Pound saluted. The Churchills touched hands. He winked at her, and then turned to his two senior military officers and said, "We have miles to go. Let us be off."

"Yes sir," they responded and boarded the train.

This time, Winston leaned forward and kissed his wife. "I shall return to you safe and sound."

"I am counting on that. *Bon voyage.*"

Prime Minister Winston Churchill had barely taken his seat in the salon car, when the train moved away. Winston looked out the window to see his wife still standing on the platform. He waved, and she returned the gesture, and then blew a kiss before she disappeared behind the train.

—

Sunday, 3.August.1941
Union Station
50 Massachusetts Avenue Northeast
Washington, District of Columbia
United States of America
10:50 hours

General Pa Watson locked the President's wheelchair at the special table in the presidential salon railcar, and then took a conventional seat across the table from Roosevelt. Watson checked his military issue wristwatch. "We should depart the station on time in 10 minutes. There are no planned stops, and we expect to arrive at New London station around 19:30 to 20:00 hours this evening. The baggage and people are all on board. We are ready to go."

"Excellent. Have the arrangements been confirmed for Elliott and Franklin Junior?"

"Yes sir. Your sons should be waiting for our arrival at New London Station."

"Thank you much for coordinating with the Army and Navy to issue their orders."

"Those were comparatively easy arrangements to make. Both of your sons are fully informed of the meeting plans as well as the expectations of them as uniformed officers during this conference."

"They shall do well and not disappoint."

"Certainly. Steve asked me to remind you that the Press has been notified and he expects a handful to a dozen or so reporters and photographers to be waiting at the rail station."

Steve was Stephen Tyree 'Steve' Early, who had been serving as White House Press Secretary since Roosevelt became president in 1933. Roosevelt trusted Early with the presentation of the President's and the administration's public face.

"Thank you, Pa. He wants light-hearted, a little humor, and a sense of the ordinary—very low key."

"Yes sir. That is what he told me. This is just a simple summer cruise on the presidential yacht for relaxation with two of your sons, a little fishing, and a tour of the New England coastline, which you have not been able to do for too many years."

Roosevelt chuckled softly. "Steve has thought to everything, hasn't he?"

"Yes sir. We want the 'impromptu' exposure to be recorded and publicized to substantiate our cover for this trip."

"I've not been to New London since I was assistant secretary of the Navy all those years ago."

"The transfer site was selected because the rail station is just across the tracks from the City Pier. The *Potomac* is already tied up there . . . arrived yesterday."

"That transfer should be easy enough."

"We have a special car, actually a small, modified bus. Once aboard *Potomac*, the plan calls for a prompt departure. We will remain within a few miles of the coast with all our lights on, so anyone who wishes to see will be able to keep sight of us. Tomorrow, we will spend most of the day on a leisurely tour of Narragansett Bay to include a slow passage of Providence harbor. By late afternoon, we will head toward Block Island, make a slow circuit around the island, and then after twilight, we will alter course to the southeast. We expect to rendezvous with *Augusta* at mid-day on Tuesday for our transfer. The cruiser's crew has practiced the transfer process numerous times, so all should be in readiness."

"Thank you, Pa. I have absolute confidence in the *Augusta*."

"I might also say that your decoy body-double has been aboard *Potomac* and out of sight for the entire repositioning process. It will be your choice entirely whether you wish to be on deck tomorrow for our . . . what shall we call it . . . our display for curious eyes to observe from shore. Once we debark, the decoy will carry on a routine of fishing, moving about the deck, lounging on deck, all activity intended to be quite visible . . . ."

The knock at the forward door stopped their conversation.

"Enter," the President said strongly.

The conductor opened the door. "Good morning, Mister President. All is ready for departure."

"Please proceed."

"Very well. The journey should take eight hours. We have no planned stops, and no delays are anticipated or forecast."

"Thank you."

The conductor departed. Within a few minutes, the characteristic gentle jolt of the initial movement tensioning up the couplings signaled their departure from Union Station. The historic journey has begun.

—

Sunday, 3.August.1941
New London Station
27 Water Street
New London, Connecticut
United States of America
20:05 hours

Steve Early boarded the train before it had fully stopped at the station and entered the salon car. "Good evening, Mister President. Welcome to New London."

"Thank you, Steve. I trust your advance coordination activities have gone well."

"Yes sir. All is as we discussed last week. There are a dozen plus reporters and photographers waiting on the platform. I would like to confine the Press exchange to no more than 10 minutes. We should not allow this exchange to become a Press conference. I would urge you to fain tiredness and desire to relax with your sons. Elliott and Junior are on the platform for a public welcome . . . a bit staged, but that is the plan."

"I know my lines, Steve. Let's get on with this."

Pa Watson knew one of his essential tasks quite well. He would wheel the President down a special ramp and to a position in front of a small podium for his support. Elliott and Junior would greet their father on the platform before his public statement. Once Roosevelt's leg braces were locked, Pa would help him stand and shuffle to the support of the podium.

Captain Elliott Roosevelt, USAAF, was the third child and second son of Eleanor and Franklin. He wore the pinks and greens service uniform and silver wings of an Army Air Forces pilot. Elliott was assigned to and on leave from the 21st Reconnaissance Squadron, based at the moment at Miami Municipal Airport in Florida, where he flew long-range search patrols with the B-17C heavy bomber or the B-18 Bolo medium bomber.

Ensign Franklin Delano Roosevelt Jr., USNR, was the fourth child and third son, who graduated from the University of Virginia Law School the previous year. Junior was called to active duty and on leave from his training as a gunnery officer.

The choreography went off without a hitch. Both sons stood behind and to either side of their famous father to ensure the best possible photographic record.

"Evening, gentlemen. I apologize for the late hour on a Sunday evening. I know how hard you work, and this is a disturbance to your beauty sleep." Several reporters chuckled as flashbulbs popped. "My sons, Elliott and Franklin," he said looking to each man behind him, "and I are going to take a nice leisurely cruise of the New England coastline to do a little fishing but mostly relaxation."

"What are your port calls?" interjected a reporter.

"None . . . to be precise. We want to see the coast from Connecticut to Maine in the time we have, so that does not leave us time for port calls. Besides, if we did dock somewhere, I would have to face your questions." Everyone laughed.

"So, there is no other purpose?" asked another reporter.

"Not that I'm aware of, do you know something I don't?" More laughter came from the Press.

"What do you think of the situation in Russia?"

"There you go, no rest for the weary." More laughter. "If you will excuse me, gentlemen, I am here as a father with two of my sons, to take a couple of weeks to recharge. I've taken off the presidential hat for the moment, and I hope you will respect that fact."

"Aren't you concerned about German submarines lurking around out there?"

This time, Roosevelt gave a hearty, robust laugh. "No, frankly, I am not. We are not at war with Germany or anyone else, and furthermore, the whole purpose of this trip is to remain close to the coastline. It is a sightseeing excursion. We have no problems with German submarines in our territorial waters."

"Have there been confrontations in international waters?"

Roosevelt laughed, again. "No matter what I say, there is always a question. That is enough for tonight, gentlemen. I have my signal from Mister Early. I need to get my boys on the *Potomac*, get them fed, and begin a little rest and relaxation. Good evening, gentlemen, and thank you for coming."

Several more questions were shouted at the President, but he ignored them. Pa and Elliott helped the President back to his wheelchair. No flashbulbs popped. Roosevelt was lowered to his wheelchair, unlocked his leg braces, and lifted his feet onto the supports.

Silently, they made their way out of the rail station, onto the special bus via a ramp, and boarded the *Potomac*. President Roosevelt and his two sons went directly to the fantail, just in case anyone wanted to photograph them during their departure. Within 10 minutes of boarding, the gangway was withdrawn and the mooring lines cast off. They were underway on the second leg of what would become an historic voyage.

—

Monday, 4.August.1941
HMS Prince of Wales
58° 53' 52" North – 3° 2' 48" West
Royal Navy Anchorage Scapa Flow
Orkney Islands, Scotland
United Kingdom
16:00 hours

The massive battleship lay peacefully at anchor in the center of the historic anchorage. Her impressive 14-inch main battery naval guns were clearly visible from miles away. The warship had swung into the light, prevailing, westerly wind.

The Prime Minister's special train had arrived at Thurso Station an hour after summer's early dawn of the northern latitudes. Several military vehicles transported the group from the rail station to the pier, where they boarded a destroyer for the three-hour, 40+ mile passage, as well as the crucial transit through the anti-submarine defenses for the historic and vital naval anchorage. The sinking of the battleship HMS *Royal Oak* at anchor, barely a month after the war in Europe had begun, shocked the Admiralty and resulted in significant changes to the Royal Navy's anti-submarine defenses and entry/departure procedures. The destroyer had moved precisely and slowly among the warships at anchor, and then moored to the starboard side of the battleship.

The transfer was equally smooth and uneventful. Captain John Catterall 'Jack' Leach, DSO, MVO, welcomed them aboard on the Quarter Deck. With him stood Harry Hopkins, who had arrived from Moscow and been welcomed aboard the warship two days earlier. Winston had wanted to talk to Hopkins then and there, but chose to wait for a better time.

After the formalities of coming aboard a ship of the line, the two senior officers and their prime minister had taken mid-day meal in the Flag Mess with Captain Leach. They did not discuss their mission. After lunch, Leach took them on a quick tour of the superstructure and the facilities available to them as well as the compartments they might need during their time aboard. Churchill wanted to see the communications compartment to assure himself that his cryptographic specialists were properly situated and ready. They were set and ready.

With their tour complete, they returned to the flag stateroom – the Prime Minister's compartment while he was embarked. As soon as they sat at the conference table, the Boatswain of the Watch knocked on the oak hatch and entered. "Captain, the Officer of the Deck has requested your presence on the bridge."

Leach nodded his head. The boatswain disappeared, closing the hatch behind him. Leach looked to Churchill and said, "Please excuse me, sir. Duty calls."

"By all means, Captain. Please tend to your command."

As Captain Leach departed, Churchill turned to Hopkins. "Harry, pray tell me how was Moscow and Uncle Joe?"

Hopkins replied, "Short answer, not good."

"How so?"

"Neither Premier Stalin, Foreign Minister Molotov, nor Marshal Shaposhnikov offered a discouraging word, but I could just sense the tension. They described their situation in as positive terms as they could, but there was no hiding the reality they have not found the means or place to stop the German advance. The rather frenetic pace of defensive preparations in Moscow is a clear sign they are expecting the Germans to reach Moscow."

"Shaposhnikov? Boris Shaposhnikov?" asked Sir John.

"Yes, General. He was introduced as Chief of the General Staff Marshal of the Soviet Union Boris Mikhailovitch Shaposhnikov."

"Well, I'll be damned."

"You know him, Sir John?" Churchill asked.

"Yes sir. Shaposhnikov was Chief of Staff of the Soviet Armed Forces from '37 until August of last year. I understood he had effectively retired. Stalin apparently brought him back . . . probably the only general officer he trusted through the purges. Rumor had it, he helped Stalin, Beria and Merkulov with the purges of generals and admirals. His survival through the worst of the purges and now his reinstatement are testaments to that reality, I am afraid." Sir John looked back to Hopkins. "When was he installed?"

"I was told two days before my departure . . . that would be the 29th by my calculation."

"He's a political animal," Dill mused.

A knock interrupted the conversation. Captain Leach entered and looked to the Prime Minister. "Sir, we are prepared to weigh anchor and make way."

"Very well, Captain. Please proceed." Leach left. Churchill looked back to Hopkins. "I would like to observe the departure from the anchorage, but before we adjourn, if it is no intrusion, I imagine your primary mission to Moscow was to assess the prospects of survival for President Roosevelt's impending decision to extend Lend-Lease to the Soviet Union."

Hopkins smiled, looked to Dill and Pound, and then back to Churchill. "Yes, that was my primary purpose. To be blunt, Prime Minister, I could find no evidence to even suggest the Red Army can or will stop the Germans. I was surprised to hear Stalin hint at making it to the protection of the Russian winter. In short, winter may well be the decider. If the Germans take Moscow before winter, it is hard to see how the Soviets will be able to sustain their fight. If the Germans reach Baku before winter, the Soviets may well be choked to death regardless of Moscow. If they do make it to winter, they may find sufficient breathing room. Stalin indicated they are moving significant units from the East."

The ship-wide loudspeaker system blared, "Shift the colors."

"The anchor is free. We are underway," said Churchill. "If you don't mind me asking, what will be your recommendation regarding the Lend-Lease extension?"

"The President took a huge risk in authorizing the arms shipment to you a year ago, July. We had reasons to believe you would be successful. There are few similar reasons in Russia. There is no Channel to cross. The Red Air Force is a shadow compared to the Royal Air Force. About the only positive factor I could find is the harsh winter. But, at the end of the day, as you say, I do not see that we have an option. We must do our best to sustain the Russians."

"Thank you for your candor and frankness. I would like to discuss this question more, later. If you will excuse me, I would like to adjourn and pick this up in a few hours, perhaps after supper."

Pound, Dill and Hopkins joined Churchill on the starboard wing of the flag bridge. The battleship was moving slowly through the myriad of anchored warships toward the western passage. Normally, the warships would be offering honors with their saluting cannon, especially with the Prime Minister on board, but there were none. Admiral Pound explained nerves were still too raw from the dramatic loss of the *Royal Oak*, nearly two years previous. The wreckage of the battleship lay on the bottom, 11 fathoms below and north of their just departed anchorage. The *Prince of Wales* made numerous unexplained turns of various leg lengths as she made her way through the hidden path of the anti-submarine defenses.

Once clear of the western passage, the *Prince of Wales* began to noticeably vibrate, as she built up speed as quickly as possible. An array of a light cruiser and half dozen destroyers were fanned out ahead of them toward the western horizon. They had probably been screening for any lurking submarine for an hour or two. The escort would remain with the *Prince of Wales* until she was up to full speed and into the night. At some point during the night, four of the six destroyers would peel off and return to convoy escort duties. Only the two fastest destroyers and the new, fast, light cruiser would remain with the battleship for the complete North Atlantic transit.

—

Tuesday, 5.August.1941
RAF North Weald
Epping, Essex, England
United Kingdom
09:15 hours

The telephonic message from the Operations Building sent most of the squadron pilots outside to watch the arrival of six, new, Spitfire Mark II fighters reported inbound and due to land at any minute. The hazy sky made observation less than ideal. Red Burns was the first to spot two sections in trail, descending on a straight-in approach. They landed smoothly, without a single bounce. These pilots had clearly handled Spitfires before. They taxied to the squadron area, and each in turn swung his tail into the designated, open, parking spot.

Brian watched Corporal Jacobs tend to the third aircraft in, so he walked toward that aircraft. All of the aircraft had basic RAF markings, roundels and fin flashes, and no tail designator letters, yet. As the pilot shut down the big Merlin engine and secured the aircraft, Brian reached Jacobs. "Is this one ours, Henry?"

"As good as any, don't you think, sir?"

"Works for me. I can't recall the last time I saw a fresh from the factory Spit."

"You will have one this afternoon, sir."

The pilot stepped out of the cockpit, jumped down from the left wing, and removed his headgear. It was only at that instant Brian noticed that the pilot was a woman with shoulder length auburn hair. From that recognition, he noticed the modest bumps on her chest under her overall flight suit. She was probably nine inches shorter than him, with an attractive face and no makeup or lipstick to alter her appearance.

"How did she fly?" was all he could think of to ask at the moment.

"As she was intended," she said and smiled.

"I'm Brian Drummond," he said and extended his right hand to her.

She shook his hand with a firm grasp. "I know who you are, Mister Drummond. I'm Jennifer Brentwood, Ferry Command." An admixture of confusion and fear must have been written across his face. "No, we have not met," she added, seeing his expression. "Some of my colleagues have mentioned you."

"Well, thank you for that . . . I guess."

"It is all good, I must say."

"Thank you, again."

"Is this bird to be yours?"

"Until Corporal Jacobs here tells me this is our bird, I cannot be so bold."

"Well, she flies like a dream." She looked to Jacobs. "No squawks, Corporal . . . not a one."

"Thank you, ma'am," Henry responded, and then stepped out smartly to do his work.

"I believe you have Hurricanes for us to move north. Is there anything else I can do for you, Mister Drummond?"

"First, I prefer Brian, in case we should ever meet again. Mister Drummond sounds like my father." *There is no reason to mention my father's passing.* "If you have a few minutes to spare, could you tell me the differences with a Mark Two?"

"Sure. Very simple. First and foremost, this one has a Merlin Mark Twelve powerplant – another 120 horsepower more than you have with the Merlin Mark Three engine. The other difference, although less significant in my opinion, is this machine was built at Castle Bromwich rather than Woolston. I guess they wanted to keep manufacturing site differences in the Mark. This one has a de Havilland airscrew, although a Rotol airscrew is perfectly compatible, so not reflected in the Mark definition. The extra horsepower gives you another fifteen miles per hour of level airspeed. Other than those differences, they are the same machines – guns, fuel, electronic kit, all the rest is the same."

"The extra power will always be appreciated. Thank you for your time. I won't keep you longer. Have a safe flight."

"Thank you, Brian . . . until the next time," Brentwood said and touched two fingers to her right temple.

Brian watched her walk away. Even in her overall flight suit, she cut a fine figure. *How did I miss that? Who told her about me?*

She disappeared into the Operations Building under the control tower and the spell was broken. Brian jumped up on the left wing to examine the cockpit. He carefully checked each item. *She was correct. Nothing is different in here.*

"Mister Drummond?" Brian pulled his head out of the cockpit to see Corporal Jacobs at the forward left wing root. "We need to move her into the hangar for an acceptance check and to apply your tail letters."

"Sure, Henry. The sooner the better." Brian jumped off the wing, stood aside and watched the crew attach the ground-handling tug. As they moved the new Spitfire fighter toward one of the hangars, Brian noted the other freshly delivered Spitfires were moving toward the hangars as well.

Brian returned to the Squadron Dispersal Hut. He found an open lawn chair and sat.

"Pretty exciting, huh," Bulldog Tolly stated.

"I suppose," responded Brian.

"Aren't you excited about getting back into Spits . . . and a new one at that?"

"Yes, sure. The Spitfire remains my favorite. I will always enjoy flying these magnificent machines, but the pleasure of flight is not why we are here."

"You don't have to be a stick in the mud," Pete Peterson interjected.

"I'm not . . . just trying to ensure we keep a sense of perspective here." *Truth be told, I am ecstatic. Nothing flies quite like a Spitfire fighter.*

Whitey stepped out of the Hut, followed by Rocket Downing, the only other 'A' Flight pilot not already outside. "'A' Flight, if I may have your attention," Whitey said, and waited for heads to turn and eyes to be on him, "the maintenance lads inform me the new aeroplanes should be ready for flight this afternoon. I am going to inform Group we are dropping to 30 Minutes," meaning their status on the tote board, "and after lunch, we will brief and launch as a flight to familiarize ourselves with the new machines. As I recall, Rocket and Bulldog have never flown Spitfires." Both pilots shook their heads to the negative. "Pardon me, Dusty, but I do not know about your experience, as I should."

"No problem, Skipper. Nope, not Spits, only Hurris so far."

"I know Red has Spit time, and we are still missing a replacement for Horse. So, after lunch, we will brief for a flight launch. We'll take some individual time while we are up, to let Bulldog and Rocket feel the machine, and then I'd like to spend some time in mock engagements. Since Red has Spit combat time, you will take the Blue Section Right Wing position for this afternoon's event."

"Will I keep the Spitfire?" asked Red.

Everyone outside listening laughed at the image of a boy losing his toy. "Yes," Whitey answered, 'you will keep your Spitfire. The rest of the squadron will transition next week. Are there any other questions?" A few shook their heads. No one spoke up. "Very well then. Corporal Harris," Whitey called.

"Yes sir," Harris responded.

"Please request our step down to Available in 30 Minutes."

"Yes sir."

"After lunch, we will brief the flight and get on with this transition. Pete, you shall retain 'B' Flight for the time being."

"You got it," Peterson responded.

"The squadron is at 30 Minutes, Skipper."

"Thank you, Corporal. Let's get the early seating, gentlemen."

No one had his flight kit. The pilots began walking to the Mess on this fine summer day.

—

Chapter 10

It does not involve the remotest surrender
of free debate in determining our position.
On the contrary, frank cooperation and
free debate are indispensable to ultimate unity.
-- Senator Arthur Henrick Vandenberg of Michigan

Saturday, 9.August.1941
USS Augusta *(CA-31)*
47° 18' 39" North – 53° 57' 16" West
Placentia Bay, Argentia, Newfoundland
Dominion of Canada
07:45 hours

"**P**a, I'd like to stand for this," President Roosevelt said. The solid steel railing on the starboard wing of the flag bridge blocked part of the President's view of the bay. Roosevelt locked his leg braces. General Pa Watson locked the wheel chair's large wheels, grasped under the President's arms, as Roosevelt grasped the railing, polished mahogany top cap, and began to pull himself up. He stood awkwardly, but he stood erect. "There she is," he announced on seeing the distinctive features of the Royal Navy battleship HMS *Prince of Wales* appear across the bow of the heavy cruiser. The warship moved slowly toward her assigned anchorage, adjacent to the *Augusta*. "Magnificent ship of the line, isn't she?"

"Yes sir, that she is."

The distinctive shrill of a boatswain's pipe broadcast on the ship's 1MC loudspeaker system. "Attention to starboard," came the spoken command. The ship's horn blasted once, indicating attention to starboard for a salute to another warship. The *Prince of Wales* answered with one blast of her horn. *Augusta* offered another single blast of her horn to signal hand salute to those officers and seamen manning the rails, followed by the report of the saluting gun 21 times. *Prince of Wales* returned the salute. Although the two warships were several hundred yards apart, Roosevelt believed he could see Churchill wave from the bridge, so he returned his wave. The boatswain's pipe signaled the end of the traditional salute, followed by "As you were."

The President watched as the British warship slowed even further to nearly stop and dropped her anchor. The light easterly breeze would eventually swing her on anchor to the same direction as the *Augusta*.

"Pa, I don't know that I can see Harry over there, but I'm certain he will have some stories for us."

"Yes sir," the general responded, with his commander-in-chief standing at the railing.

"Have all the arrangements been made?"

"Yes sir. The Prime Minister is scheduled to arrive at eleven hundred. We will be on the Quarter Deck to greet him after honors. As requested and agreed, only Mister Churchill will come aboard for this first visit. Harry should come across with him, although I have no positive confirmation of that. No one else is expected at that time. We will move you both to the Flag Mess for lunch and your private discussions. Unless you and Mister Churchill decide to change the plan, the two military staff groups will gather aboard Augusta for introductions and the initial plenary meeting in the Officer's Wardroom. You will host a joint group evening meal in the Wardroom at nineteen hundred hours."

"Very well. I think our arrival honors are complete. If you would be kind enough to assist me, I'll go to my stateroom to get some work done before Mister Churchill arrives. Would you kindly inform Elliott and Franklin to be ready for the arrival ceremony? "

"Yes sir."

10:50 hours

President Roosevelt had moved to the cruiser's Quarter Deck early. Dressed in a light, off-white, summer suit, he had arranged himself to be standing and leaning slightly against the superstructure bulkhead for stability. The President insisted upon standing with Elliott and Franklin, both sons in fresh uniforms. General Watson stood close by in case he was needed. Roosevelt had plainly seen the admiral's barge depart from the British battleship toward the American cruiser.

"Lay to the quarterdeck the sideboys," announced the 1MC.

Eight sailors in freshly pressed, blue dress uniforms arranged themselves in two, precisely straight lines, shortest to tallest, facing each other on either side of the deckside accommodation ladder. This was not the first time these men had performed this duty. The Officer of the Deck, a lieutenant, in his Dress Blue uniform with medals and a polished sword on his left hip, took his position at the end of the sideboy formation.

The duty boatswain's mate monitored the approach of the admiral's barge. The warship's 1MC loudspeaker system announced "Prime Minister, United Kingdom arriving." As the boat reached the water level platform of the accommodation ladder, the boatswain piped 'Alongside.' A minute later, as the Prime Minister's shallow peaked hat reach the deck edge, the boatswain piped 'Over the Side." The sideboys, boatswain and officer of the deck came to attention. The military men on the quarterdeck saluted. Prime Minister Churchill passed between the sideboys and stopped, standing at attention. The ship's small band played four ruffles and flourishes, and then immediately transitioned to the "Admiral's March."

11:00 hours

Once honors were rendered, Churchill dressed in pseudo-naval attire, slowly stepped toward the President. He extended his right hand before they were within arm's reach, and then shook hands. "At long last, Mister President, we meet."

Roosevelt replied, "Glad to have you aboard, Mister Churchill."

"An honor, sir."

Roosevelt introduced his middle sons to Prime Minister Churchill, who had more than a passing familiarity with their respective stories. Churchill's knowledge impressed them all.

"Did you bring Harry with you?" asked the President.

"Yes, certainly. He came across with me. He should board shortly, now that honors are complete."

"Excellent. I hope and trust he has been most useful to you, and thank you very much for giving him a ride home. I understand his trip to England was not so comfortable."

Churchill laughed heartily. "I am afraid that may be an understatement. Only Harry can properly describe his experience. Nonetheless, I was grateful to have him with us."

"I anxiously await his report. As agreed, we shall have a private lunch and discussion before the first plenary session with our military staffs."

"Excellent. Our chiefs shall arrive on board at three bells," using the traditional naval time reference, in this instance meaning 13:30, "in plenty of time for the beginning of the plenary session. I presume we shall make the appropriate introductions at that time."

"Yes, that is the plan. Now, I think lunch is ready for us. May I introduce my senior military aid, Major General Edwin Watson." Pa saluted. Churchill returned the salute and extended his hand. They shook hands vigorously.

"A pleasure to meet you, Pa." Churchill said.

"So, you know his nickname," Roosevelt observed.

"Harry has been a very gracious guest."

"Very well, then. Pa, if you will escort the Prime Minister to the Flag Mess, I shall have the lads cart me up to join you shortly."

"Yes sir," Watson responded and gestured for Churchill to precede him.

The President signaled for the small team of four seamen to bring his narrow wheel chair, which they would carry up the ladderway, several decks, with him in it. They had become quite adept at moving their commander-in-chief despite his disability.

11:20 hours

One of the able seamen pushed Roosevelt's chair into the Flag Mess. The modest table was exquisitely set with gold-leafed china, silverware and crystal glasses. "Thank you, Pa," Roosevelt said, discreetly dismissing his senior military aide, who departed promptly and closed the hatch behind him. "Winston, please be seated." Churchill took the only chair with a place setting, while Roosevelt positioned himself opposite the Prime Minister. "It is a bit early for lunch, but I know the stewards are eager to impress us with their cuisine, and frankly the sooner we have the dishes cleared, we can focus on our discussions."

"Very well. I bow to your wisdom."

The stewards expertly served first a delightful New England clam chowder, followed by a mixed green salad with light virgin, olive oil and balsamic vinegar dressing. The main course was roast duck with a tangy orange glaze, mixed vegetable medley, and a lightly curried, saffron rice, and an excellent French Sauvignon Blanc wine. The stewards also offered an ample serving of a rich, freshly churned, chocolate ice cream for dessert.

The President thanked the stewards for their excellent service and the delicious meal. The dishes were cleared. The table straightened and a fresh, white tablecloth spread smoothly across the table. The two national leaders waited patiently for the galley hatch to close. They were now alone.

"If I may," began Churchill. Roosevelt nodded his consent. "It is a genuine honor to finally meet you . . . may I call you, Franklin."

"By all means, Winston. I dare say we shall be bosom buddies through this storm and tumult. Thanks to your generosity and candor, we have been communicating like pen pals for nearly two years. Yes, Winston, the familiar is quite appropriate between us."

"Thank you for that and for agreeing to meet with me. I know you have taken considerable risk in meeting one of the belligerent nations in the current conflict. I do not presume to know your feelings, but I assume you appreciate the threat before us. Thanks in no small measure to the serious risk you took last year to supply us with your surplus small arms to make up our losses in France, and of course the vital destroyers you lent to us. You have done so much already." Roosevelt nodded and chose to listen. "I surmise you recognize that the U.S. entry into the war against the Axis is inevitable and only a matter of time."

Roosevelt waited, studying Churchill's eyes. "I appreciate your recognition of the risks we take. I will confess my considerable apprehension and trepidation with the situation just one year ago. We saw you dangling by a mere thread – a very thin thread. I have always admired your spunk, Winston, as we Americans like to say. However, you did it, you managed to make Hitler blink. I dare say, you are far stronger today."

"Yes, we are," interjected Churchill.

"What is your assessment of the situation in the Soviet Union?"

"Not good, I'm afraid. The Germans continue to make steady progress with little sign the Russians can or will mount sufficient resistance to stop their advance. Harry and I have enjoyed several long chats regarding his journey to Moscow, once we were reunited aboard Prince of Wales, five days ago. I believe he will provide you with an accurate perspective."

"Yes, but, will they stand?"

"Humph! That is the question, isn't it?" Churchill considered his words. "The collective opinion of His Majesty's Government, of which I agree, is that the critical milestone is the onset of the Russian winter. If they can hold on until winter, they will likely survive and gain strength, as we did after the summer of 1940. If Moscow falls before winter, then we believe they will not remain viable as a government or resistance. There is general resentment of the Communist government, but we should never understate the power of *Rodina*—the Motherland."

"Their will to fight . . . for the Motherland."

"Rather than the government?"

"Exactly."

"So, your thinking is, supply the Russians as best we can, to sustain them to and through the winter, and beyond."

"That is the notional proposal."

"After your message and before departing on this little sojourn, I directed my Commerce Secretary to develop a plan for fulfilling your requirements."

"Thank you, Franklin. I would be remiss if I did not speak up for my people. Will U.S. industry be able to redouble their production efforts?"

Roosevelt chuckled softly. "Yes, Winston, that was my instruction. In addition, I have directed the Navy to extend our convoy escort operations to the Icelandic meridian. I have also directed two battalions of Marines to be deployed to Iceland, to protect our interests. If the presence of my Marines might relieve some of your security personnel for other assignments, as it were, we are agreeable to protecting your interests in Iceland, while we protect ours.

The Marines are expected to arrive early next week. I also brought Admiral King, who will manage our escort operations."

"Excellent. When I was informed of your plans, I amended our return. It is my intention to waylay at Iceland. I would be honored to inspect your Marines and convey our gratitude for their sacrifice."

"We will make that happen, Winston. The information provided to me by the intelligence fellows indicates the invasion threat for England has passed."

"For the time being, I should think, which is precisely why events in Russia are so vitally important and of keen interest. There is little doubt, in our view, if the Germans manage to subdue Russia, they will turn their full fury upon us, again. That evil corporal cannot feel safe with a free Great Britain. He must realize we shall gather our strength, as we are doing, and ultimately defeat him. The longer Russia can remain viable in terms of occupying his attention and resources, the closer his defeat comes."

"I do agree, Winston. We shall do our part."

"Excellent. Franklin, I do not know what intelligence you have from the East. MI6 has terribly fragmentary information from the Baltic countries, Poland and Ukraine that the German special units, they apparently call them, may be carrying out mass executions of those local inhabitants who do not find favor under the German archetype. Have you heard anything like this?"

"No, I have not, but I will certainly ask Donovan to see what he can find."

"MI6 has a standing intelligence objective to develop a better, clearer picture of what is going on in the East, but these signs are not encouraging."

"Agreed."

The two looked into each other's eyes for several seconds, as Winston considered his words. "May I speak plainly and privately with a good friend?"

Roosevelt's infectious smile beamed. "By all means."

"I believe you can see reality as well as I can. That reality is perhaps a little more intimate for my countrymen and me, but you feel the substance of world events just as vividly. Truth be told, defeating Nawzee Germany by ourselves, even with your industrial and peripheral support, will take a very long time. I remain confident that vile corporal will not be able to sustain his Third Reich. The oil embargo of Japan cannot be suffered and the prevailing junta will not abandon their hegemonic enterprise." Churchill paused to give the President an opportunity to comment. Roosevelt remained sternly focused and unflinching, but did not take the opportunity to speak. Churchill continued, "I think we both know entry of the United States into this *mêlée* is inevitable. It is only a matter of time. I asked for this private discussion, so we could speak frankly and plainly." Roosevelt nodded his agreement. "I sought the opportunity to plead the case of all freedom-loving peoples, and the British case . . . the sooner the United States joins the fight, the sooner we shall vanquish the Axis belligerents."

"As you say, between friends and in private," began Roosevelt, "I have no disagreement with your assessment. Yes, I think our entry into this war is inevitable. The heroic stand of your pilots . . . ,"

"And, some of yours, Franklin," interjected Churchill.

Roosevelt chuckled. "Yes, including some American volunteers . . . the heroic stand last summer has turned popular opinion. The America Firsters have far less influence than they did a year ago, and what influence they have today is slipping through their grasp. Speaking plainly, the United States is far less prepared for war than the British were in 1939. It was only a short time ago, our armor brigades have been practicing their battlefield tactics with Jeeps covered in cardboard in some comical semblance of an actual tank."

"You signed into law several massive rearmament acts last year," Churchill noted.

"Yes, we did, but those building programs take years to achieve rate production. We also implemented a serious conscription program last year, but training new troops takes time as well. We tried to negotiate with Imperial Japan without success and the oil embargo was our last shot short of war. The military council that now dominates Japanese political systems has shown no signs of altering course."

"We are at war with Germany . . . not yet, with Japan."

"We are one step short of war with Japan, I'm afraid."

"To be blunt, Franklin, what will it take for you to declare war on Germany and join us on the field of battle?"

Roosevelt smiled, again. "I am fairly certain you are familiar with our laws, Winston, and you know perfectly well that only Congress can declare war. I do not have that authority. Let me say, I am very keenly aware how tightly bound congressional opinion is to the will of the people. Whatever happens, I must convince the people that war is necessary to protect them and protect our way of life. The people will trust Congress. Today, we are fairly certain the votes in Congress do not exist, and if we made the attempt, just the process would turn public opinion away. Again, to speak plainly and privately, I need an incident or a series of incidents to convince the American people."

This time, Churchill smiled, perhaps more with nervousness than humor. "Thank you for your candor, Franklin. Yes, I am familiar with your laws. I am a politician and an author. I am not a lawyer, as you are, so I claim no expertise in American common law. The extension of your escort operation will provide the opportunity by the very nature of their work alone. I have little doubt the Germans will make the mistake we need."

"I suspect you are correct, Winston. When they do make a mistake, I shall not miss the occasion to present the case to the American people."

"Very well, then. I am mindful of the hour."

"Yes, your military staff should be aboard by now. One last item before we stop for the plenary session . . . I have taken the liberty to draft a declaration, preferably a joint declaration, that we may wish to issue at the conclusion of this conference." Roosevelt offered a single sheet of typewritten paper to Churchill. The Prime Minister glanced at the paper as he took it from the President and placed it on the table.

"I will study your words tonight and have it vetted. I must pass any such document through my colleagues in the War Cabinet. We can discuss it tomorrow after our worship service aboard *Prince of Wales*, if that is acceptable to you."

"Yes, of course. And, your battleship is a magnificent ship of the line, I must tell you."

"Yes, she is. Our shipyards hurried the battle damage repairs from her engagement with *Bismarck* last spring, but she is still a beautiful ship."

"And, a glorious victory, as well."

"Indeed. The thought of a German capital ship marauding through our convoys was a chilling nightmare. I did not sleep well that week."

"Our military staffs are waiting. Shall we make our way to join them?"

"By all means, Franklin. Thank you for taking the time with me." Churchill had second thoughts before they left the first private session. The April 17th ULTRA message still haunted him. He was uncertain about timing. If he opened the issue at the beginning of their conference, the President would have time to think things through before they parted. If he waited until the end, he would not have time to say no. Or, maybe HMG should continue to wait for an instigating event.

Franklin sensed turmoil rumbling through Winston's thoughts. "At difficult times like these, I have always found it best to just spit it out."

Winston smiled. "Certainly! This is not an easy topic to open. For the moment, I must ask for this part of our conversation to remain strictly between us, as colleagues . . . as friends."

"That sounds rather grave."

"Well, in a form, it is. Last year, we took a difficult unilateral action to transfer vital military and scientific secrets for a variety of mortal reasons. That exchange has paid and will pay enormous dividends to our joint war effort to defeat Germany."

"Agreed . . . benefits beyond our wildest imagination."

"Yet, at considerable risk, none of those items were our most precious secret."

"Yes," he paused, "and, I presume this may be the moment to transcend."

Again, Winston smiled. "As good as any! The story is long. However, in the interest of our short time available, please allow me to summarize. You can ask whatever you wish, whenever you wish." Churchill waited for Roosevelt's consenting head-nod. "We are not quite routinely deciphering German sensitive communications." Roosevelt did not react. "We refer to the program under the code word ULTRA. We occasionally refer to by a decoy reference 'Boniface.'" Again, no reaction. "On the 17th of April this year, we decoded a message from the German chargé in Washington to Foreign Minister Ribbentrop informing him, based on information provided directly from Soviet Ambassador Urmansky, that the Japanese diplomatic code had been broken by the Americans."

Roosevelt felt the blood draining from his head. He felt dizzy and nauseous.

"Are you OK? Are you well?"

Roosevelt waved his hand to give himself a moment to regain his composure.

"Should I call for a physician?"

"No, no, I am OK. I did not expect that revelation. It was a bit of a shock. So, this is the moment we open the kimono." Roosevelt smiled. The color had returned to his face. "We have achieved sporadic success deciphering Japanese diplomatic codes and less success with the active naval command codes. The code name for our decryption program is MAGIC. Allow me to ask, do you have any evidence the Germans notified the Japanese of the disclosure from the Soviets?"

"Yes. A week later, we decoded a message from the German Foreign Minister to his Japanese counterpart. We have yet to detect even an acknowledgment by the Japanese. We have no indication the Japanese reacted to the disclosure."

"I am not aware of any change in the productivity of MAGIC. I would like to make very discreet inquiries."

"That would be my desire as well. However, unless you intentionally disclosed your MAGIC yield to Ambassador Urmansky, the very real potential may exist that you have a Soviet mole very high in your government, or at least with access to MAGIC." Churchill paused to allow Roosevelt a contemplative break. "To be direct," Churchill paused again for Roosevelt's consenting head nod, "we cannot risk ULTRA being exposed by a mole, or any other perhaps well-intentioned individual, to the Soviets, and thus the Germans. ULTRA has been an absolutely vital intelligence source."

"Winston, I am stunned, embarrassed and perhaps even ashamed that we must have this conversation. I am immensely grateful for your trust in me. With your consent, I would like to ask Bill Donovan to discreetly evaluate whether there have been any changes in the code or the yield of MAGIC that might indicate the Japanese have reacted. I would like to ask him for his assessment of how Urmansky became aware of our decryption program."

"Bill is a good man. I trust him as well. However, until we understand what happened to generate Urmansky's disclosure to the Germans, I would ask you not to refer to ULTRA or our decryption program."

"Agreed. The burden remains squarely upon us to answer the security questions. We will get to the bottom of this. Once Bill has the answers we need, I will send him to you . . . to brief you directly and answer whatever questions you or your specialists might wish to know."

"Very well," answered Churchill. "Now, I'm afraid we may have missed the luncheon with our staffs."

"We have a few minutes before the first plenary session. I'll ask the duty steward to get us a sandwich."

The two national leaders talked about more mundane and routine topics as they ate a quick bite. Harry Hopkins arrived to notify both leaders they needed to move to the cruiser's wardroom.

14:00 hours

The Wardroom had been commandeered for the first formal meeting. The staff introductions had been gregariously completed with the calm professionalism of the senior military officers.

With Prime Minister Churchill were:

-- Admiral Sir Dudley Pound;

-- General Sir John Dill.

With President Roosevelt were:

-- Chief of Naval Operations Admiral Harold Rainsford 'Betty' Stark, USN [USNA 1899];

-- General George Marshall;

-- Commander-in-Chief, Atlantic Fleet, Admiral Ernest Joseph 'Ernie' King, USN [USNA 1901];

-- William Averell Harriman—Special Envoy to Europe for the Lend-Lease Program;

-- Harry Hopkins;

-- and of course, General Pa Watson as well as Captain Elliott Roosevelt and Ensign Franklin Roosevelt, Jr., who attended the plenary sessions.

Hopkins clinked his water glass with a small butter knife.

Roosevelt waited for everyone to be seated and quiet. "As host for this meeting," began President Roosevelt, "I welcome our distinguished guests from across the Atlantic." The group clapped their hands. "The Prime Minister and I have agreed on the agenda for this conference. However, before we begin the substantive discussions, Harry, would you please give us the administrative details, to get that out of the way?"

"By all means, Mister President," Hopkins responded and chose to stand. "This is our first plenary session. The President will host dinner here, aboard *Augusta* this evening at seven. Tomorrow, we have Divine worship services at eleven aboard *Prince of Wales*, followed by lunch hosted by the Prime Minister and the second plenary session at two o'clock in the afternoon. The Prime Minister will host Sunday's dinner at seven, aboard *Prince of Wales*. The remainder will be determined as necessary after the first two plenary sessions. That is all we have at the moment, Mister President."

"Very well, thank you, Harry. As indicated, the Prime Minister and I have agreed upon the agenda. First priority is the Battle of the Atlantic. I have already informed the Prime Minister of our decision to extend American escort operations to the Icelandic meridian. For that, we brought Admiral King, who will have direct responsibility for that action. There is much to coordinate to ensure seamless protection for the supply convoys to Great Britain. Second priority is the Lend-Lease supply effort, for which we brought Averell Harriman, who is our point man for your support. We want to ensure our fullest commitment to meeting your needs. Third priority is the request for supply support to the Soviet Union. We must mutually agree upon a means of fulfillment before we can commit to that requirement. Fourth priority is the establishment of an informal committee of military chiefs to begin coordination of our strategic level planning, as initiated by our staff conversations last winter. Also, as I informed the Prime Minister, the United States cannot subscribe to any combat operations beyond the escort duties of the Navy for the time being. Mister Churchill, do you have preface words you would care to offer?"

"Thank you, Mister President, for your hospitality and gracious welcome. You have set the stage quite well, I should think. We look forward with eager anticipation to working together to the ultimate defeat of Germany and Italy, and the restoration of peace in Europe. It is the intent and commitment of His Majesty's Government to work within the constraints by which you are bound, and we shall persevere until the inevitable victory is won. With that and mindful of the agenda, Sir Dudley, if you will, please give this august audience a thorough overview of the Battle of the Atlantic."

The Royal Navy's First Sea Lord opened the substantive portion of the first plenary session. They would discuss the convoy and anti-submarine operations at sea for the next two hours with questions and answers, and a few take away actions on both sides. The detailed military coordination between the two nations had begun and would set the tone for the military staffs throughout the ensuing hostilities.

—

Sunday, 10.August.1941
HMS Prince of Wales
47° 18' 47" North – 53° 57' 05" West
Placentia Bay, Argentia, Newfoundland
Dominion of Canada
14:30 hours

The Sunday service on the main deck fantail had gone exceptionally well, after being carefully choreographed by the Prime Minister himself, down to the very hymns they sang. The massive 14-inch naval guns of the aft main turret established a poignant venue for the joint worship ceremony.

The elegant luncheon had served as a nutritive environment for their social interaction. The specially prepared meal, complete with wine, an unusual addition for the American officers, served its purpose. Laughter actually complimented the meal. At the conclusion, the Prime Minister and the President excused themselves, retiring to the flag stateroom that served as Churchill's quarters for their continued private discussions. Harry Hopkins would remain with the President, along with Pa Watson, who excused himself once the President was situated. The chiefs continued their practical discussions of coordination.

Churchill opened their private talk. "I have studied your draft declaration, Franklin, and I have received the comments from my colleagues of the War Cabinet." Roosevelt nodded his head slightly. "To speak plainly, you are asking me . . . us . . . to consent to the dissolution of our empire. That is a very tall order. Our Empire has existed for more than a century . . . more like two or three centuries. No servant of the Crown can agree to such an outcome."

Roosevelt raised his hand, to gesture his desire to speak. "I recognize, acknowledge and fully appreciate the concern you have expressed. I respect your reluctance to agree *prima facie* with the spirit of clause three of our draft. Let it suffice to say that the United States is committed to the free choice and expression of the indigenous peoples."

"They may well choose to withdraw . . . to opt for independence."

"They may well indeed, and if so, then so be it."

"The majority of your draft is acceptable. However, we fear many of the indigenous peoples are ill-prepared to execute self-governance in contemporary times among the assembly of nations."

"That may well be, Winston, but the United States cannot condone anything less. I cannot defend to the Congress or to the American people our extraordinary resource and industrial support, if that support contributes to the sustainment of unilateral, dictated, governance of colonial administration. Winston, to be blunt and direct, the colonial days are over . . . or will be soon. I think in your heart you know this must be the course."

"Perhaps so, Franklin, but that does not make the pill any easier to swallow. My entire life, my father's entire life, has been in service to the Crown and the Empire. Nonetheless, we, the War Cabinet and me, can support a joint declaration. We should like to adjust the wording in a few places. This may take a few iterations to achieve mutually acceptable words."

"Understood. We can edit as much as necessary to achieve a joint declaration. We joined here under strict secrecy. We will leave here in secrecy. Yet, it is time for us to make a worldwide public statement of our aim. As we discussed yesterday, our joining you in the fight against rabid fascism is only a matter of time. This declaration will be part of our collective effort to mobilize the American people."

Churchill smiled, but it was not a comfortable smile. "If the price of American contribution to victory is our empire, then so it shall be. The freedom of choice of which you speak is a worthy principle. However, I must say, our colonies may choose to remain under His Majesty's dominion."

"Yes, they might. If that is their free choice, then all Americans can support that outcome. We have no intention of dictating to you or any other nation. We are only interested in free choice."

"I shall do my best to convince my colleagues on the War Cabinet, and of course, we must also convince His Majesty."

"If I may be of any assistance to that end, I stand ready to assist. I am at your service."

"Thank you, Franklin. We will hammer out the proper words over the next few days. While that process is on-going, I wanted to raise with you personally an overarching strategy to set the proper tone for the details our military chiefs of staff are presently discussing."

"What may that be?"

"His Majesty's Government remains quite concerned about Japanese aggression in the Pacific region."

"We share that concern."

"Your oil embargo was appropriate and necessary. They, the Japanese, cannot continue their military domination of the region with impunity. As you know, we have joined you in that oil and resource embargo. It is our opinion that given the oil embargo, war with Imperial Japan is likely inevitable."

"We continue diplomatic negotiations to find a peaceful solution, yet at the bottom line and between us only, I am convinced of that inevitability as well. We are ill prepared for war, Winston, even more so a two-theater war. We have moved a substantial portion of our Atlantic Fleet into the Pacific Ocean as a precaution and warning to the Japanese, but we cannot and will not disregard or discount the enormous submarine threat faced by the Royal Navy everyday in the Atlantic. We will do our part to help with convoy protection operations, but we are presently incapable of general naval combat operations, and even more so land or air operations."

"We were in the same position two years ago. You have sacrificed to bridge the gap in our forces. Yet, I suspect the Japanese will not wait for us to become ready."

"You are most likely correct."

"I thought it appropriate to float the notion of Germany first." Roosevelt nodded his head in tacit agreement. "Germany is the lynchpin among the Axis countries. Italy cannot stand without Germany, and the Germans have been instrumental in the formulation of their so-called Tripartite Pact, last year. Even if we face war with all three, the fall of Germany will undoubtedly take down Italy in the process and weaken Japan's position."

"In general, broad principle, I would agree with your assessment. However, we cannot predict how Japan will react to the sanctions we have imposed upon them. They are not in a strong position. Their conquests have gained valuable resources, yet they are dependent upon maritime transport to do anything with those resources. That makes them vulnerable."

"Certainly. So, just to be plain, you do agree with a Germany first approach?"

"That is our position today . . . with the only proviso being any offensive action by Japan mitigating that strategic position."

"Thank you, Franklin. That is what I needed to say to you personally. Should we join the chiefs?"

"Yes, I would like to hear more of their discussions."

Surprisingly, as if by some secret signal, General Pa Watson knocked and opened the door. He moved quickly to assist the President in moving to his narrow wheeled chair. Four sailors appeared shortly after the President was situated, and lifted the wheelchair and the President. Churchill followed the small entourage down several decks to the Wardroom. Two armed Marines protecting the entry came to attention and saluted as the President approached. Roosevelt wanted Churchill to proceed ahead of him to alert the chiefs their political leaders had arrived, and asked that they sit together away from the main conference table. After some commotion and recognition that their respective commanders-in-chief intended to listen rather than participate, the senior military officers returned to their discussions and negotiations.

—

Tuesday, 12.August.1941
HMS Prince of Wales
47° 18' 47" North – 53° 57' 05" West
Placentia Bay, Argentia, Newfoundland
Dominion of Canada
08:15 hours

The Prime Minister stood at the starboard flying bridge railing, aft of the lookout, and watched the USS *Augusta* slowly and precisely depart the anchorage. The RIVIERA conference was officially concluded. He had sent his last few mark-ups of their joint communiqué across to *Augusta*, late last night as Draft 7. Churchill was expecting a confirmation message. The conference had been a good and generally productive set of meetings. Pound and Dill concurred in the general success. Several open action items needed more intelligence information, which both sets of chiefs committed to collect as best they could and share the results. The Prime Minister rated the first meeting as broadly successful, as well. He knew Roosevelt was correct; however, the clause three statement still bothered him for more than a few reasons. They had managed to soften the words in small measure in several places. The negotiations had been illuminating and informative.

The *Augusta*'s wake appeared to boil. She was picking up speed quickly as she cleared the harbor into the broader bay. Her destroyer escorts awaited her in open waters, most likely having already screened the area for any lurking ambitious U-boat captain. In another few minutes, she would disappear behind the buildings of the Argentia air station.

"Excuse me, Prime Minister." John Martin's voice caused both the lookout and Churchill to turn to the speaker. "This open transmission message just arrived." The Prime Minister's duty private secretary handed the single sheet of yellow paper to Churchill.

```
PPUS
MMTB RQSP NR 1752
U 121154Z AUG 41
FM POTUS ABD USS AUGUSTA
TO PM ABD HMS PRINCE OF WALES
SUBJ JOINT DECLARATION
URGENT PROMPT DELIVERY
BT
UNCLAS
TEXT OF DRAFT 7 YESTERDAY EVENING AGREED BREAK
EXPECT JOINT PUBLIC RELEASE 14 AUGUST 1941
BT
NNNN
```

"Excellent," said Churchill. "Now, we have only the public release on Thursday. We will be at sea. Please resend the President's message to the War Cabinet and Foreign Office with my simple statement that Draft 7 is confirmed. Please release the document as approved on the 14th to our ministries and news agencies."

"Yes sir. I also brought a copy of Draft Seven, in case you wanted to have one last read through before release."

"Thank you, John," he answered and took the proffered white sheet of paper.

```
DRAFT 7, August 11, 1941; 19:05 (Z)
            JOINT DECLARATION
    The President of the United States of
America and the Prime Minister, Mr. Churchill,
representing His Majesty's Government in the
United Kingdom, being met together, deem it
right to make known certain common principles
in the national policies of their respective
countries on which they base their hopes for a
better future for the world.
```

First, their countries seek no
aggrandizement, territorial or other;

Second, they desire to see no territorial
changes that do not accord with the freely
expressed wishes of the peoples concerned;

Third, they respect the right of all
peoples to choose the form of government under
which they will live; and they wish to see
sovereign rights and self-government restored
to those who have been forcibly deprived of
them;

Fourth, they will endeavor, with due
respect for their existing obligations, to
further the enjoyment by all States, great
or small, victor or vanquished, of access,
on equal terms, to the trade and to the raw
materials of the world which are needed for
their economic prosperity;

Fifth, they desire to bring about the
fullest collaboration between all nations in
the economic field with the object of securing,
for all, improved labor standards, economic
advancement and social security;

Sixth, after the final destruction of the
Nazi tyranny, they hope to see established
a peace which will afford to all nations the
means of dwelling in safety within their own
boundaries, and which will afford assurance
that all the men in all lands may live out
their lives in freedom from fear and want;

Seventh, such a peace should enable all
men to traverse the high seas and oceans
without hindrance;

Eighth, they believe that all of the
nations of the world, for realistic as well as
spiritual reasons must come to the abandonment
of the use of force. Since no future peace can
be maintained if land, sea or air armaments
continue to be employed by nations which

threaten, or may threaten, aggression outside
of their frontiers, they believe, pending the
establishment of a wider and permanent system
of general security, that the disarmament of
such nations is essential. They will likewise
aid and encourage all other practicable
measures, which will lighten for peace-loving
peoples the crushing burden of armaments.
                AUGUST 14, 1941
      Franklin D. Roosevelt - Winston S. Churchill

"It will soon be official and a public notice in the eyes of citizens and our enemies," Churchill said and paused for a moment, staring at the paper, "and, a worthy document despite my misgivings. I pray I am not judged too harshly by history for contributing to this statement."

Martin chuckled softly. "After what you have already done for the Empire and for freedom itself, I highly doubt that, sir. If you will excuse me, I shall get these sent per your instructions. Do the chiefs need to advise on this?"

"No. Not necessary. This is a political statement, not a military one. You can certainly provide them a courtesy copy, but no need for counsel on this one."

"Very well, sir," Martin said and departed to the interior.

Churchill turned back just in time to see the stern of *Augusta*, with a bloom of white foam behind her, disappear behind a hangar on the Argentia airfield. Winston waved with no one to see. He took a deep breath and left the flying bridge. There was work to be done before the *Prince of Wales* sailed as well, later this afternoon.

———

Thursday, 14.August.1941
HMS Prince of Wales
58° 14' North – 35° 21' West

John Martin knocked on the Flag Stateroom hatch, and then entered. Prime Minister Churchill stopped his dictation and gestured to Martin to let him have it – whatever it was that disturbed his work. "Excuse me, sir, the duty special communications man is here with material for you."

Churchill gestured for Martin to let the man in, and then for the others to clear the compartment. The man in an Army corporal's uniform entered and stood aside by the door, until he was alone with the Prime Minister and the hatch was closed. He then approached Churchill at his desk, saluted crisply, and extended his left hand, holding a single folder with a bright red border.

The Prime Minister returned the man's salute, took the folder and placed it on his desk. He opened the folder. A familiar, single sheet of pink paper sat on top.

MOST SECRET - ULTRA

```
RESEND
DATE: 14 AUGUST 1941
TO: PM
FROM: C
FINAL SOLUTION
SECRET
DATE: 11 AUGUST 1941
TO: ESG1 ESG2 ESG3 ESG4
FROM: RSHA
SPECIAL UNIT ACTION
BEGIN
ESG UNITS FALLING BEHIND OBJECTIVES BREAK
MUST EXPAND TO ENGAGE WITH APPROPRIATE LOCAL
PERSONNEL TO IMPROVE PRODUCTION BREAK FINAL
SOLUTION REMAINS TOP EMPIRE OBJECTIVE BREAK
REPORT REVISED PLANS TO THIS HEADQUARTERS
END
SECRET
```

MOST SECRET - ULTRA

Well, look what Boniface turned up. Winston read the message several more times. It apparently took GC&CS several days to decode the SS message. He recognized RSHA – *ReichsSicherheitsHauptAmt* (Empire Main Security Office), the dominion of *SS-Gruppenführer* Reinhard Tristan Eugen Heydrich, principal deputy of *Reichsführer-SS* Heinrich Luitpold Himmler, who in turn was one of the essential deputies of *Reichskanzler* Adolf Hitler and chief of the SS. The other designations were not familiar to him and he knew he could not ask the communications man, but there was little doubt the word 'production' in an RSHA message was probably not a good sign. Churchill turned to the next page on blue paper.

MOST SECRET – PM EYES ONLY

```
MOST SECRET - PM EYES ONLY
DATE: 14 AUGUST 1941
TO: PM
FROM: C
FINAL SOLUTION
BEGIN
BONIFACE PRODUCT YESTERDAY BREAK POINTS OF
EXPLANATION MIGHT BE USEFUL BREAK ESG ARE
SPECIAL ACTION GROUPS IN GERMAN EINSATZGRUPPEN
BREAK ROUGHLY BRIGADE SIZE UNITS OF SPECIALLY
TRAINED SS PERSONNEL BREAK FINAL SOLUTION
BELIEVED TO BE PREFERED NDSAP SS SD TERM
FOR LIQUIDATION OF JEWS COMMUNISTS AND
OTHER UNDESIREABLES LIKE GYPSIES DISABLED
AND MENTALLY ILL BREAK THIS IS FIRST
COMMUNICATIONS REFERENCE TO FINAL SOLUTION
BREAK DO NOT PRESENTLY KNOW EXTENT AND SCOPE
OF ESG OPERATIONS BREAK APPEAR TO BE RELATED
TO OPERATION BARBAROSSA BREAK IF SO MAY BE
CONFINED TO EAST BREAK DIRECTED SIS FIELD
AGENTS TO COLLECT ANY AND ALL INTELLIGENCE ON
SPECIAL UNIT ACTIVITIES AND SPECIFICALLY FINAL
SOLUTION TO BETTER UNDERSTAND IMPLICATIONS
OF BONIFACE INFORMATION BREAK NEED TECHNICAL
REVIEW BY APPROPRIATE PERSONNEL UPON YOUR
RETURN
END
MOST SECRET - PM EYES ONLY
```

MOST SECRET – PM EYES ONLY

'Final Solution,' ay, that cannot be a positive term. I've heard that term before . . . but where? I wish I had had this message for my private discussions with Franklin. I cannot just resend this to the President. I will wait until after the technical review Menzies has called for in this matter, and then inform the President what we have found.

Churchill looked to make sure there were no other papers in the folder. There were none. He closed the folder and returned it to the courier. The Prime Minister said, "Acknowledge receipt. No reply necessary. That will be all. Thank you, corporal."

The corporal took the folder back, saluted, and turned to leave.

Churchill gazed out the open porthole to the blue sky beyond, as the massive battleship swayed gently in the ocean swells. He could feel the deep, low amplitude vibration of the propellers as they were making good speed. They should arrive in Iceland tomorrow.

John Martin returned. "As you requested, Prime Minister, the navigator has confirmed our arrival at Reykjavik tomorrow morning at eight AM local time. The captain expects four hours to refuel, resupply foodstuffs and retrieve the post. Your agenda has been confirmed. We should be able to debark at eight thirty. A detachment of American Marines will be formed at dockside for your inspection. The commanding officer will be available to escort you and answer whatever questions you may have. As you requested, HMS *Churchill* will be tied up astern of us and also available for your inspection."

"Very good, John. That should fill our available time. I wanted to pay my respects to the newly arrived Marine contingent . . . a big step for the Americans, and a necessary one for us. The *Churchill* is the former USS *Herndon*, one of the first destroyers lent to us by President Roosevelt last year. This will be my first opportunity to make a direct assessment of these old, surplus, American destroyers the Navy has been complaining about."

"At least they are working and on patrol," Martin added.

"Precisely the point, John. They are not brand new warships, but I do believe they are functional, and at the very least might cause a U-boat captain to hesitate, so useful in that sense alone. Thank you for confirming tomorrow's plans. Perhaps not a stop the captain needs to make, but a necessary one for me."

"Captain Leach asked me to remind you that he would like to sail by noon tomorrow, so we can make the Scapa Flow anchorage before sunset on Saturday."

"Yes, yes. I am quite sensitive to that requirement. Now, let's get the stenographers back in here and get the remainder of the Despatch Box dealt with. While I am working through the daily papers, would you be so kind to research the term 'Final Solution,' probably associated with Hitler and his National Socialist thugs. If the term has been used before, I would like to know when, by whom, and in what context? Also, evening meal is approaching and I would like to sup with the ship's officers."

"Yes sir. They are expecting you in the Officer's Mess for evening meal. I will see what I can find regarding the term 'Final Solution.' How soon do you need the information?"

"The appropriate research facilities may not exist aboard this ship of the line. Try to find it as soon as you can upon our return to London. You may need to ask the Foreign Office or MI6 for assistance."

"Very well, sir. I shall retrieve the duty stenographers, while I take a quick look-see, and then I shall rejoin you to clear the Despatch Box." Churchill nodded his head and turned his attention to the papers before him.

Martin recalled the duty stenographers and the Prime Minister picked up exactly where he left off without skipping a beat. He always impressed his clerical staff with his memory and grasp on facts and events, and his relentless work ethic. The stenographers usually worked in pairs, and in three or four relays, when he was in England. The clerical staff on overseas travels was often half or less of what he usually employed.

—

Chapter 11

Action is transitory—a step, a blow,
The motion of a muscle, this way or that—
'Tis done, and in the after-vacancy
We wonder at ourselves like men betrayed:
Suffering is permanent, obscure and dark,
And shares the nature of infinity.
                          -- William Wordsworth

Sunday, 17.August.1941
RAF North Weald
Epping, Essex, England
United Kingdom

The Spitfire Mark II had handled quite like his old Mark IA, less than a year ago. He felt so at home and comfortable in the confining cockpit. The extra horsepower from the upgraded engine helped at a technical level, but was not noticeable on a practical air combat basis.

The latest mission had involved a heavy section of Messerschmitt 110 Destroyer twin-engine fighters on a determined attack of a fast, six-ship convoy that had turned north out of the Thames Estuary. Not quite a squadron of Bf109s had escorted the attackers. To Brian, the Germans appeared to be focused on one ship in particular within the convoy, as if they knew what was on that ship and they were determined to destroy it. Brian had wondered what was so special on that particular ship. He reminded himself to mention his observation to the debriefer.

'A' Flight had engaged the fighters, while 'B' Flight in their Hurricanes took on the Destroyers. The squadron had successfully defended the convoy. No aircraft had been clearly shot down during the engagement, although hits had been achieved. The first air combat mission in their new Spitfire Mark IIs had been successful – the convoy had sustained no hits . . . several near misses, but no hits.

Brian's Green Section landed second and smoothly, followed by the 'B' Flight sections in trail. They taxied back to their assigned revetments. Brian switched off his radios, oxygen and secured the Merlin engine.

"Shots fired, ay, Mister Drummond?" said Corporal Jacobs.

"Yep . . . about half a bag, by my guess . . . hits but no victories I could observe."

"You can't get them every time."

"Nope, but the mission was successful. We protected the convoy."

"Congratulations, sir. Any squawks?"

"She's a perfect machine."

"We'll get her reloaded and refueled . . . in case you are called up later."

"Thank you, Henry."

The debriefings of each pilot were completed in 45 minutes. As each pilot completed their portion, he joined the group on the grass outside the Dispersal Hut.

Dusty Langford leaned over to Hunter, "While you were finishing your debriefing, Whitey told us the Air Ministry confirmed that 'Tin Legs' Bader was shot down over France on the 9th, and bailed out." Bader had been leading the Tangmere Wing – Nos.145, 610 and 616 Squadrons – on a CIRCUS Sweep. "Several of his buddies saw his parachute descent and saw him land."

Wing Commander Douglas Robert Steuart 'Tin Legs' Bader, DSO, DFC, had commanded No.242 Squadron at RAF Duxford, flying Hawker Hurricanes during the Battle of Britain, and was the first commander of a fighter wing when he was shot down. He had lost both his lower legs as a consequence of crashing his Bristol Bulldog Mark IIA, during an impromptu airshow on 14.December.1931. He fought hard both physically and bureaucratically to return to full flight status, which he achieved in November 1939.

"Was he captured?" Brian asked.

"Don't know, yet . . . according to Whitey," Dusty answered.

"Well, hopefully, he made it, and he's alive and not seriously injured." Brian turned to Tolly. "So, Bulldog, that was your first combat in Spits. What did you think?"

"Like you told me, she feels more agile and sporty than the Hurri. I also know now why you like Spits so much. Magnificent machine."

"Yes, it is," Brian responded. "But, Spitfires do have their weaknesses, like armament. We have three-oh-three's and the Germans have 20-millimeter cannons. We have lead bullets. They have exploding and incendiary projectiles. Their machines keep running in negative 'g'; ours do not. They have better high altitude performance than we do. Never forget your weaknesses, and fly to minimize those weaknesses."

"Understood, Hunter. I have not forgotten your words."

Whitey joined them. "We are at 30 minutes. It's lunchtime. Let's get our sustenance and get back down here lickety-split, as you Yanks like to say."

"Hear, hear," several of the pilots declared. Those who still had some or all of their flying kit on, doffed their gear, stowed the equipment on their assigned pegs, and joined the others on the short walk to the Officer's Mess for lunch.

—

Tuesday, 19.August.1941
No.10 Downing Street
Whitehall, London, England
United Kingdom
16:30 hours

"Lord Cherwell for his scheduled appointment, sir," announced Jock Colville, the Prime Minister's duty private secretary.

The venerable scientific advisor entered. They sat in adjacent, large, leather chairs.

"Welcome back, Winston. I trust your journey was productive and successful," the Prof said.

"Yes, very good on both counts. It was good to finally get my first face-to-face meeting with Franklin Roosevelt . . . good man, good mind, keen understanding."

"I presume you saw the final report from Mister Bensusan-Butt in your Despatch Box for your railway journey back to London."

When the investigative statistical analysis task was envisioned in May, the Prime Minister's science advisor Lord Cherwell, Professor Frederick Lindemann, had known precisely who should do the work. While Lindemann had served Churchill as his science advisor at the Admiralty, the Prof had hired David Miles Bensusan-Butt to form a statistical analysis group for better understanding of Royal Navy operations. Bensusan-Butt had been educated at Cambridge, a student of John Maynard Keynes, with whom he worked before joining The Economist, and then the civil service before the war began.

"Yes, I read the report . . . very interesting . . . as we suspected, I should say."

"Based on our discussions last spring, I thought we would find optimistic reporting from the Bomber Command aircrews, but to be frank, I did not expect the results Butt developed. Here is the page of the report's conclusions." Lindemann handed the single page to Churchill.

The Prime Minister read through the proffered page.

SUMMARY.

STATISTICAL CONCLUSIONS.

An examination of night photographs taken during night bombing in June and July points to the following conclusions:

1. Of those aircraft recorded as attacking their target, only one in three got within five miles.

2.   Over the French ports, the proportion was two in three; over Germany as a Whole, the proportion was one in four; over the Ruhr, it was only one in ten.

3.   In the Full Moon, the proportion was two in five; in the new moon it was only one in fifteen.

4.   In the absence of haze, the proportion is over one half, whereas over thick haze it is only one in fifteen.

5.   An increase in the intensity of A.A. fire reduces the number of aircraft getting within 5 miles of their target in the ratio three to two.

6.   All these figures relate only to aircraft recorded as attacking the target; the proportion of the total sorties, which reached within five miles, is less by one third.

Thus, for example, of the total sorties only one in five get within five miles of the target, i.e. within the 75 square miles surrounding the target.

RECOMMENDATIONS.

1.   These results though fairly reliable should be checked by a thorough expert study of the day photographs, and by a comparative study of photographs of German and British towns.

2.   In order to keep these figures up to date, and to obtain continuous records of the success of our navigation, staff should be set up to maintain statistical records of night photographs and any other evidence that may be available.

3. This staff should consist of at least one trained statistician, with a sufficient clerical staff. He should have authority to modify forms and questionnaires in order to make sample enquiries, e.g., to replace some existing questions for a certain period by

others designed to elucidate some particular
point.

———————

"Yes, that is what I read. Not a very encouraging result."

"No . . . worse than I expected."

"I would like to have your unvarnished view of Butt's work," Churchill directed his science advisor.

"To be blunt, I am disturbed by the definition of the target for this analysis. A five-mile radius around the specified target – the object of the mission – is a 78-square-mile area. That hardly seems like a precise definition of what the target was for this analysis. I asked David why he chose a five-mile radius, and his response . . . they had to start somewhere to yield results. Even with this inordinately large definition of the intended target, only something like five *per centum* of the released bombs impacted or had effect within the target area. Although David chose not to report his casual observations, there were more than a few events where not a single bomb hit the prescribed target. In a typical 50-ship raid, 350 tons of high explosive and incendiary bombs had no effect on the actual designated target facility."

"How is 'effect' determined in this context?"

"A high explosive bomb produces blast – a shock wave – and shrapnel damage, with some heat effects. A typical 500-pound high explosive bomb has a blast radius of roughly 400 yards, or 16 one hundreds of a square mile in area. From another perspective, again in a 50-ship raid, roughly 1,400 bombs are dropped. On some missions, not one of those bombs hit the target . . . had any discernible bomb damage effect on the planned target, even as more broadly defined by Butt."

"Aren't you being a bit harsh, Prof?"

"Perhaps. The worst performance was against targets in the Ruhr Valley, in Germany proper."

"Where the defenses are deep, multi-layered and determined."

"Yes, that is true. They did better against less well-defended targets. However, the issue we agreed to address was the difference between what Bomber Command was reporting on mission results versus the post-raid, aerial photographic documentation that actually established the bomb damage on the assigned targets."

"Our suspicions validated," mumbled Churchill. "Has Butt's report been distributed to the Air Ministry?"

"Yes . . . to the Air Ministry and Bomber Command."

"There is no point dropping bombs, if we cannot hit the targets we need to destroy." Winston paused to think. The Prof nodded his head in agreement. "How long has the Air Ministry had the Butt report?"

"Two days . . . yesterday morning."

"I will ask Jock to schedule a Defence Committee meeting tomorrow afternoon to discuss the Butt Report."

"The whole committee, or just the Air Ministry, or even just Bomber Command?"

Churchill considered the question. "Let's make it the whole Defence Committee. The other ministers and chiefs should hear this, and contribute to the solution."

"Would you like me to attend?" asked the Prof.

"Yes, of course . . . and Bensusan-Butt, now that I think of it. We must get a grip on this. Bomber Command has rejected daylight operations against defended targets due to excessive loss rates. Now, we see they are unable to hit their assigned targets during nighttime bombing operations, and we still sustain losses . . . just not as high as during the day. Dropping tons of high explosive bombs on countryside or residential areas, and not military targets, will only waste valuable resources and not achieve victory."

Lord Cherwell connected with Churchill's eyes. "I am sure you will recall our discussion about advances in electronics, several months ago."

"Yes, but will that new Wizard War equipment solve these discouraging results?" the Prime Minister asked.

"They have not reached the demonstration stage, as yet, so performance is only theoretical. The results of component testing along with the calculated . . . expected . . . anticipated performance give the scientists and engineers optimism their new kit will help."

". . . but not solve the night bombing accuracy problem."

"There are no guarantees, Winston. This is war."

Churchill sneered and stared at the Prof. "I shall not bite. Nonetheless, I am inclined to suspend our general night bombing campaign until the night accuracy can be improved."

"That seems a bit extreme, Winston, but that is your choice."

"Indeed." Churchill stood and walked to the window, overlooking the garden. The bomb damage from The Blitz was slowly being erased. "Is there anything else," he said without turning around, "I need to absorb Butt's work?"

"I think you have the essentials, although I must close with the observation . . . Bomber Command can and does hit the target at night, just not to the degree claimed by the aircrews in their mission debriefing reports. Let us not be too hard on the bomber crews."

"Yes, yes." Churchill instinctively looked to the sky above the capital city. Puffy cumulous clouds obscured half of the blue sky. Less than a year ago, German bombers filled the daylight sky above the city in a desperate aerial battle that has now abated . . . at least for the time being. The Germans abandoned the daytime as well. However, they managed to achieve far better night bombing accuracy, at least based on the results on the ground, than what the British had achieved. The Prime Minister knew they had to get ahead of that problem. "Anything else, Prof?"

"No, Winston."

Churchill turned back to his trusted science advisor. "Very well, then. Thank you for organizing this work . . . very important. Please convey my gratitude to Mister Bensusan-Butt for his exceptional work. Lastly, I would like you to work with the Air Ministry to institutionalize these analytical processes to improve operational performance."

"Already in work, Winston."

They concluded their discussion. Before Lord Cherwell departed, the Prime Minister instructed his private secretary to schedule the Defence Committee meeting to address the Butt Report. Churchill moved on to the next of myriad issues before His Majesty's Government.

—

Sunday, 24.August.1941
RAF North Weald
Epping, Essex, England
United Kingdom
14:20 hours

A rather dashing, chiseled, medium build, squadron leader walked alone toward the No.71 Squadron Dispersal Hut. He had a Distinguished Flying Cross ribbon under his pilot wings. "I presume this is Seventy-One Squadron," he said, in a decidedly British accent.

"Yes sir," answered Salt Morton.

"Is Flight Lieutenant Whittington about?"

"Yes sir . . . in his office."

The visitor disappeared into the Dispersal Hut. Salt shrugged his shoulders, as if he was saying, what's up with that?

Brian lay back in his lawn chair, closed his eyes to enjoy the sun's warmth on the summer day, only to have a large group of clouds obscure the sun. The peacefulness of the mid-morning in East Anglia struck Brian. *A year ago, conditions were monumentally different and worse. This is almost lazy compared to those crazy days.* The sun would feel better, but it is still warm. His thoughts drifted off, as he allowed himself to take a short nap.

"Gentlemen," the voice of Corporal Harris brought Brian back to consciousness after an indeterminate period of time, "the Skipper would like you to assemble inside."

Grumbles and grunts punctuated the rise of the pilots from the grass or chairs. Once everyone was inside, Corporal Harris notified Whitey. He appeared with the unnamed visiting squadron leader.

"Your attention please,' Whittington began. "I am honored to introduce our new commanding officer Squadron Leader 'Tug' Meares. He comes to us from Seven Four Squadron and Six Double One Squadron before that. Welcome to the Eagle Squadron, Mister Meares."

Squadron Leader Stanley Thomas 'Tug' Meares, DFC, had served in both No.74 and No.611 Squadrons during the Battle of Britain. He was an accomplished fighter pilot.

"Thank you, Whitey. It is an honor to join you, gentlemen. Before we jump into our routine business, I understand from Whitey that you have some concerns for the welfare of Doug Bader. I have come from the Air Ministry, yesterday. Bader was captured and is a prisoner of war. Apparently, when he bailed out, one of his prosthetic legs was caught in the aircraft. He is now a guest of the Germans. The Air Ministry was all a-twitter with an open message from none other than Adolf Galland, offering safe passage to drop a spare leg for Doug. That mission was accomplished on the 19th to the amazement of many in Whitehall."

"Well, how about that . . . ," Dusty said and added, "like the Christmas ceasefire."

"Christmas ceasefire?" asked Bulldog.

"Good point, Dusty," Whitey responded. "Yes, indeed, Christmas 1914, on the Western Front. There was an unofficial truce and fraternization between Allied and German troops. The war was suspended for a few hours, and they acted like human beings."

"Thanks, Whitey. I did not know that," Bulldog said.

"Then, they returned to another four years of meat-grinder war in the trenches," added Whitey.

"Now that the history lesson is over," Meares brought them back, "we are removed from combat duty for a few days for two reasons. One, I need to learn about you, each of you, and how you fly; and, you need to adapt to me as your skipper. Two, the Air Ministry informed me upon this assignment that Sunday, we will receive the first of our brand new Mark Five 'B's."

"We just transitioned to Mark Two's," Red Burns observed.

"Do you not want the Five's?"

"No, no, I didn't mean it that way, Skipper. It is just that we have only had Mark Two's for a couple of weeks.

Meares smiled, "How about let us not look a gift horse in the mouth."

"Yes sir."

"The 'B' wing on the Mark Five gives us cannons."

"Oh, yeah, baby," proclaimed Red Burns.

"And, a more powerful engine . . . ."

"Wow!" Red added.

"This one has some bite," Pete Peterson offered.

"Indeed," responded Meares. "Whitey told me you transitioned in stages as the new aircraft were delivered. I expect we will do the same with this move. I do not know how many will be delivered when, so we shall have to be flexible. I do not want to mix types within a given section, so that may affect our assignments. We shall be off the line for at least a week . . . perhaps more, depending upon delivery numbers and rate."

Squadron Leader Meares took the squadron up in the afternoon for some basic maneuvering drills, and then mock combat engagements of 'A' Flight versus 'B' Flight. The squadron seemed more on point after the news of the new aircraft soon to arrive.

—

Sunday, 24.August.1941
Chequers Court
Ellesborough, Buckinghamshire, England
United Kingdom
17:15 hours

Commander Alastair Ignatius Denniston, CMG CBE, arrived early from the notification telephone call to the Prime Minister's duty private secretary two hours ago. Denniston was chief of the Government Code and Cypher School at Bletchley Park, often referred to as Station X—the decryption and analysis group that processed communications intercepted and collected by Station Y. The specialists at Bletchley were the source of the ULTRA decryption of German command messages.

The Prime Minister waited for Denniston in the large study. These sudden visits always meant something important from ULTRA. Denniston was announced and entered the study wearing the uniform of a Navy Volunteer Reserve commander with the Buff Box manacled to his left wrist.

"Good afternoon, Prime Minister," Denniston said.

"Good afternoon, Commander."

Denniston placed the Buff Box on the table between two chairs. Churchill retrieved and unlocked the case and removed the candy-striped folder.

"'C' asked me to bring these messages directly to you. They are four days old. We just managed to decipher them this afternoon."

Churchill removed the first message.

MOST SECRET - ULTRA

```
SECRET
DATE: 20 AUGUST 1941
TO: SUPREME COMMANDER NORTH AFRICA
FROM: COMMANDER AFRICA CORPS
BREAK
REINFORCEMENT UNITS ARRIVED SAFELY BREAK
DEPLOYMENT HAS BEGUN EXPECT TO BE FULLY
INTEGRATED BY SUNDAY BREAK ONE SUPPLY SHIP LOST
ENROUTE
END
SECRET
```

MOST SECRET - ULTRA

"Rommel, again."

"Yes sir. The significance is not so much what the message says, but what it does not say." Churchill waited for Denniston's explanation. "We believe the mentioned reinforcement units are an armor division and an infantry division transferred from the Eastern Front."

"Why is that significant? We move divisions around quite a lot ourselves, trying to match resources to situations."

"Taken with the second message, it suggests the Germans are progressing in the East better than planned and that Rommel may be having some difficulties."

"That judgment is a bit of a stretch, it seems to me," the Prime Minister said and retrieved the second message.

MOST SECRET - ULTRA

```
SECRET
DATE: 23 AUGUST 1941
TO: COMMANDER ARMY GROUP CENTER
FROM: ARMY HEADQUARTERS
INFORMATION: ARMY GROUP SOUTH
BREAK
THE LEADER ORDERS SECOND ARMOUR GROUP DETACHED
AND TRANSFERRED BY FASTEST AVAILABLE MEANS TO
ARMY GROUP SOUTH BREAK HAIL HITLER
END
SECRET
```

MOST SECRET - ULTRA

"Oh my, this seems rather extreme, doesn't it? Do we know the composition of this Second Armor Group?"

"This group has grown since it was created last year. To the best of our knowledge, the Second Armor Group is comprised of six, first line, armored divisions and five mobile infantry divisions with assorted attachments."

"Commanded by . . . ?"

"Colonel General Heinz Guderian."

"Oh my . . . one of their best armor generals along with a good chunk of resources, and on orders from Hitler himself. I see your point. Why would he diminish the advance to Moscow?"

"We are asking the same question. We have no direct information from ULTRA or any other source. So, from a purely conjectural standpoint, we see two primary explanations. One, the Germans are more concerned about taking Kiev and perhaps Crimea than they are about Moscow."

"That does not make sense. Moscow is the capital of the entire Soviet Union, not just a region."

"They might also be more confident with the progress of Army Group Center than with Group South or Group North for that matter. Or, perhaps, they felt Center was advancing too fast and creating a salient with exposed flanks."

"That did not stop them in France. They took the risk and dealt with what threats appeared on their flanks that confronted them. This does not make sense. I would have taken forces from North and South to strengthen

Center and take Moscow before winter. Perhaps we are missing some essential contributing factors."

"Yes, well, unfortunately, we do not have any indications of the reasons. We are just guessing."

"You said earlier, your estimate is Rommel's reinforcements came from the Eastern Front as well?"

"Again, an assessment without direct evidence to substantiate it."

"This moving of Guderian may well prove to be another strategic blunder by that foul corporal. I know you will keep a close eye on things, as you always do. Please convey my thoughts to 'C,' if you see him before me."

"Yes sir, I will."

"Is there anything else?"

"No sir, that's it for this afternoon."

"Would you care to stay for supper before you head back to Bletchley?"

"Thank you very much for your generous offer, Prime Minister. I shall beg your forgiveness. I really must get back as soon as possible."

"Very well, Commander. Thank you for making the journey. Quite appropriate. If you or 'C' have no objection, I would like to visit your facility next month."

"I cannot imagine an objection, Prime Minister. We would be honored. I will confer with 'C' and notify your private secretary for arrangements."

"Thank you."

Denniston returned the messages to the candy-striped folder, inserted them in the proper place, and locked the Buff Box case. He nodded his head and departed.

—

Wednesday, 27.August.1941
RAF North Weald
Epping, Essex, England
United Kingdom

What remained of No.71 Squadron waited in eager anticipation for the arrival, once the inbound message had been received. Rocket, Dusty and Sweet were still away on their 48-hour pass. They still did not have a replacement pilot for Horse Harrow, who had been killed the previous month, for Rusty Bateman, who was shot down and became a POW. It was taking longer to get replacement pilots, since they were now competing with their sister Eagle squadrons – Nos. 121 and 133 Squadrons – although Squadron Leader Meares thought replacements to bring them back to a full complement of pilots could be arriving any day.

The message did not specify the number of aircraft inbound, only that a ferry flight was to arrive shortly. Even Meares and Whittington sat outside in the morning sun. They did not have to wait long. Three aircraft in a nice 'V' formation appeared from behind a cloud. They maintained a smooth descent for a straight in approach to landing.

The pilots stood and walked to the first open parking spot, as the three, freshly painted Spitfires taxied to them in trail. The long cannon barrel protruding from the inboard portion of each wing leading edge offered the most distinctive visual difference from the previous versions of the aircraft. One cannon replaced the two inboard machine guns of the previous versions. They gathered behind the left wing, as the lead pilot shutdown and secured his aircraft. The other two aircraft followed in sequence, as they parked in adjacent spots.

When the lead pilot took off his helmet, goggles and oxygen mask, he appeared to be a middle-aged man, with partially grey hair, which seemed a bit odd for wartime. Then again, he may well have been a pilot from the Great War, who was now too old to fly on active duty, but certainly capable of flying for Ferry Command.

Meares talked to the ferry pilot, while the other squadron pilots began their inspection of the new aircraft. Brian looked across the tails to the third aircraft to see Jennifer Brentwood, the ferry pilot he met three weeks prior, notice him and wave in an enthusiastic and delicate feminine manner. Brian returned her wave. Brentwood and another female pilot walked off toward the Operations Building. They were onto the next leg of their mission. Meares finished talking to the pilot, and the older man began his walk to join his comrades.

The Supermarine Spitfire Mark V incorporated a Merlin 45 engine rated at 1,470 shaft horsepower – a substantial increase over the Mark II's 1,175 horsepower Merlin XII engine, with a slightly larger diameter, wide blade, Rotol propeller to take the increased horsepower. The new engine included a better single-stage supercharger for improved high altitude performance and more importantly a unique carburetor that kept fuel feed to the engine during negative 'g' maneuvering. While the increased engine power would help, it was the 'B' wing variant the pilots were most eager to get their hands on. Two of the four 0.303 Browning machineguns in each wing were replaced by a single Hispano Mark II 20mm cannon with a 60-round ammunition drum per gun.

Brian did his own walk around the aircraft. Each wing had a blister on the top skin to accommodate the new cannons. The cannon barrels looked big. Brian did a quick mental calculation . . . they were not quite three times the bore as their normal 0.303 machine guns. *I wonder if we will have explosive projectiles for these cannons like the Germans do?* The barrels protruded nearly a foot from the leading edge. *I wonder if they will put protective tape over the muzzles like they do for our gun ports?* Brian wanted to see the engine, but doubted there would be any external differences. Brian waited his turn and jumped up on the left wing root to examine the cockpit. Everything he was familiar with was just as it had always been, except freshly painted. The most noticeable difference Brian saw was the control column spade circular red machine gun firing button had been replaced by a silver, rectangular button. He wanted to press it, to get a feel for the mechanics of the switch, but he knew better than to play with things he did not know fully regarding their function. From the appearance, the switch had a thumb depression on the top and another on the bottom, and a ridged middle section. *One of those thumb positions is probably for cannons and the other for machine guns. If the ridged section was depressible, then that would probably be for both types of armament. I cannot see a selector panel. I need to learn more.*

When Brian jumped off the wing, Squadron Leader Meares was waiting for him. "Sorry, sir, I did not realize you were waiting for me."

"What did you think of her?" asked Meares.

"She's fresh."

"Indeed."

"Still not sure of the firing button."

"Top gives you cannons only. Bottom guns only. The middle section gives you everything. The gents that have flown the 'B' wing Spits, they call the middle the 'All Hell' selection, or some call it the 'Heaven Can Wait' position."

"Wow. Nice to have some bite. What about the ammunition for the cannons?"

"Even better. We have a full range of projectile choices: ball, high explosive, incendiary, armor piercing, and a fancy combination high explosive / incendiary projectile, with and without tracer elements. The tracer element is yellow, I am told. I have not fired these 'B' wing cannons before."

"Me either."

"We have not had a chance to talk, as yet. I am well aware of your accomplishments and reputation, Brian. My thinking with only three new aircraft, it is probably not wise to have one section or a mixed section. I would like to take you and Whitey up with me, once the aircraft are out of the acceptance checks and the tail designators applied. We will cycle each of the pilots in two ship sections with one of the three of us leading for their familiarization and training flight. We will hold onto the Mark Two's in case we happen to get called out, but I expect we will be in training for a week or two. Once we get the other aeroplanes in, we will move through transition faster. We will need gunnery training to get used to the cannons and the new three-way firing button."

"Sounds like a plan to me, sir."

"Just to be clear, Brian, Pete is senior to you. By rank, he should have priority over you. But, you have more combat experience than any of us. I intend to have a private chat with him, but I did not want to cause any conflict between the two of you."

"I don't think he will object, sir."

"Very well. Let's rejoin our comrades. I will brief the plan before lunch. We should get the Mark Five's back by mid-afternoon."

Meares patted Brian on the back, and then walked to the Dispersal Hut together. Brian eagerly anticipated the afternoon flight in the new fighter.

—

Friday, 29.August.1941
Standing Oak Farm
Winchester, Hampshire, England
United Kingdom

"Great to have you home, Brian," Mrs. Charlotte Drummond said lovingly to her husband, as he extricated himself from the small taxi. She held their sleeping, 10-week-old son Ian in her arms.

Brian kissed Charlotte without an embrace, not wanting to disturb Ian. "Great to be home, my dear."

The taxi drove away slowly. The crunch of the gravel diminished as the vehicle climbed the ridge and disappeared over the crest.

"How is our boy?" Brian asked, but did not wait for her answer, as he leaned to smell and kiss Ian's forehead. The baby stirred but did not wake.

"He is not quite sleeping through the night. I hired a nanny to help me tend to our son. She is inside. I will introduce you."

"I am so glad you got help, especially without me here."

Charlotte motioned for them to sit on the long, heavy, wooden bench outside, to savor the late summer's day. "Would you like to hold him?"

"Of course."

Charlotte cradled the infant in his arms before they sat down. "How is the war?"

Brian gave her a smirk, as he knew she had little interest in the war; however, he appreciated her gesture. "We started our transition to the new Spitfire Mark Five. The engineers have done a fine job. Anyway, we are back off-line in a training status, again."

"Well, then, at least you won't be shot at for a few days or weeks, correct?"

"Correct."

"Then, I am happy. How long do I have the pleasure of your company this time?"

"I need to be back at North Weald by noon, Sunday, and ready to fly."

"We can work with that. Would you like to meet our son's nanny and my right hand in raising our first born?"

"Sure."

Charlotte gestured for Brian to give her the sleeping baby, but he shook his head and followed her into the house. The blackout curtains had been pulled back and draped, clearing the entryway. A woman in a light green, smock dress stood at the sink with her back to them, washing dishes. "Edith," Charlotte said. The woman turned, saw them, smiled and dried her hands. She was an attractive young woman, perhaps Brian's age or less, with modest, medium build, and light auburn hair and brilliant, emerald green eyes. "This is my husband, Brian." She turned to look at Brian, "This is Edith Hanscom. She has been with me for a month and an enormous help."

"Thank you Mrs. Drummond. Nice to finally meet you, Mister Drummond," she said, extending her right hand. "I have heard so much about you, sir, and thank you so much for your service to the kingdom."

Brian gently squeezed her proffered hand with his left hand, not wanting to disturb the baby. "A pleasure to meet you, Miss Hanscom. Thank you for helping Charlotte. She has a lot to handle on the farm."

"Would you like me to take Ian, so you two can catch up?" asked Edith.

"That would be most welcome," Charlotte answered before Brian was able to decline. Edith carefully took Ian from his daddy's arms and took him upstairs, presumably to her quarters. "We need to talk," Charlotte said without any detectable emotion and led her husband to the living room. She sat on the couch and patted the seat next to her. As Brian sat next to her, Charlotte picked an off-white, square card. "I received this in the post, day before yesterday," she said and handed Brian the card.

Air Commodore and Mrs. John Spencer
cordially request the presence of
Flying Officer and Mrs. Brian Drummond
for dinner
and an evening of good cheer
19:30; Saturday, September 20th

at their residence:
No.417 Sudbury Hill, Harrow
Regrets only please:
Bushey Heath 2471

———————

"A handwritten, formal, dinner invitation . . . Brian, what is going on? Good cheer? There is a bloody war on, hard to be of good cheer when my husband gets shot at by bloody Germans. Has the war ended and I have not heard the news?"

Brian turned to face her as much as he could, sitting next to her. "Air Commodore Spencer called me earlier in the week to tell me they had sent a dinner invitation to you, for us. This is the soonest I could get a pass to come talk to you about this."

"Why? Is there a problem or something serious?" Charlotte asked.

"No . . . nothing like that. They wanted us to bring Ian for dinner, and if possible, spend the weekend with them. While the children are too young to know, they want to build a more personal relationship with us and for our children. They want the boys to grow up together, as friends."

Charlotte displayed a rather puzzled expression. "While our children are a few months apart in age, John is quite a few ranks above you, Brian. Doesn't that sound a little strange?"

"In that sense, perhaps so, but I never really thought of it like that. After all, we all know each other."

"And," she said and stopped to smile broadly, "you have fathered both children."

"Charlotte . . . please."

"I just love teasing you about that little fact of life. Anyway," she stopped when she heard Edith coming down the stairs and Ian was crying. Ian was swaddled and definitely unhappy.

"Excuse me, ma'am. It is feeding time for Baby Ian."

By the time Brian turned to look back at Charlotte, she was sitting in the separate chair with her right breast exposed and ready to accept their son. Edith placed Ian in Charlotte's right arm, resting on the stuffed chair arm. Ian found her nipple quickly and began to suckle. Brian looked back to Edith, who simply smiled.

"Would you like me to wait for Ian to finish?" asked Edith.

"No, that's OK. He'll sleep after his tummy is full."

"Very well, ma'am," she said and actually went outside.

Brian knelt in front of Charlotte. He formed the words – I love you – without speaking the words and leaned forward to kiss her lips.

"I love you, Brian," she said softly. "I need to stay calm while Ian is at breast. Now, would you be a dear and retrieve a fresh hand towel. I need to sop up my left breast before I make a mess of this blouse."

Brian did as she requested. By the time he returned to her, Charlotte had her left breast uncovered and was trying to wipe away the milk leaking from her as-yet unused breast. Brian wrapped her left breast with the towel.

"Thank you," she said softly.

"You are most welcome," Brian responded and sat on the couch, to watch Charlotte tend to their first-born.

—

Chapter 12

Make no little plans;
they have no magic to stir men's blood.
-- Daniel Hudson Burnham

Saturday, 6.September.1941
Bletchley Park
Sherwood Drive
Milton Keynes, Buckinghamshire, England
United Kingdom

The comparatively short ride from Chequers Court had been smooth and easy in the Prime Minister's 1939 Rolls-Royce Phantom III, black limousine. The roadways in the country remained undamaged by war, in stark contrast to the repairs upon repairs of the city streets in London. Prime Minister Churchill was alone in the back of the vehicle, not common but this was a special visit. Only Detective-Inspector Thompson and Churchill's driver Sergeant Stanley 'Stan' Carrick occupied the front seats.

The sign at the closed, black, wrought iron gate simply read GOVERNMENT FACILITY – KEEP OUT in large, bright red, capital letters. The heavily armed Army guards conveyed the seriousness of the sign and were not casual as the limousine approached. The guard sergeant stood at Inspector Thomson's door. Walter lowered his window. The sergeant said, "Papers, please."

Thompson said, "This is the Prime Minister," motioning to the back compartment, "for a scheduled visit."

The sergeant leaned forward to visually see all three faces. He gave no indication of recognition and simply repeated, "Papers please."

Thompson displayed some irritation, retrieved the identification documents for all three of them and handed them to the stern sergeant. The man examined each set of papers, compared the photographs to the vehicle's occupants, and then apparently ticked off each name on a clipboard list. Once satisfied, he said, "The Prime Minister is cleared to enter the manor house. Inspector Thompson and Sergeant Carrick must remain with the vehicle in the manor house courtyard. Welcome to the facility, Prime Minister," said the sergeant and saluted sharply. Churchill returned the salute, although the sergeant could not see his salute. The gate opened. All six of the guards saluted as the limousine passed. They drove through what seemed like a small forest, along a continuously curved gravel path, so that no one outside could see anything except trees inside the gate. The roadway became a circular drive with a well-manicured lawn and several mature conifer trees within the driveway. Sergeant Carrick took the clockwise direction to place the passenger doors toward the house.

Churchill recognized Commander Denniston standing alone at the main entrance of the intricate, Victorian manor house. No identifying signs could be seen, yet Churchill knew this was the headquarters building of the Government Code and Cypher School (GC&CS, in some circles also known as the Golf, Cheese and Chess Society). Thompson opened his door and started to get out before the car stopped. As soon as the limousine came to a complete stop, Walter had the passenger door open. Prime Minister Churchill stepped out.

"Welcome to Bletchley, Prime Minister," said Denniston.

"I have finally made it to the magic kingdom," Churchill responded and extended his right hand to the naval officer and shook his hand.

"Sergeant Carrick and I," interjected Thompson, "shall be here when you are done." Detective-Inspector Thompson and Sergeant Carrick knew the location by name, but neither of them knew what the purpose of this unusual facility was . . . other than it had to be important with the security present at the gate alone, and undoubtedly surrounding the grounds.

"Thank you, Walter. I shall not dally."

Commander Denniston gestured for the Prime Minister to precede him into the manor house. Denniston guided the Prime Minister up the ornate central stairway, down a hallway to the left, and into a small conference room. He closed the door behind them and gestured for the prime minister to sit at the small table. A wooden box with a black enamel label riveted to it was the only object on the table. The label had one word – Enigma. Churchill knew what the box was, but he had never seen what was in the box.

"First," began Denniston, "Colonel Menzies sends his regards, sir, and felt his presence for your requested briefing was not necessary. We have made all the arrangements you requested, and 'C' was briefed and approved of those arrangements.

"Second, 'C' asked me to ask you if you were aware of the U-boat attack on the American destroyer *Greer*, two days ago?"

"Yes, I was briefed by the First Sea Lord, yesterday. Do you have new information?"

"Only that she was apparently en route solo, carrying mail to American personnel stationed in Iceland. An aircraft spotted a submarine, alerted *Greer*, and dropped four aerial depth charges. Two torpedoes were fired at *Greer*. The destroyer in turn attacked the submarine in a running battle that lasted more than three hours. The *Greer* was not hit, and apparently the U-boat made good her escape."

"Thank you. I had not heard that additional information."

"You are most welcome, sir. Now, with the administrative bits handled, let us get to the purpose of your visit." Denniston placed his hand on an oak box. "This is Enigma. In fact, this is the box captured in Poland before the war, and as I understand the history, you were involved in its acquisition."

"How much of that story do you know?" Churchill asked.

"Only that you were essential to the approval of the mission."

"I do not know if I was essential, but the intelligence chiefs along with Agent Diamond, sought my counsel. That was March of '39, at my home. It was a huge risk, but there it is . . . the reward for that risk."

"Yes sir. We have two field boxes, of which this was the first. We also received an operations manual from the Poles via the French. In the early days of our decryption efforts, this box was essential to what success we achieved." Denniston opened the box revealing what appeared to be a small typewriter. "The type keys are used to input the letters of a message received. The lights above the keys indicate the decoded output letter, corresponding to each input letter. The genius of the device is the electro-mechanical encryption-decryption process. The plug board on the front," he said, pointing to a set of plugs and wires, "is set generally once a month. The operator of each box is instructed to the specific setup for a given month. The three wheels at the top," pointing to the thumbwheels, "are each uniquely wired internal to each wheel. We believe each unit was supplied with five unique rotors. Wheels are interchangeable, and the specific wheel and location are part of the settings." Denniston lifted the black metal cover and removed one of the wheels. "Each has its own identifier marking," he said, pointing to the large Roman numeral 'IV' on the side, "and the letters on the edge establish the unique one-time connection." Churchill took the freed rotor and examined it closely. "Each rotor has a slip ring to place a particular letter under this index mark," he said, again pointing to the mark. "Each keystroke advances the right rotor one place. One full rotation advances the next wheel one place, and when the center has turned through one full cycle, the last wheel is advanced one place. If the last wheel advances through one full cycle as well, then they start over and just continue rotating like that."

Churchill handed the rotor back to Denniston, who returned it to the box and closed the access cover. "We do not yet have a codebook, but we have deduced that the daily settings for each box change at midnight Berlin time, worldwide. The codebook must contain the day, month and year, along with the index ring setting for each assigned rotor, the position of each specific rotor in the wheel sequence, and the precise plug board setup – 'A' to 'N', 'T' to 'F', and so on. The mathematicians among us calculated the possible combinations of any given day's settings as 2.6 times 10 to the 25th power, or 26 sextillion combinations."

"That is a big number," Churchill observed matter-of-factly.

"Yes sir, indeed it is, and the settings change every single day without fail."

"How on earth can you possibly find the one setting for a particular day and a particular message?"

"Answering that question properly would take us several days . . . time I suspect you do not have. I will say, in the early days, our success was blind luck. As we have gained experience with decoding Enigma messages, we achieved more reliable success. There were several key elements to each German message. The first key group is always a clear text sequence that defined the initial rotor settings used by each individual operator. Thus, the receiver can ensure his machine settings match the sender's settings. Another key element . . . every message, especially from SS units or other NSDAP organizations, always concluded with 'Hail Hitler,' or *Heil Hitler* in German. Repeat letters give us a significant leg up in the decoding process. The armed forces units were not as reliable users of the concluding statement. Also, the encrypted portion of all Navy at sea messages, especially from the submarine service, begin with a local weather observation in specific letters that have been in use by the German Navy for several decades now. Before we finish this afternoon, I would like to show you the latest work of our computing team."

"That would be my pleasure."

"Excellent. I would like to cover a little of our message decoding process before we take our short tour."

"By all means," Churchill responded.

"We have a card we use as an illustrative example of the decoding process," Denniston said, produced a letter-sized card from a drawer below the tabletop, and passed it to the Prime Minister.

```
Line A:   Angriff im morgengrauen auf drei ebenen
Line B:   QXESI  JELXR  BNTYD  AASLG  CVGOP  ZXUEF  HDWN
                        Transmission
Line 1:   QXESI  JELXR  BNTYD  AASLG  CVGOP  ZXUEF  HDWN
Line 2:   ANGRI  FFIMM  ORGEN  GRAUE  NAUFD  REIEB  ENEN
Line 3:   ANGRIFFIMMORGENGRAUENAUFDREIEBENEN
```

```
Line 4:   ANGRIFF IM MORGENGRAUEN AUF DREI EBENEN
Line 5:   Attack at dawn on three fronts
```

"Using this simple example," Denniston began, "the text of the desired German message is Line A. This example does not include the clear text key group – a sub-message between cryptographers, like a checksum in accounting. The cryptographer assigned to the unit confirms his box settings, and then keys into his Enigma box the Line 'A' message. His assistant writes down each encrypted letter at it appears in the lights," he said, pointing to the light array above the Enigma box keyboard. "Then, the encrypted message is probably handed to a communicator, who adds the administrative transmission data. Messages are transmitted in five-character groups using Morse code – a series of dots and dashes as we say, or short and long touch strokes – which is represented by Line 'B.'"

"Why five-character groups, out of curiosity?" asked Churchill.

"In simple form, the technique has been used for decades for ease of transmission by the operator. This technique reduces errors and makes the message accounting more straight forward."

"Thank you."

"Certainly. Our listening stations – the Station 'Y' sites – copy down the message groups as sent as well as record the receipt time and direction cut, if they can obtain it. The hard work occurs in breaking down the code settings for each day from Line 1 to Line 2; this is the decryption step. The Line 2 message goes to the German translators who sometimes group the letters, like Line 3, to better see the German words. The translators produce Line 5, which then goes to the analysts. Some are here, but most are at SISHQ. We keep the listeners completely separate from the code breakers, who are in turn separate from the translators and analysts. Very few people are privy to the whole process. Each individual is instructed to discuss nothing about their work outside their specific hut . . . with no one. We have had to dismiss a few even for minor infractions. We take the security here very seriously."

Churchill nodded his head in recognition. "I see Line 5?" he asked, changing the subject.

"Yes, when the raw message is appropriate. Mostly you see the analyst's version, which has the administrative details removed and may include additional information from other sources. The message we discussed a fortnight ago were what we refer to as the analyst's versions. If the Line 5 message, or anything resembling the Line 5 message, is contained in a document, it must be classified MOST SECRET – ULTRA to identify the special handling necessary for an Enigma decrypt. As you well know, yourself . . . MI6 and other cleared individuals often use the code word 'Boniface' to further mask the origins of the information."

"Well, now I know. I have known pieces of this for some time now, but it is much clearer to me and emphasizes the vital need for limited access. We keep a tight hold on that list, and we certainly make sure anyone given access understands they will not likely survive even the slightest compromise of ULTRA."

"As it should be, sir. We are doing our part here to ensure security of what we do here. MI5 pays close attention to all of our personnel, including me."

"This is a very sensitive business and I cannot exaggerate the importance to our war effort. 'C' knows, and you should know, President Roosevelt was informed by me of ULTRA, although he has not yet seen any of the products of ULTRA. I met with him last month. We both feel the American entry into the war is inevitable and only a matter of time . . . and instigation. My point here is, you should prepare for American cryptographers and intelligence specialists joining this effort, at least at the Line 2 and subsequent level."

"Yes sir. We will be ready. I am mindful of your time, Prime Minister. If you have no more questions on the Enigma unit and the process, I would suggest we continue our tour."

"Yes, certainly, thank you for your concern. I should say at this stage, the Americans have a similar decryption campaign on Japanese command communications under the code name MAGIC. President Roosevelt and I intend to exchange cryptographers and analysts to integrate these two programs." Denniston nodded his acknowledgment.

Both men rose. Churchill followed Denniston back downstairs and out the rear entrance of the manor house. They walked slowly along a series of hastily constructed, simple buildings.

"We have built a series of huts to focus the work. In the early days, Hut 1 was for our listeners. As war approached, they were all distributed to multiple listening stations scattered around the kingdom and collectively called 'Y' Station, where the reception details, including the direction to the transmission source are recorded. The 'Y' Station items are received in Hut 1 and logged in here. Essentially, Hut 1 delivers the Line 1 message." They passed another building with a sign Hut 2. Denniston gestured toward the building, as they continued to walk. "This is Hut 2. We call it our Relaxation Hut. We provide tea, beer, sometimes wine, biscuits, sandwiches and such, to give our personnel a break from the rigors of their work. We have more than a few personnel that have been known to work in excess of 24, sometimes 30 plus hours, on a particular problem." The Prime Minister stepped inside, followed by Denniston. Churchill shook hands, said hello, and told the half dozen, surprised personnel inside that he was very proud of their work at Bletchley Park. They continued their tour. "Huts 3, 4 and 5 house our general intelligence analysis personnel. They see only Line 5 data, although such data has a fairly clear implied source. Huts 6, 7 and 8 are our specialized cryptanalysis sections . . . our first line code breakers, using our developed, and perhaps I must say, traditional techniques . . . very hard, frustrating work. Each hut specializes in operating groups. Hut 6 works Army and Air Force codes. Hut 7 works Japanese, SS, SD and intelligence codes. Hut 8 handles the naval messages."

"The U-boats."

"Yes sir. This is where we learn what we can about the German submarine service operations."

"May I go inside?"

"Yes, of course, Prime Minister."

As they did in Hut 2, Churchill greeted the surprised people, shook hands and congratulated everyone on their vital work. Most of them thanked the Prime Minister for his visit. Others could only nod their heads. They returned to their walkabout.

"Over there," Denniston pointed to a noticeably separate building, "is Hut 14, our communications center, to carry out our indigenous communications tasks." Denniston gestured to the closest building. "Hut 10 is 'C's hub, our primary interface with MI6 and Broadway House. The next and last hut I wanted you to see is this one," he said, motioning to the green, clapboard building with a simple square, black on white sign – Hut 11. "This is our future. I wanted you to see this, especially because of what the small team is doing to improve our performance."

"Do tell, Commander."

"Well, let me show you." They entered the building. No people were in the building. It was a larger, single room, more like a large workshop with tools, parts and papers scattered about various tables. A large panel of dials dominated the room. The clicking, whirring and clattering noise was overpowering, as the dials rotated . . . some continuously, others at a slower, intermittent or staggered rate. Denniston motioned for them to step outside. Once the door was closed, again, the sound of the machine was barely perceptible. "My apologies, sir. I expected the team to be present, and I am not sure why the bombe is running."

"Bombe?"

"It is the name we use in honor of our Polish colleagues who developed an earlier version several years ago. They called their device a *bomba kryptologiczna*, if I pronounced it correctly, which translates into cryptologic bomb. There are conflicting explanations for the genesis of the name, but that is what they called their machine, nonetheless." Denniston looked around the area outside the hut, presumably to see if anyone could be listening. "That device is an electro-mechanical computational machine designed by Alan Turing and Gordon Welchman that solves Enigma box settings for any given day."

William Gordon Welchman was a Cambridge-educated mathematician recruited by Denniston before the war began to help solve the problem presented by the German coded messages. Alan Mathison Turing was six years younger than Welchman and followed in a similar path as a Cambridge-educated mathematician also recruited by Denniston before the war as a code breaker. Turing gained some renown with the 12.November.1936, publication of his scholarly paper titled: "On Computable Numbers, with an Application to the *Entscheidungsproblem*" that described the basis for devices he called "computing machines." The two mathematicians along with several other colleagues took the Polish work to produce the Bombe as a deciphering machine.

"We generally run the Bombe in the early morning hours," Denniston continued in a muted voice, "when the Germans transmit their first messages with the day's settings. The Bombe helps us to quickly establish the setting . . . in a matter of hours, compared to days and weeks using our old manual methods." A young man approached. Denniston noticed the man and said, "Ah, there you are Turing. I was wondering where you and your lads were off to."

"We took a break in Hut 2, while the machine was running."

"May I intro . . ."

"I recognize Prime Minister Churchill, Commander," Turing said and extended his hand to the Prime Minister.

Churchill grasped Turing's proffered hand and shook it. "I understand you are quite the *wunderkin*, Mister Turing."

The young mathematician ignored the Prime Minister. "Do you know computational machines?"

"I have heard of such devices from the Prof."

"The Prof?"

"Lord Cherwell, my trusted science advisor . . . we call him the Prof. He has taught me about your paper and advocated for us to use one of your computational machines for organizing and analyzing Battle of the Atlantic data."

"Why is the Bombe running, Turing?" interjected Denniston.

"We thought we had the day's settings as usual this morning, but something happened in late morning. More than a few messages would not decipher. We needed to run at the settings, again. We re-ran this morning's information and a new pass after the change occurred. We are not sure what happened. They did not appear to have changed the settings. Traffic analysis is looking at the background of the affected messages to see what consistency can be found."

"I must say, Mister Turing, we cannot overstate the importance of the work you and your colleagues are doing. Is there anything I can do to help you?"

"Turing . . . ," cautioned Denniston.

Turing verged upon a sneer. "We always need more resources, Prime Minister."

"Turing, please, the Prime Minister has more than enough on his plate," protested Denniston.

"It is quite alright, Commander. Do you have some proposal to present to me?"

"No sir, but I am certain we can generate a few in short order," Turing responded.

"Excellent. I eagerly await my opportunity to contribute."

"Thank you, Prime Minister. Now, I am very conscious of the time and your schedule," Denniston added.

"Great to meet you, young man," Churchill said and shook Turing's hand. "Shall we?" he said to Denniston and did not wait for a response. Commander Denniston jumped to catch up. Churchill waved to the few people he saw, whether they recognized the Prime Minister or not. He moved quickly through the manor house and stopped just outside the main entrance door. He noticed both Detective-Inspector Thompson and Sergeant Carrick extricate themselves from the limousine, before he turned to face Commander Denniston. He thanked the Director, Government Code and Cypher School, for his patience, tutelage, forbearance and hospitality. They were soon off to Ditchley Park to join his wife and two of their three daughters, rather than back to Chequers Court, where he had spent last night virtually alone.

—

Sunday, 7.September.1941
RAF North Weald
Epping, Essex, England
United Kingdom

The debriefing for the disaster of a mission had taken far longer than any single mission since the worst days of the Battle of Britain. Brian's intelligence debriefer, a male corporal he had not seen before, offered no illuminating information. He was anxious to compare observations and impressions of what happened.

The squadron mission had been a large 'Rodeo' fighter sweep of the enemy airfields near Calais. They had been assigned to the ground attack portion of the mission, while two Hurricane squadrons each attacked the other two airfields in the area. No.71 Squadron's mission target was the Calais-Coquelles airfield – a fighter and bomber air base. Four Spitfire squadrons had been assigned to high cover for the ground attack element . . . at least that was the plan. Yet, once they crossed the French coastline, nothing had gone to plan.

Brian joined his brethren, found an open lounge chair and settled in, leaning his head back against the chair. "Where's the Skipper?" he asked.

"He was called to Group," Whitey answered without raising his head or opening his eyes, "before he even finished his debriefing."

"What for?"

"He did not say, but our guess is today's calamity."

"What the hell happened?" asked Dusty.

"We got jumped by fighters at treetop height. Tracers crossed me just before I saw the flash of Bulldog's fireball. I have no idea if he was hit by the Germans, or struck trees or something else, but there is no way he could have survived that."

"We never got a call from the cover guys," added Salt.

"Apparently, the Germans snuck in under our cover," Brian said. "They have been preparing this trap for a while. It was too precise. The cover lads had their hands full, so no wonder they got through unnoticed." Brian paused. "Did we lose anyone else?" he asked.

"Yeah," answered Pete from the other side of Salt. "We also lost Rocket and Red."

"Damn! That was a costly mission," Brian added.

"And, that is probably why Tug was called to Watnall so fast," Whitey said, referring to No.12 Group Headquarters at Watnall, Nottinghamshire.

"What happened to Rocket and Red?" Brian asked.

Salt sat up and turned toward Brian. "When the Germans jumped us, Whitey abandoned the ground attack, turned hard to engage the attackers. I pulled up and rolled over Whitey to cover him. As I pulled through my turn, the Germans were among us. Red collided with one of the German One Oh Nines. Both of them spiraled in and impacted buildings, trees and other ground objects . . . must've hit a refueler on the ground." Salt Morton lowered his voice, "No way he could have survived that either."

"And, Rocket?"

"We're not sure," Whitey answered. "The Skipper took off before we could talk to him. None of the rest of us saw what happened to Rocket. It was a helluva mess down there."

"Did anyone get hits on our target?"

"We don't know that either," said Whitey. "Perhaps Tug did, since he was leading, but none of the rest of us saw a thing, other than Rocket did not rejoin after we disengaged. Corporal Harris got instructions from the Skipper upon arrival to report all three of them missing, no chutes, and presumed dead."

"Well, that is not good." Brian said. "I was lined up on my first target when they jumped us and had to engage. I never fired a shot on the target."

"Nope," added Pete. "Not good!"

"I didn't ask Jimmy when I finished, what is our status?"

"Unavailable," answered Whitey.

"Is that a valid status?" Hunter asked.

"Nope, not to my knowledge," Whitey responded. "It probably means we are not available and not released. My guess, with a loss like we had this morning, Fighter Command wants us to stand down, but they want us to remain here until the Skipper returns. We're down five pilots and no replacements inbound, yet."

The pilots retreated to their thoughts. *How did the Germans get so damn lucky? Have they got some new detection device or procedure? This is going to be a very long war with days like today.* Brian stood and began to walk slowly down the line.

"Don't go far," shouted Whitey, "the Skipper could be back any minute."

Brian looked over his shoulder, nodded his head once, and continued on his walkabout. *Damn, three lost in a single mission.* He stopped to talk to his ground crew. They were working on fixing a coolant leak only discovered upon landing. They asked about the morning's event, and Brian told them in general what had happened. Losses like the day's tragedy were hard on everyone. The crew returned to their task, and Brian continued his saunter. At the end of the line of fighters, he stood motionless, staring into the far tree line and thicket of trees and underbrush. *At least I got some shots off at the bastards. I wonder if the gun camera film will show any hits . . . nothing close to another victory I could tell; but, at least I did not come back with the gun port tapes still intact. How on earth did they get past our covering wing?*

Brian heard the idle protests of a single Merlin engine. He turned to see a single Spitfire landing, but the angle prevented him from seeing the tail designator letters. He started to slowly walk back to the Dispersal Hut. The aircraft had turned and was taxiing, when he finally saw the letters – 'XR-A' – that was the Skipper. Brian picked up his pace, considered whether to jog back, but he still made it back before Meares shutdown and extricated himself from his aircraft. Brian stood by the door and was joined by Corporal Harris.

"Inside gentlemen, if you please," Squadron Leader Meares said and gestured to the door. The remaining six fighter pilots took their seats, as the Skipper hung up his flight equipment. "I am sure everyone has compared notes by now. We lost three precious pilots today. It was not a good day. As you know, I was called to Group, for what I thought would be a dressing down for losing a third of our available pilots. Group Intelligence indicated this was the second such loss within Fighter Command and an assessment is underway to evaluate the changing German tactics and our response to those changes. Anyway, the purpose of my call up to Group was not about today's events. We have two replacements due to report in tomorrow . . . two fresh pilots straight from the OTU. They apparently do not have a lot of operational time, but they are qualified. With today's losses and the newbies arriving, we are back in training status. That said, the big news is, I finally received a decent explanation for what happened to the new aircraft we were supposed to receive last week. At the last minute, they were diverted to Ten Group, by the Aircraft Production Ministry without notification of the Air Ministry . . . very unusual. Nonetheless, I was assured that early next week we will receive the remainder of our new aircraft and finally have a full complement of Spitfire Mark Five 'B's, as we were informed a week and a half ago."

"Oh yeah." "Fuckin' A." "Hear hear," came the spontaneous acclamations.

Meares eventually held up both hands for silence. "I am as excited as the rest of you to get the 'B' wing into combat at parity with the Germans. They have outgunned us for too long. We need to get a feel for the engine improvements, but more importantly, we need to adapt ourselves to the 'B' wing gun combinations. The spade will have a three-position toggle in place of the single firing button. The top will be cannons only. The bottom will be guns only. And . . . the center position fires everything simultaneously, like all hell is breaking loose, from what I am told."

"Oh yeah, we can use some of that!" exclaimed Sweet Sweeny.

"Indeed! We have a lot to adjust to with these new machines, so everything taken together, a week or so of training seems quite appropriate. We are released for the day. I was promised half the aircraft would arrive in the morning and the other half in the afternoon. Let us gather at the Mess and toss back a pint or two to celebrate our lost lads. We will reconvene at eight straight up tomorrow. Dismissed. I shall be along shortly."

The pilots decided to walk to the Officer's Mess on the late summer's day. They were probably going to consume more than a pint or two on this night. The mixture of good and bad news was simply too much to be sober much longer. Before he took his first swallow, Brian thought, *I'd better call Charlotte before I get tanked. She might hear about today's . . . event. I'll call after this first beer.*

—

Tuesday, 9.September.1941
RAF Biggin Hill
Biggin Hill, Bromley, London, England
United Kingdom

The mission had gone surprisingly well for No.609 Squadron. There was something rewarding about expending their ammunition into all manner of enemy equipment on the ground from locomotives and railcars, to fuel tanks and even an ammunition storage area that nearly clipped the wings of several squadron mates.

Jonathan felt the same sense of accomplishment as the rest of his mates. He was one of the first pilots to complete his intelligence debriefing. The overcast and light breeze offered a refreshing respite to the uncomfortably warm interior and the sweat of intensity in their strafing mission. His Right Wing joined him.

Pilot Officer Victor Hanson 'Vicky' Clegg of Lancashire sat in the chair next to his leader. "That was fun."

"Don't get to thinking that is normal."

"It's not?"

"In fact, it is quite rare that we get away with expending our entire capacity of ammunition without being shot at."

Vicky opened an envelope he apparently had been handed by Corporal Warren after completing his debriefing. He read the single sheet message. "Wow!" he exclaimed. "Read this," he added and handed the paper to Harness.

```
                    High Flight
       Oh! I have slipped the surly bonds of Earth
       And danced the skies on laughter-silvered wings;
       Sunward I've climbed, and joined the tumbling mirth
       Of sun-split clouds--and done a hundred things
       You have not dreamed of--wheeled and soared and swung
```

High in the sunlit silence. Hov'ring there,
I've chased the shouting wind along, and flung
My eager craft through footless halls of air.
Up, up the long, delirious burning blue
I've topped the wind-swept heights with easy grace
Where never lark, or ever eagle flew —
And, while with silent, lifting mind I've trod
The high untrespassed sanctity of space,
Put out my hand, and touched the face of God.
                Pilot Officer John Gillespie Magee, Jr., RCAF;
                        No.412 Squadron, RAF Digby

"That was sent to me by a classmate of mine in the Training Unit—Canadian fellow," Clegg said.

"He seems to have captured the emotions of flight rather well, if you ask me."

"I'll pass it around."

"Maybe we should ask Corporal Warren to type us up a few copies. I'd like to send a copy to Hunter and Red."

Flight Lieutenant Robert Gates 'Sparky' Morrow walked past and must have heard Jonathan's statement. "You apparently missed the latest casualty report. Red bought it two days ago on a mission into France. Seventy-One Squadron lost three pilots that day, Red being one."

"Hunter?" asked Harness with hesitation.

"Not on the list. In addition to Red, they lost two other Americans, Tolly and Downing, as I recall."

"That leaves Hunter as the only American survivor remaining from our squadron."

"Indeed," Morrow added, as he settled into an outdoor lounge chair.

Jonathan asked for and received the poem. He took it inside to request Corporal Warren's assistance with the administrative task of making a half dozen copies or so. The poem was too good not to share. The poem gave him an easy excuse to share with Brian and ask him what happened. Hopefully, he would also ascertain how his friend was doing.

Friday, 12.September.1941
No.10 Downing Street
Whitehall, London, England
United Kingdom
04:35 hours

Churchill sat motionless and silent for several minutes after the radio was switched off. His only companion for President Roosevelt's latest fireside chat was his principal military adviser beyond the chiefs of staff – General Pug Ismay. The President's worldwide broadcast lasted just 30 minutes. Ismay stood . . . undoubtedly thinking the day was finally done, as dawn was approaching. Churchill looked up, as he saw the general move.

"What did you think of the President's speech?" the Prime Minister asked.

Ismay sat back down. "It sounded like he was presenting his rationale for declaring war. We know of most of the incidents he cited, but a few were new to me. Are the Americans going to declare war against the Germans?"

"I think not . . . at least not yet. He is preparing the ground, yes, as we discussed and agreed last month at Placentia Bay. We both believe the U.S. entry into this war is inevitable and only a matter of time. This latest incident with the destroyer Greer is exactly what we expected. The Germans are doing their part to bring the Americans in."

"He spoke very clearly to his people, to us . . . and to the Germans."

"Yes, he did . . . quite clearly . . . 'rattlesnakes of the Atlantic,' indeed."

"An apropos analogy."

"That it was . . . that it was. Freedom of the seas and commerce is precisely the correct argument." The Prime Minister paused to think, clearly not done with the conversation. Ismay did not move and remained motionless. "Pug, I know I am not putting you in a comfortable position, but I'm afraid I must." Ismay nodded his head in acknowledgment rather than consent. "We are going to move Sir John Dill."

"I have suspected as much for several weeks now. He is a very good man . . . a fine and capable general. "

Churchill nodded his head. "Yes, he is, but he is not the chief we need."

"He could retire."

"Yes, he could, but I shall do my best to retain him on active service. He is too valuable to lose to retirement. I wanted your opinion for an appropriate assignment for Sir John."

"You sent Sir Hugh Dowding to America. Perhaps that is a worthy assignment."

"That is my opinion precisely, Pug. Sir Hugh will return home in a month or two. I see Sir John as our chief of mission, our primary coordinator with the American chiefs of staff. He is a great administrator, but he is not the combat leader we need, as we prepare to take the fight to the Germans."

"If I may be so bold, sir, who do you see replacing Sir John as CIGS?"

"I am leaning toward General Brooke, but I am not convinced my colleagues on the War Cabinet will agree."

Lieutenant General Sir Alan Francis Brooke, KCB, DSO & Bar, had assumed the position as Commander-in-Chief, Home Forces in July 1940. During the difficult and tumultuous weeks during the summer of 1940, when the United Kingdom was under direct threat of invasion by the Germans, Churchill interacted repeatedly with Brooke as they struggled to prepare the defenses against the impending invasion.

"Also a good man."

"Yes, he has done an exemplary job with Home Forces."

"Yes sir."

"I do not need to say this, Pug, but this was a private and personal discussion between us. I do not want any of this repeated."

"Yes sir. My lips are sealed."

"Thank you, Pug. You are a worthy counsellor, and I appreciate your perspective on things. Now, it is late. We have a full day ahead. Have a good night's sleep, or as much as you can achieve. Also, thank you for listening to the President's speech with me."

"You are quite welcome, Prime Minister. Have a good . . . what is left of the night."

"I will. I shall retire as well and knock off some of the Dispatch Box before sleep claims my consciousness. I will see you later this morning."

"Yes sir. Good night." Ismay stood and departed.

Churchill rose as well, switched off the lights, and walked to the stairs and the master bedroom. Winston smiled to himself. *Yes, that was a very encouraging speech.*

Friday, 12.September.1941
Oval Office
The White House
Washington, District of Columbia
United States of America
17:15 hours

Harry Hopkins entered the Oval Office for the umpteenth time on this particularly busy day. He knew the President was working his way through the incessant and neglected stream of paper—reports, proposals, recommendations, requests and even simple communications between friends. Roosevelt looked up from his reading material.

"We received confirmation from Henry Stimson that they broke ground on the new headquarters building in Arlington. I think they are calling the project the Pentagon for the design."

"None too soon," Roosevelt responded with fatigue in his voice. "Those wooden buildings on The Mall are just not going to cut it with the expansion that is coming."

"I think everyone is agreed on that part. The Army Corps of Engineers has assigned a hard-charging West Pointer to supervise the construction—Colonel Groves as I recall."

Colonel Leslie Richard 'Dick' Groves Jr., USA CE [USMA 1918] had a well-deserved reputation as a no-holds-barred prosecutor of engineering building programs. The massive Pentagon construction project was simply the latest in this portfolio. The recently approved plan called for the five-sided, multi-story headquarters building for the Departments of the Army and Navy to be completed in two years. No one believes such an ambitious plan could be done in that amount of time. However, some in the Corps of Engineers instinctively knew that if it could be done, Groves was the man who could deliver.

"I don't know him, but Henry and George seem to think he can do it."

"That is my understanding as well. Henry believes they will have one wing done and ready for occupancy next year. His office and staff will be the first to move in, along with the Chief of Staff and his staff."

"What are they going to do with the Munitions and Main Navy buildings?"

"As I have been informed, they intend to use those buildings as annex facilities to cover the personnel expansions. I've not seen a detailed plan, yet, but I know they are working on it."

"So, we will have those ugly buildings around for a while?"

"For the foreseeable future, I'm afraid. Also, Frank Knox called to inform us that convoy escort operations have been extended to the Icelandic meridian and have taken full control of Iceland's defense. The last of the British departed the island earlier this week and were redeployed."

"I'm sure Winston will be pleased."

"Certainly, although he have not heard from him, as yet."

"We will."

"Frank indicated they expect to have the first mid-Atlantic handoff during an eastbound convoy later this month."

"And then, we wait for the Germans to take the bait." Harry nodded his head and chose not to speak. "I have not seen anything else from Winston on the Russian support effort."

"I believe the British are working through Averell Harriman now. They made their first Arctic convoy run from Iceland to Archangel late last month, with another scheduled for later this month."

"No one will ever accuse Winston of not being bold," Roosevelt observed. "The challenge of meeting the British and Russia supply demand is daunting at a minimum. Didn't the Foreign Office inform us of British armed forces now operating in Russia in support of the Murmans?"

"That's correct. Lord Halifax notified us that a fighter wing of four Hurricane squadrons along with support personnel was successfully deployed to Murmansk three days ago and expects to join combat operations in support of the Soviet defense of Leningrad within a week."

"That must be a measure of their confidence in their own defense as well as the Red Army."

"It would sure seem so, and, they . . ." The knock at the door stopped his thought. Grace opened the large door. "Excuse me, Mister President. Mr. Hopkins, if you please . . ."

Roosevelt nodded his head and returned to his reading.

Hopkins stepped out of the office for a few minutes and returned. He closed the door behind him. "Bill Donovan is available to brief you on the situation in Russia," he said.

"No. I'm tired, Harry," Roosevelt announced. "Please ask him to see me tomorrow morning. I would appreciate a thorough briefing on what we know."

"Very well, I'll set it up," Harry said and turned to take care of the task.

"Wait, before you go," Franklin paused for Harry to stop and turn to face him, "I want to visit Missy in the hospital tomorrow at say 11:30, so if Bill can complete his briefing by around 10:45, that would work. I'd rather do the intelligence before visiting Missy, and don't schedule anything after my visit with her."

"She is stable, but still quite diminished," Hopkins said.

"There is always hope, Harry."

"Yes sir. I will see to the arrangements."

"Very well. Also, please inform Grace I am done for the day. I want to leave for the residence, now."

"Allow me to get her started on the arrangements for tomorrow, and I'll be right back to escort you upstairs," he said, referring to the location rather than the means of transport.

President Roosevelt organized the papers on his desk, while Hopkins talked to Grace Tully. Within minutes, Hopkins returned to push the President's wheelchair out of the Oval Office and West Wing to the Main House elevator to lift them to the second floor residence. The electric elevator had been operating in the White House since 1898 and the McKinley administration. The electric version had replaced the first hydraulic elevator installed during the Chester Arthur administration in 1881, after the assassination of President Garfield. The electric elevator had been upgraded several times by the time Franklin Roosevelt needed it.

—

Chapter 13

Honest, unaffected distrust of the power of man
is the surest sign of intelligence.
-- Georg Christoph Lichtenberg

Saturday, 20.September.1941
Chequers Court
Ellesborough, Buckinghamshire, England
United Kingdom
12:30 hours

Churchill entered the expansive study/library of the Prime Minister's country residence, dressed in only his bedroom slippers and robe. Business suit bedecked, Colonel Stewart 'C' Menzies stood patiently by the desk. When Jock Colville announced the unscheduled arrival of Colonel Menzies, Winston had chosen not to take the time to dress, since he knew the head of MI6 usually only carried the "Buff Box" for something important.

The two men did not speak. Churchill retrieved his key from around his neck and opened the cover on the beige case manacled to 'C's left wrist. He removed three pieces of paper from the 'Boniface' folder and read the top paper.

MOST SECRET - ULTRA

```
SECRET
DATE: 17 SEPTEMBER 1941
TO: ARMY HEADQUARTERS
FROM: COMMANDER ARMY GROUP CENTER
BREAK
REPLACEMENT FORCES IN PLACE BREAK EXPECT TO
COMMENSE FINAL ASSAULT ON PRIMARY OBJECTIVE NO
LATER THAN 2 OCTOBER BREAK COORDINATION WITH
GROUPS NORTH AND SOUTH COMPLETE BREAK ALL IN
READINESS HAIL HITLER
END
SECRET
```

MOST SECRET - ULTRA

"Have you notified the Military Mission in Moscow?" Churchill asked.

"No sir. How much can we tell them and what should we pass along to the Soviets?"

"You must mask the source appropriately, as we have done before. I am still quite mindful of that April 17th message that the Soviets and Germans know more than they should. Reliable sources, plural, should suffice. What do we know about Russian preparations for the attack?"

"What we know has come from our Military Mission and the Embassy. Our personnel have witnessed fresh troops – infantry, artillery and armor – arriving by rail from the east and marching west. Morale of these new troops appears to be high. Reports from usually reliable sources suggest the government has prepared to evacuate the capital city. However, NKGB and NKVD units are not allowing withdrawal or refugee departures. Those reports also indicate summary street executions for violators."

"They are going to defend the city," the Prime Minister said with a tone more questioning than statement.

Menzies nodded his head. "That is our assessment, sir."

"Center, that is Field Marshal von Bock, is it not?" asked the Prime Minister.

"Yes sir, it is."

"Very capable field officer by my knowledge."

"Ours as well, sir."

"What delayed their assault? Winter is rapidly approaching for them."

"We have fragmentary information, even from Boniface, but our analysis suggests Hitler himself pulled several corps from von Bock's Group Center to support Group North's assault on Leningrad."

"He has those forces back, now?"

"Apparently, at least that is our interpretation of the Group Center message from von Bock himself."

"Perhaps the infamous Russian winter will stall their attack, as it consumed Napoleon."

"They are not to winter, yet. The Met Office forecasts their first sustained snowfall in mid to late October. If the Germans are not prepared for such winter operations, they will learn a very hard lesson, as the French did over a century ago."

"What happens if the Germans take Moscow?"

"Our opinion depends upon whether the Soviet government and security infrastructure remain intact during their withdrawal. If they stand and fight, and they are defeated, then Russia will fall. If they can reconstitute the government in the Urals, then their capacity to continue resistance could be maintained.

If the Murmansk-Moscow supply line is broken, we will have few comparable logistical lines. The possibilities diverge from there."

"Humm," Churchill muttered, lapsed into thought for several seconds, and then he turned his attention to the next message.

MOST SECRET - ULTRA

```
SECRET
DATE: 17 SEPTEMBER 1941
TO: ARMY HEADQUARTERS
FROM: COMMANDER ARMY GROUP SOUTH
BREAK
EXPECT TO ENTER KIEV NEXT WEEK BREAK
INDICATIONS SUGGEST RED ARMY MAY HAVE ABANDONED
CITY BREAK SPECIAL ACTION GROUP ASSIGNED
IN SUPPORT THIS COMMAND REPORTS SURPRISING
COOPERATION BY UKRAINIAN POPULACE BREAK GROUP
SOUTH OPERATIONS WILL BE COORDINATED WITH GROUP
CENTER OFFENSIVE BREAK EXPECT TO MAKE MARISPOL
AND CUT OFF CRIMEA WITHIN TWO WEEKS
END
SECRET
```

MOST SECRET - ULTRA

"Now, this one, Group South, is Rundstedt – Field Marshal Gerd von Rundstedt, correct?"

"Yes sir."

"There is that name, again – Special Action Group."

"Quite correct . . . a trained special SS group assigned to each Army Group area of operations. These are the killing squads we identified last month. These groups have been very busy in the Baltic States already, and now they are working their dastardly deeds in Ukraine . . . and apparently with Ukrainian complicity. The regular army appears to be turning a blind eye to the work of these special action groups."

"Or, maybe, the army has its hands full with combat operations and no capacity or stomach to take on the political challenge." Churchill paused to think. "We need to keep a close eye on those special action groups, as best we can. Now, what do you think the objective is for Group South? Clearly, the objective for Group Center is Moscow."

"We do not know precisely, but both the Germany and Russia desks agree, Group South is pointed to the Baku/Caspian oil fields."

"What is your prognosis, if the Germans achieve their objectives?"

"Well, that is the penultimate question, isn't it? If the Germans take Leningrad, Moscow and Baku, they will likely consolidate their holdings. There are no indications that Hitler has interests beyond the Ural Mountains . . . at least none we have detected in his writings or speeches. The Caucuses, on the other hand, could be just a waypoint. Once they solidify their hold on European Russia, they will have substantial resources to sustain their territory and war machine. It does not take much imagination to see an advance on the Arabian Gulf oil fields of Persia, the Mandate territories and even the Arabian Peninsula. Which brings us to the third message."

Churchill put the second message at the back of the short stack and read the third message.

MOST SECRET - ULTRA

```
LSECRET
DATE: 19 SEPTEMBER 1941
TO: ARMY LOGISTICS COMMAND
FROM: ARMOUR GROUP AFRICA
BREAK
URGENTLY NEED MOST ACCURATE MAPS OF TOBRUK AND
VICINITY TO SUPPORT SPECIAL OPERATION PENDING
BREAK GREATEST DETAIL VALUABLE BREAK SEND BY
MOST EXPEDITIOUS MEANS AVAILABLE HAIL HITLER
END
SECRET
```

MOST SECRET - ULTRA

"So, Rommel's vaunted *Akrikacorps* is up to something in Tobruk. We must hold Egypt!" the Prime Minister exclaimed. "Without Egypt and the Suez, the whole Middle East could fall."

"Hitler and perhaps even his General Staff expected to have Russia subdued by now," Menzies said. "Resistance by the Red Army is stiffening.

BARBAROSSA has already proven to be more difficult than the Germans expected. While the Russians are still withdrawing and have shown no sign of stopping the German advance, they are bleeding the Germans. These gains for Hitler are not coming cheaply. Let us assume the worst-case scenario; the Germans take Leningrad, Moscow and Baku before winter sets in. They will need at least six months, and more likely a year or more, to refit, rearm and resupply their armed forces. We have seen no references beyond BARBAROSSA."

"Yes, well, that foul bastard seems to be feeding on the blood of others."

"Yes, and we remain a clear threat to his hegemony. If the Soviets are subdued, he must eliminate the threat we represent, and his offensive to cripple our interests and our ability to wage war will quite likely be a two-pronged attack on Suez and Egypt. If he defeats us there, he will control the Mediterranean and have hold of our jugular."

Churchill grunted and started pacing as he thought. He looked at the third message, again. "Something is afoot at Tobruk."

"Yes, but we do not know exactly what. I have taken the liberty to inform Middle East Command and specifically General Wavell. We will keep looking for details and clues."

"Very well. Thank you, 'C.'" Churchill paused. "As an interesting and perhaps intriguing side note, I cannot fail to notice the absence of the mandatory 'Heil Hitler' from the close of von Runstedt's message. Is he tempting fate?"

"Hard to say, but we have detected no reaction, and this is not his first time at omission."

"One more item, the joint Anglo-American military team to the Red Army agreed to at Placentia Bay is due to depart tomorrow for Scotland, Scapa Flow and Moscow via Archangel. They cannot see these messages or know the source, but they should have the essence of the Russia content, properly masked."

"Yes sir. I shall see to it upon my return to London."

"Can you stay for lunch, Stewart?"

"No, thank you, sir. If you will excuse me, I would like to get back to London."

"Very well," the Prime Minister responded. He returned the messages to the proper folder and locked the case. "Thank you for bringing these," he said patting the Buff Box. "Safe journey."

The Director, Secret Intelligence Service, turned and departed the Prime Minister's study and country residence.

Churchill stared out the window at the meticulously manicured garden. "They must not succeed," he said aloud to no one. A knock at the door broke his thoughts.

His valet Sawyers looked in. "Would you like to dress before lunch, Prime Minister?"

"It is just me for lunch, today, isn't it?"

"Yes sir."

"Then, no. Is lunch ready?"

"Shortly, sir."

"Sawyers, please find Colville and ask him to join me for lunch."

"Very well, sir."

Churchill looked around the room for nothing in particular, and then made his way to the dining room.

———

Saturday, 20.September.1941
Waterloo Station
Lambeth, London, England
United Kingdom
18:30 hours

Dressed in a fresh, properly festooned uniform, Brian had been waiting on the platform for not quite 20 minutes, when Charlotte arrived. Her elegant, medium blue, dress-suit with small, conservative matching hat gave her a distinguished, verging on elegant appearance, and carrying Ian swaddled in her left arm and sleeping. She had a small travel case in her right hand that she literally dropped. Brian marveled at how she looked—angelic was the word that came to mind. The broad smile on his face broadcast his appreciation. She had smiled in return, and they had walked toward each other, embraced and kissed, to several cheers from passing travelers.

Three-month-old Ian complicated travel for Charlotte. Brian took the peaceful child from his mother's arms, cradled him in his arms, and smelled and kissed his forehead.

"How was the journey with the baby?" asked Brian with genuine curiosity. After all, to his knowledge, it was Charlotte's first rail travel with Baby Ian.

"Much better than I expected," Charlotte answered. "Edith has been a dream come true at helping me with Ian. She instinctively knows when it is time for her to take him and most of all she allows me to sleep. Frankly, I was more concerned about not having Edith with me."

"You are absolutely gorgeous."

Charlotte stopped and faced her husband. She smiled broadly. "Thank you, my darling."

"I could just stand here and stare at you all afternoon."

"That will make me feel a bit awkward."

"Well, I do not want that."

"And, we have an engagement to attend and we best not be late."

Brian picked up the case. Charlotte placed her right hand underneath Ian, and grasped Brian's elbow as they walked to the Underground railway system at Waterloo Station. The Drummond family boarded the Jubilee Line, switching to the westbound Piccadilly Line at Green Park Station, and arriving at Sudbury Hill Station 30 minutes before their appointed appearance time. The late summer evening, two days prior to the autumnal equinox, remained overcast but unusually mild and dry. They decided to walk up the hill to the Spencer residence. Brian had respected Charlotte's contemplative mood both on the Underground and as they had walked up the small hill toward the Spencer's residence.

—

Saturday, 20.September.1941
No.417 Sudbury Hill
Harrow, London, England
United Kingdom
19:30 hours

Brian stopped, facing the large, black, heavy looking door with large, polished brass numerals '417' above the equally polished brass knocker, still holding Charlotte's warm, soft hand. He faced his wife without releasing her hand. "Are you ready for this?"

Charlotte actually smiled, nodded her head, and said, "As ready as I will ever be. Shall we?" she asked, gesturing toward the Spencer's front door.

They walked up to the door. Charlotte grasped the knocker and firmly knocked twice. They did not have to wait long. The door opened. Air Commodore Spencer was in his uniform with quite a few more ribbons than Brian had. "Welcome to Sudbury Hill, Charlotte," he said and gently kissed the back of her proffered hand.

"Thank you for inviting us, sir."

"Please, for both of you, this is a social evening, not official. Let us agree on our familiar given names. I prefer John."

"Very well, John. Thank you."

John nodded his head, stood aside and gestured for the Drummonds to enter. "Great to see you, again, Brian."

"Likewise, sir." Brian saw John's frown. "Sorry. Thank you for inviting us to your home."

John smiled and nodded his head. He took their hats and hung them on open pegs of a wall mounted hat rack. John led them into the living room, where they found Mary breastfeeding five-month-old Malcolm. She was wearing a nice pair of grey wool slacks and a light pink blouse that was open to facilitate Malcolm's feeding. She had no cover for modesty.

"Great to see you both, again," Mary said with a smile. "Welcome to our home. Please excuse me, it is nearly Malcolm's bedtime and I am trying to fill his belly."

"Quite all right, Mary," Charlotte responded. "I may start leaking myself. It has been several hours since I last fed Ian."

"Perhaps Malcolm can help, but he's nearly done . . . slowing down quite a bit."

"Thank you, Mary."

"Would you care for a cocktail?" John asked his guests.

"Seltzer water, if you have it, please, John," answered Charlotte.

"I'll have what she is having, dear," Mary added.

"Brian?"

"A pint of bitter would be fine by me."

"Very well, then, seltzer for the ladies and a pint for our hero."

Hero . . . what the hell is that?

Malcolm apparently finished and detached on his own. He was sound asleep. Charlotte moved to Mary and took the baby, cradling Malcolm in her left arm.

Ian began to move and make the first sounds of what would soon become a cry.

Mary reattached her brassiere cup, buttoned her blouse, and then stood to retrieve her son. "I'll put him to bed and be right back. Come with me, Charlotte. I'll show you your room. We've prepared a crib for Ian." Charlotte lifted Ian from Brian's arms and followed Mary.

As Mary and Charlotte left the living room, John returned with a tray of drinks – sparkling water for the ladies and dark beer for the gentlemen.

John turned to Brian. "I suppose the ladies had baby business to tend to."

"Yes."

"How is the squadron recovering from that dreadful day a fortnight ago?" John asked Brian.

"It was a rough few days, but we received two replacement pilots, already inbound, within days and the remaining three we were short, earlier this week. Considering . . . I would say we are doing rather well, thank you."

"Excellent. I read through the after-action reports with keen interest. It was very unusual for us to lose three pilots like that. How do you like your new Mark Fives?"

"Helluva machine, I must say. We practically giggle with excitement with the added power, speed, high altitude performance, and of course the 'B' wing cannons . . ."

An older woman, clearly dressed in a cook's working outfit entered the living room. "Dinner is ready to be served, Mister Spencer," the cook announced.

"Very well, Harriett," John responded. "The ladies are tending the babies, so we should wait for their return."

"Very well," Harriett Peterman answered. "Should I send Miss Perkins up?"

"That would be quite nice, even if for nothing more than to meet our other mother, Charlotte Drummond."

"Very well. I'll send her up straight away."

"You were going to tell me about the 'B' wing 20 millimeter cannons," John said to Brian, once Harriett disappeared.

"I've only used them on ground attack so far, but judging by how they chew things up, I suspect they will be a godsend in aerial combat."

"That is what they are designed for. We were concerned early on about mixing machine guns and cannons."

"Not a problem. We adapted to the firing switch quite easily. I've heard no complaints from any of the guys and certainly none from me."

The ladies returned sans children and were giggling, as if they had shared some inside joke. They went directly to the well-appointed dinning table. The meal was indeed a simple but deliciously prepared meal – pork chops in a wine sauce, mushy peas in the British tradition, and scalloped potatoes, served with a very nice French chardonnay. Their talk meticulously avoided the war in any form or its consequences to their lives. The topics included the nuances of the Harrow community, Standing Oak Farm, and mostly the boys. Laughter and lighthearted joking, common to enduring friends, complemented their conversations.

The remainder of the evening, along with two more bottles of chilled champagne, was occupied with thoughts of children and the effects on their lives. It was after midnight when they retired. Neither John nor Brian had duty the following day, so they would sleep in and have a nice country breakfast in the morning.

—

Monday, 22.September.1941
No.10 Downing Street
Whitehall, London, England
United Kingdom
17:45 hours

General Ismay entered the Prime Minister's office unannounced. "Excuse me, Prime Minister."

"Yes, Pug. What is it?"

"I've collated various bits of information from the service departments that I think you will appreciate."

Churchill looked up from his papers and gestured impatiently for his senior military aide to proceed.

"One Fifty-One Wing saw its first aerial combat action two days ago and shot down three One Oh Nines near Leningrad and damaged a dozen others. That had to be a surprise to Germans seeing British Hurricanes defending a Russian city."

"Plenty of Germans there—a rich hunting ground. They've nearly surrounded the city and cut off all of their logistic lines. The city is under siege."

"Quite right, so I imagine our fighters over there are having an impact."

"Pug, four squadrons of Hurricanes are hardly a drop in the bucket on the Russian scale."

"Well, then, perhaps they are of more symbolic value than military significance."

"They are a visible contribution, which at this stage is important. What else have you?" Winston asked impatiently.

"The Joint Anglo-American Mission to Russia has departed by ship. They are expected to arrive in Murmansk by the end of this week. They should make Moscow next week."

"Hopefully ahead of the Germans," Churchill mumbled. Ismay did not react. "And, the Germans have taken Kiev."

"Yes. MI6 confirmed that fact this morning. They took the city on the 19th and quickly consolidated their gain. Signs continue to mount that the Ukrainians are helping the Germans . . . apparently see them as liberators from the oppression of the Russians."

"That is certainly an active dynamic in the surrounding countries. No love lost on the Russians, or more properly the Soviets."

"I think Russians would suffice . . . long history of animosity, there. The German's Army Group South literally passed through Kiev and left control of the city to the SS. They have moved so fast of late that they are on the verge of overrunning another Soviet army."

"They have got to find the means to stop the damn Germans," Churchill said with some force. "The Joint Mission must convince Stalin that help is on the way."

"How much help can we afford to offer?" asked Ismay.

"My flippant answer would be, whatever it takes to sustain them and keep them in the fight. They are tying down several scores of German armor and infantry divisions. Those are forces that cannot be used against us. Realistically however, we are walking a very fine line in what we can spare to give to the Russians."

"How will we know the proper amount?"

Churchill smiled. "When victory is achieved."

"That is a long way off."

"Indeed, it is Pug . . . a very long way off . . . with many disappointments to suffer until the day arrives, as it most assuredly will arrive."

General Ismay nodded this head in acknowledgment and departed the office, leaving the Prime Minister to return to his endless paperwork.

—

Wednesday, 24.September.1941
Office of Naval Intelligence (N-2)
Main Navy Building
Constitution Avenue & 18ᵗʰ Street Northwest
Washington, District of Columbia
United States of America

Newly appointed Director of Naval Intelligence Captain Alan Kirk was still adjusting after returning from his assignment in England and sat in his office in anxious anticipation. The special and highly classified MAGIC cryptography office asked for the unscheduled meeting, which meant they had something important. There were no other clues as to the topic, however a MAGIC-implied topic had something to do with the Japanese and the Pacific Theater. Kirk did not respond to the double knock on the door. He knew the courier would be opening the door shortly, and he was correct.

A Navy lieutenant dressed in khaki working uniform entered the director's office. He closed the door behind him. "Sir," he spoke softly, as he approached the director's desk, "we deciphered a Japanese Foreign Office message using the Green code," he checked his wrist watch, "ten minutes ago. I think you will see the urgency."

Kirk nodded and extended his right hand. The lieutenant placed his brief case on the director's desk, unlocked it, retrieved a folder with slanted red border stripes, and handed it to the director.

TOP SECRET - MAGIC

```
TOP SECRET
MSG DATE: 22 SEPTEMBER 1941
TO: CONSULATE HONOLULU COMMERCIAL DESK
FROM: FOREIGN OFFICE LIAISON COORDINATOR
SUBJ: REQUEST FOR INFORMATION
BEGIN
THIS OFFICE REQUESTS DETAILED INFORMATION ON
UNITED STATES NAVAL STATION PEARL HARBOR HAWAII
AND ASSOCIATED AREAS BREAK THE FOLLOWING AREAS
OF INTEREST SHOULD BE REFERRED TO BY LETTER
DESIGNATOR ONLY
A FORD ISLAND MOORINGS AND HARBOR HYDROGRAPHIC
SURVEY IF AVAILABLE
B SUBMARINE BASE
C AIR BASE SUPPORT FORD ISLAND HICKAM WHEELER
KANEHOE
D PETROLEUM AND AMMUNITION STORAGE
E HARBOR ENTRANCE DEFENSES
INFORMATION REQUIRED PROMPTLY BREAK PERIODIC
UPDATES TO THIS OFFICE REQUIRED SHORT OF
VIOLATING LOCAL LAW BREAK CARE SHOULD BE TAKEN
TO AVOID DETECTION
END
TOP SECRET
DECRYPT DTG: OP20GZ 241602Z SEP 41
```

TOP SECRET - MAGIC

"I'll be damned," pronounced Kirk. "Well done." Kirk looked up from the pink paper. "I have not seen you before, lieutenant. Who are you and where are you assigned?"

"Lieutenant Jeremy Ryan, sir. I am the duty crypto officer in the MAGIC room."

MAGIC was a highly classified, compartmented, restricted access program involved in breaking Japanese encrypted messages. The Japanese employed a series of code forms that were given color names to identify the specific cryptologic code set. The United States had been routinely deciphering Red and Blue codes since the early 1920's, and Green diplomatic code since the mid 1930's. After the Tripartite Pact and the opening of the war in Europe, the Japanese began using a sophisticated encryption-decryption device the Americans called Purple. At the time, the Americans did not know the Purple device was built on the design and process of the German Enigma device.

"This is the most definitive information we have to date of Japanese intentions," said Kirk.

"I must caution against jumping to conclusions, sir. They are clearly interested in our naval and military resources in Hawaii, but there is no mention of offensive military operations or intentions."

"This is a target list," protested Kirk, "or at least the makings of a target list."

"Every warship in Pearl Harbor is a target by definition, sir. This message changes nothing. There could be many reasons for such interest. We need and are certainly looking for corroborating information. We obviously thought this message was important enough to notify you promptly."

"Who has seen this?"

"Other than the cryptanalyst who completed the decipher and myself, only you, sir."

"We need to notify Admiral Kimmel and General Short."

"Sir, I'm obligated to remind you that neither Admiral Kimmel or General Short are on the MAGIC access list."

"Yes, right." Kirk handed the message back to Ryan, who in turn took the message, returned it to and locked the brief case. "Will there be anything else, sir?"

"No, Lieutenant Ryan. Thank you. That will be all." Ryan departed.

Kirk swiveled his chair to look out the window and the rain continuing to fall. *They have to be warned. Ryan is correct. The message was not definitive. Although coupled with other clues, it certainly takes on an ominous tone.* Kirk knew he could not refer to the deciphered Japanese diplomatic message, but they had to do something. The correct person to talk to on the Navy staff was the head of the War Plans Division.

Unfortunately, the Director, War Plans Division – at the moment, was Rear Admiral Richmond Kelly 'Kelly' Turner, USN [USNA 1908] – insisted upon all intelligence material intended for the fleet had to be filtered and sanctioned by War Plans, to ensure current operational plans were consistent with the intelligence being sent to the fleet. As a consequence, Turner had made it his personal mission to oppose all of the changes Kirk proposed to reform the Office of Naval Intelligence, to improve the efficiency and productivity along the lines of Admiral Jumper Pike's Naval Intelligence Department of the Royal Navy. Kirk's stint as naval attaché in London for the two years prior to his assignment to ONI had given him a stark perspective on the traditional intelligence bureaus. Roosevelt's appointment of Colonel Bill Donovan as this new Coordinator of Information three months earlier had amplified the need to upgrade ONI, or potentially be left behind by Donovan's innovation in intelligence operations. He was already picking up clandestine field operations.

This is just too important.

Kirk left his office and made his way through the maze of the building to the Operations wing of the building and the War Plans Division. The name and title in gold leaf on the door – RAdm R.K. Turner – Director – marked his destination. Kirk entered the outer office.

"Good afternoon, Captain Kirk," announced his secretary in an exceptionally cheery voice.

"Good afternoon to you, Jane. Is the Director available?"

"Yes sir. He is wading through the incessant flow of paper."

Kirk nodded his head, knocked on the inner office door and entered. "Good afternoon, Admiral, sorry to interrupt your favorite pastime."

Turner laughed. "Quite alright, Alan. What can I do for you?"

Kirk sat in a sturdy wooden chair across the desk from Kelly Turner. "What plans do we have for a major fleet action with the Japanese in the vicinity of Hawaii?"

Turner laughed, again. "Do you know something I don't?"

"I'm just curious, sir. What if the Japanese made a major fleet attack on Pearl Harbor?"

"Alan, this is not more conspiracy nonsense, is it?"

"Humpf. Do we have defense plans for Pearl Harbor?"

"Yes. Admiral Kimmel has re-positioned most of the Pacific Fleet to Pearl, as a show of force, to discourage further Japanese expansion eastward. Now, I have answered your question, Alan; so, you should answer mine."

"Which is?"

"What have you learned that we don't know?"

"We have indications from reliable sources that the Japanese may be developing information useful for targeting purposes . . . as if they are planning an attack . . . specifically at Pearl Harbor and the Oahu military facilities. The oil embargo has to be hurting them; however, they are not likely to bend to the President's will. They are quite capable of striking out with a sizable fleet at the largest naval opposition force in the Pacific region. In my opinion, we should notify Admiral Kimmel and General Short to prepare them for the potential of major fleet action in Hawaii territorial waters."

"Are you serious, Alan?"

"Yes."

"Have you briefed the Chief?"

"No."

"Don't you think you should . . . before we take such an . . . alarmist act. Dramatically increased naval activity could be seen as even more threatening to the Japanese than the oil embargo. I cannot support stirring up the Pacific Fleet, and I would not encourage you going to the Chief, unless you have unequivocal information."

"Rarely happens in the realm of intelligence," answered Kirk. "Plus, you made it very clear that intelligence distribution goes through you."

Turner did not take the bait. "Admiral Stark has been very clear that we must avoid making matters worse than they already are."

"Yes, I understand that Kelly, but the Japanese interest in Pearl should be threatening to us."

"The Japanese would not be so foolish to attack us."

"I certainly hope you are correct, Kelly, yet my job is to be suspicious and vigilant."

"Yes, well, Alan, we have had our differences on the place and use of intelligence. I sincerely hope our debates are not amplifying our disagreements."

"If I cannot convince you, then I doubt I can convince anyone else. However, my inclination is to at least make Kimmel and his staff aware of what the Japanese are doing."

"Understood, Alan," Turner said, and then added rather dismissively, "anything else?"

"No sir. Thank you for your time."

Captain Kirk left the Plans Division dissatisfied with the conversation. He clearly needed to acquire more substantiating information, but his suspicions regarding Japanese intentions in the Pacific were all too real. Something was going on, and it was his job to figure it out.

—

Thursday, 25.September.1941
RAF Manston
Ramsgate, Kent, England
United Kingdom

The morning had gone well. Prime Minister Churchill felt a sense of history. The King appointed him to replace Major Freeman Freeman-Thomas, 1ˢᵗ Marquess of Willingdon, who passed last month, to be Lord Warden of the Confederation of Cinque Ports. The five Southeast ports of Kent and Sussex – Hastings, New Romney, Hyth, Dover, and Sandwich – had been associated since the 12ᵗʰ century. As part of his new ceremonial position, Churchill visited

Walmer Castle, built near Deal late in the reign of King Henry VIII, which continued to serve as the official residence of the Lord Warden.

The Prime Minister and his wife Clementine, along with the remainder of his entourage – Ismay, Thompson and Colville – were then driven to the Royal Air Force aerodrome he last visited at the height of the Battle of Britain, for another ceremonial visit, having accepted another honor. He was now an honorary commodore for No.615 (County of Surrey) Squadron. The planned visit included reviewing the pilots, maintenance crews and their Hurricanes.

The Prime Minister's limousine stopped in front of the squadron Dispersal Hut. The squadron pilots and ground crews in formation came to attention. The partly cloudy skies and moderate, early autumn temperature made the outside formation comfortable. Churchill stepped out of the car door opened by Detective-Inspector Thompson. The Prime Minister reviewed the troops, complimented the commanding officer on his fine looking men, and for the honor and visitation invitation.

The squadron leader commanding officer dismissed the formation of men. The ground crews moved away, presumably to change uniforms and return to their duties. The pilots broke formation but did not move. The Prime Minister was ushered into the squadron Dispersal Hut. A rather unusual full table of sandwich quarters, cookies, and other pastries had been laid out for their distinguished visitors.

An RAF corporal asked the Prime Minister, "Would you care for some tea, sir."

"Good God, no!" Churchill protested. "My wife drinks that. I'll have a brandy."

The corporal appeared caught in the headlights for a few seconds, and then answered, "Straight away, sir." The corporal whispered to the squadron leader, who nodded his head to an unheard question. The corporal disappeared into the commander's office, returned with a bottle of brandy, poured a couple of fingers in an ordinary glass, and handed it to the Prime Minister.

"Thank you, young man. Well done!"

The pilots sat in arranged chairs. Churchill's companions sat in a row of chairs orthogonal to the pilots. The commander tapped a glass for quiet that came promptly. The Prime Minister was the last to sit and took an open chair next to Clementine.

"On behalf of Six One Five Squadron," began the commander, "we welcome you, Prime Minister, Mrs. Churchill, General Ismay and guests. We are honored by your presence, sir. We would like to show you a short film of our most recent mission over France."

"By all means," responded the Prime Minister.

The window shades were pulled down to darken the room. A film projector was switched on. The film displayed gun camera footage from each Hurricane fighter. They shot up several small merchant ships just off the coast and trying to transit the English Channel. Aircraft were hit on the ground and even one in the process of taking off. Fuel storage tanks were ignited. The film ended with a rather dramatic explosion of what must have been an ammunition storage bunker.

"That is what we have for you, sir." The projector was switched off and the window shades raised. "Would you care to say a few words, Prime Minister?"

Churchill finished off his brandy, stood and placed the empty glass on the food table. He turned to face the pilots. "I must say, that is rather dramatic film of your exploits. It warms the cockles of my heart to see the fight being taken to the bloody Huns." He looked to his wife. "Excuse my language, darling." Clementine simply smiled. "I think you will agree what you accomplished on this mission," he said gesturing to the movie screen, "is quite different from where we were a year ago. The tide of war is changing. We shall know success and certainly failure. This is not an easy path we are on, or a short journey we have embarked upon. Yet, with the stout hearts of young men such as yourselves, we shall ultimately prevail in this affair. Thank you for the honor of being your commodore and for inviting me to visit you. Good hunting, gentlemen."

"Attention!" commanded the squadron leader. The pilots sprang to their feet and maintained the position of attention out of respect for their prime minister.

"Carry on," Churchill said, and began shaking each pilot's hand and thanking him for his service to the Crown. Churchill conveyed his gratitude to their commanding officer, and then stepped outside.

As Winston approached the limousine, Detective-Inspector Thompson stepped out of the car and opened the passenger door for his charge. Just then, a dozen Spitfires began landing. Churchill turned to Ismay, who was a few steps behind, and asked, "Who are they?"

"'XR' is Seventy-One Squadron, I do believe, sir. The American Eagle Squadron, if memory serves," answered General Ismay.

"What are they doing here . . . at Manston?"

"I do not know that, sir. Please allow me to check." Ismay returned to the Squadron Dispersal Hut, presumably to ask or perhaps call base operations for the requested information. Winston stood behind the car, watching the Spitfires taxi to their parking spots. Ground crews swarmed over each aircraft as the propeller stopped. Ismay returned. "It is indeed Seventy-One Squadron. They have just returned from France. They landed to refuel and rearm before returning to North Weald."

"I know one of their pilots. I would like to go say hello."

"Are you certain, sir? They are not expecting you."

"I know, but what harm can I cause just to convey a greeting."

"Very well, sir," Ismay responded.

As Prime Minister and Mrs. Churchill took their seats in the limousine, Ismay conveyed instructions to Thompson. The driver moved the automobile to the line of fighters. "There he is," announced Churchill, "the 'XR-G' aeroplane." The pilot stood alone at the tail of his aircraft, as the ground crew filled his fuel tank and reloaded his ammunition magazines. When the pilot noticed the large black car with small Union Jacks fluttering on each front fender, he looked around, probably confused as to why a large black car was approaching the line of fighters. The pilot still wore all of his flight equipment, leather flight helmet with goggles on his forehead and his oxygen mask dangling from the left of his helmet. His service pistol in a shoulder holster could just be seen under his yellow floatation vest. The car stopped and Churchill did not wait for the door to be opened to step out. As soon as Flying Officer Brian Drummond recognized the passenger, he removed his flight helmet, came to attention and saluted crisply, even though he was not wearing his service hat. The Prime Minister returned his salute. "How fortuitous," Churchill said and extended his right hand to Brian. They shook hands. "I happened to be here on other business. I understand you and your mates are just returning from France. How did your mission go?"

"Excellent sir. We shot up our assigned airfield and supply area quite well. Our new 20-millimeter cannons do a masterful job of chewing things up. I know we destroyed at least a half dozen fighters and achieved a handful of secondary explosions that I was aware of, so a very good day for us."

"The last time I saw you . . . ," he stopped, as Clementine joined them.

"Great to see you, again, Mister Drummond."

"The pleasure is mine, Mrs. Churchill."

"The last time I saw you, in London, not quite a year ago, you were near the end of your recovery. I presume you are back to full, robust health."

"Yes sir. Thank you for asking."

The Prime Minister looked around to the adjacent fighters. Brian followed Churchill's scan. All activity had virtually stopped, with men frozen in place, staring at the extraordinary scene before them. "Excellent." Churchill smiled at Hunter Drummond. "I'm afraid I have disturbed the flow of your endeavor."

Brian laughed. "I don't think anyone will mind, sir. We completed our mission, with good success, I might add, and we are headed back to our base. I suspect we will be released once we complete our mission intelligence debriefings. If I may be so bold, sir, I think the lads would really appreciate shaking your hand."

"Done," the Prime Minister answered without hesitation. He looked around and waved his arms for the men to come forward. Mrs. Churchill retreated to the limousine before the pilots and ground crews surrounded the Prime Minister.

Churchill shook every hand offered to him and offered words of encouragement, as only he could do. Brian interrupted long enough to introduce Squadron Leader Meares. After the Prime Minister had shaken all the hands that wanted a connection, he told the gathering that he had just seen the most recent gun camera film from No.615 Squadron. He congratulated the squadron on a successful mission and told everyone he looked forward to watching their gun camera film. He also joked about being half American himself. There were cheers and words of encouragement for the Prime Minister as well. As the impromptu events seemed to wane, Squadron Leader Meares asked for the Prime Minister to excuse them. He promised a low pass on departure, and the Prime Minister consented.

Flying Officer Drummond was the last to say good-bye and thank the Prime Minister for saying hello.

The pilots took to their aircraft.

Churchill joined Detective Thompson and General Ismay at the limousine. Clementine was already inside. The men watched and listened to twelve Spitfire Mark V's start up, taxi and take off. They turned to the south, out over the Channel as they formed up as a 'V' of 'V's, and then turned directly toward where they knew the Prime Minister was standing and watching. They came across the airfield at high speed, passing just a hundred feet above the Prime Minister, and then they began a slow climb to the north. The sound of twelve powerful engines at full throttle was awe-inspiring by itself. Prime Minister Churchill gleefully applauded. He also noticed that they were not the only ones outside watching the low pass and departure.

Churchill and his entourage were soon on their way to the Minster Railway Station on the Southeastern Line, for their journey to Chequers for a long weekend. Two of their daughters would join them to say good-bye to Jock Colville, who was scheduled to begin two years of active service in the Royal Air Force, next week. They wanted a fitting send-off for an appreciated and loyal assistant private secretary. U.S. Ambassador Gilbert Winant would also be joining them at the country residence.

—

Tuesday, 30.September.1941
RAF North Weald
Saffron Walden, Essex, England
United Kingdom

The mission had been simple and uneventful. 'A' Flight had been assigned a protection of shipping patrol for what must have been a comparatively high-value cargo in one or more of a half-dozen merchant ships departing from the Thames Estuary. They spent a long couple of hours orbiting over the convoy at 5,000 feet with their throttles pulled way back to maximum endurance power.

They noticed but did not broadcast about flights of bombers and fighters taking off from eastern airfields. The only radio calls they received were routine "are you still there" calls from the sector controllers until the "return to base" call was received.

The six 'XR' Spitfire Mark V's skirted the few scattered, fair-weather cumulous clouds between their patrol altitude and RAF North Weald. They landed as a flight of six on the large grass landing area and taxied to their assigned parking spots.

The 'B' Flight pilots all sat outside in the advancing, pleasant, early autumn afternoon air. Whitey had his eyes shut in an apparent attempt to remain out of what seemed to be rather heated conversation. The 'A' Flight arrivals did not have to ask what they were arguing about.

The recently assigned newbie Pilot Officer Henry Carl 'Buddy' Courtland of Bridgeport, Connecticut, spoke up, "The Series is set."

"What series?" asked Pilot Officer Joshua David 'Frog' Forcier, the newest of the replacement pilots and flying Brian's Left Wing position.

"The World Series, you idiot," barked Buddy.

The newest of the 'B' Flight newbies Pilot Officer Michael Raines 'Sloppy' Butterfield of Seattle, Washington, chimed in, "Buddy's all worked up because it's another subway series."

"Yankees and Dodgers?"

Buddy ignored the question and pressed home is argument. "It should be the Red Sox instead of the damn Yankees. The 'Splendid Splinter' set an all time batting record for the whole damn season. The Sox deserve to be in the Series."

"So you say," interjected another one of the replacement pilots—Pilot Officer James Edward 'Jimmy' Stonestreet of Charlottesville, Virginia, and Tug's Right Wing. "Sounds like New York and Brooklyn earned the spot as the best teams in their leagues."

"Yeah, but, Ted Williams was the best in all of baseball."

"He's one guy . . . not the whole team," Sloppy added.

The fifth of the recently arrived replacement pilots—Pilot Officer Bradley Thomas 'Hick' Hickerson of Macon, Georgia—stayed out of the conversation, as did Brian and the other senior pilots.

Tug appeared at the Dispersal Hut door. "Have the 'A' Flight pilots finished their debriefing?" he asked what had to be a rhetorical question, since he knew the answer.

"No sir," responded Brian.

"Then let's knock off all this babble and get your duty done first."

Brian did not wait and proceeded directly to the adjacent intelligence hut. The other 'A' Flight pilots followed suit without complaint. As they walked away, Buddy picked up his argument, unwilling to let go of his opinion that Ted Williams and the Boston Red Sox should have been in and won the World Series. These pilots were ex-patriot Americans who had not lost their taste for the fall sports spectacular.

—

Chapter 14

Like one that on a lonesome road
Doth walk in fear and dread,
And having once turned round walks on,
And turns no more his head,
Because he knows a frightful fiend
Doth close behind him tread.

-- Samuel Taylor Coleridge

Thursday, 2.October.1941
RAF North Weald
North Weald, Essex, England
United Kingdom

The entire No.71 'Eagle' Squadron landed safely; they took off with twelve; they landed with twelve fighters. This particular day had been another very good day – the best day Brian could remember. They clipped five bad guys, one of those belonging solely to Flying Officer Brian Drummond – his 20th victory, and his first combat success since his crash a year ago. Brian was now an ace four times over. He achieved hits on several others with his smaller caliber machineguns, but no kill shot that he was able to observe. He liked the results the 20mm cannons in his 'B' wing provided, and the spade rocker switch selector had become second nature, an easy three-position switch to use in highly fluid aerial combat. His only complaint was the 60-round feed drums just did not hold enough cannon ammunition.

Even the newbies had proven themselves in fine form. All of them had joined the squadron last month. They were all college graduates with some knowledge of flight before they had arrived in England and all were commissioned as pilot officers upon completion of the advanced training unit. Stonestreet and Hickerson had joined the squadron and the Skipper's Blue Section on the 9th of September. Courtland had joined on the 11th, been assigned to Whitey's Red Section, and achieved his first aerial victory on this day – the first of the newbies to do so. Butterfield had joined the squadron on the 15th and been assigned to Peterson's Yellow Section. Brian laughed to himself every time he heard Sloppy's callsign, as he was reminded of how he acquired the moniker in his operational training unit. The 24-year-old pilot was not particularly fastidious about his attire, his eating and drinking habits, or even his flying precision; his saving grace was his tactical instincts and aggressiveness. Brian finally picked up his left wing replacement pilot in the person of Frog Forcier, who joined the squadron on the 17th of last month. Today's fighter sweep over

Northeastern France was Forcier's first 'shots fired' combat mission, and he did get several hits on target, but no confirmed victories. Frog was three years older than Brian with barely one seventh of Brian's flight time and a fraction of his combat experience.

The American volunteer squadron had been assigned as the covering unit for Mud Morrison's No.32 Squadron of Hurricane fighters. Brian and the other American pilots had been rather busy, so he did not actually witness Mud's success. The columns of black smoke rising from their target airfield seemed to be testament to their success, and no Hurricanes were lost either. They would gain confirmation in a day or so, after the post-mission aerial photography could be analyzed.

They would genuinely celebrate this evening, a dramatic change from their losses of a month ago, but tomorrow was a duty day, so they could not celebrate too much. The mood outside the Dispersal Hut, even on this overcast day, was decidedly effervescent. Yes, this was a good day.

—

Monday, 9.October.1941
Cabinet Room
No.10 Downing Street
Whitehall, London, England
United Kingdom
15:30 hours

The War Cabinet only gathered early at the Prime Minister's request. The latest intelligence from MI6 had not been encouraging. The prospects of German victory in Russia mounted by the day. The relief of thwarting the German invasion attempt and of ending The Blitz had dissipated. The news from Russia had brought back the unease and tension of the worst days of the battle for their very survival of a year ago.

"Let us get started," announced Cabinet Secretary Sir Edward Bridges. He waited for quiet, and then looked to Churchill. "Prime Minister."

"Thank you, Sir Edward. The Secret Intelligence Service acquired fresh information on the situation in Russia that I consider to be of the utmost urgency, and thank you for amending your busy schedules to assemble early. I have also invited 'C' to join us and brief us on their findings. Sir Stewart, if you would be so kind . . ."

"As the War Cabinet should recall, the German Army Group Center initiated what they likely believe will be their final assault on Moscow one week ago. The first indications came from our joint military mission to Moscow, yesterday. We received two separate, reliable confirmations early this morning that the Red Army has withdrawn again, toward Moscow and is feverishly taking up a new defensive line just 140 miles from the capital city."

"What is your prognosis?" asked Clement Attlee with solemnity.

"I'm afraid, not good," the SIS Director-General answered succinctly. "We have seen a stiffening of resistance by the Red Army in the last few months, but clearly, the latest information confirms that stiffening has not been sufficient to stop the Germans. They have slowed the rate of the German advance but not stopped it. Stalin has moved nearly a dozen fresh Siberian divisions and continues to move his forces east to west. German Army Group South continues to make steady progress and their axis of advance remains pointed directly at the Caspian Sea. Army Group North has made the least progress in the last month, but it is still advancing toward Leningrad. I should add that our One Fifty-One Wing based near Murmansk has appointed itself well in their operations with the Red Air Force."

"This latest information," interjected Churchill, "brings us closer to the collapse of Russia. We must face that potential. If the Soviets do fall, the prospect of a renewed invasion threat grows substantially for us. The Russian winter is not forecast to arrive soon enough to help them.

"I must interject here, Prime Minister," said Menzies. Churchill nodded his head. "Several reliable sources reported the first snowfall began three days ago. Our latest report from yesterday reported that snow was still falling, however there was no significant accumulation, as yet. So the Met Office may be a bit off on their forecast."

Churchill nodded his head, again. "That is earlier than the Met Office forecasted, so then . . . there is hope. We are in a race to winter. This appears to be truth or consequences time on the Eastern Front. We are sending everything we can spare by convoy to sustain the Red Army through these days of trial. The hard debate we had in June of last year, surrounding Sir Hugh Dowding's resistance to sending more fighter squadrons to France, has returned. Just yesterday, I received a personal message from President Roosevelt articulating his rapidly growing concern about our supplies being diverted to the Soviet Union and falling into German hands. His concerns are real. I share his worry. We are in a multi-stage race against time. I would be remiss if I did not raise the question to the War Cabinet, do we stop supplies to Russia?"

"That would send a very negative message to the Soviets, to the Americans, and to the world," Attlee responded. "Even if Moscow falls that does not mean Soviet resistance ceases. Russia is a very big country."

"It does, if the Russians sue for peace," added Anthony Eden, "as the French did after the fall of Paris. We know such discussions have occurred and are on-going in the Kremlin. On top of that, if the Germans got lucky and eliminated Stalin, the likelihood of an armistice, or worse surrender, goes up substantially."

"Rationing is already biting deep into the standard of living for our people," contributed Arthur Greenwood. "The supplies we send to Russia are taken from the mouths of our people, Winston. We must not forget that reality in our deliberations."

"No one is forgetting that reality, Arthur," Churchill responded. "So, to the question before us, what say you?"

Eden answered first. "The risk of stopping while the Russians are still fighting for their very existence is far too great. The consequences beyond the Russian situation are incalculable. I shall advise pressing on."

"I agree with Anthony," Attlee announced.

"Against my better judgment," said Greenwood, "I cannot disagree . . . at least at this stage. The Russians have not given up. They continue to fight valiantly. We need them to stay in the fight."

"I must pick up the lance dropped by Arthur," said Minister of Labour & National Service Ernest 'Ernie' Bevin, Member of Parliament for Wandsworth Central and current General Secretary of the Transport and General Workers Union. "I think we have gone too far in demanding sacrifices by our people. I share Roosevelt's apprehension. The Soviets are so close to collapse. It seems to me, we are wasting precious supplies intended for our people on a lost cause. I say stop."

Churchill nodded his head and looked to Lord President of the Council Sir John Anderson.

"I say we stay the course," Sir John said.

"I'm with Sir John," added Sir Kingsley Wood, "stay the course."

Churchill nodded his head, again. "I say stay the course as well. That puts the War Cabinet at six to one to maintain support to Russia." Winston paused to hear any objections. "I wish to acknowledge the concerns of Arthur and Ernie," Winston added, "but, we are decided. I shall craft a reply to the President, and we shall issue appropriate instructions to the Admiralty and Ministry of Supply. I will say for the record," he added, looking and nodding to Sir Edward, "we are taking considerable risk, and asking our people to redouble their sacrifices, to support our brothers in Russia during their mortal engagement. However, I think Anthony struck resonance. The consequences of withdrawing material support from the Red Army at this critical moment are indeed incalculable. Time shall tell the tale as to whether our decision this evening was good. Sir Stewart, please keep a close eye on events in Russia, as I know you will, and keep the Cabinet informed as quickly as possible.

"Now, let us move to our agenda. We have much to cover," Churchill said.

Sir Stewart 'C' Menzies was excused. The War Cabinet turned to the more mundane of their deliberations, dealing with the rebuilding process to restore residual damage from The Blitz and the seemingly perpetual bomb disposal process. The ever-present Battle of the Atlantic had to be addressed, but more specifically the on-going effects on food supplies, military equipment, and even coal for heating and industry. The business of the War Cabinet remained incessant and perpetual.

—

Monday, 9. October. 1941
Oval Office
The White House
Washington, District of Columbia
United States of America
10:50 hours

"**W**here are we on the paragraph six effort?" President Roosevelt asked his special assistant Harry Hopkins.

The administration and specifically the President sought the repeal of the paragraph six provisions of the Neutrality Act of 1939, to remove any legal interference to his expansion of war support to the United Kingdom and its Commonwealth countries, and to the Republic of China. It was another step back from the isolationism so prevalent in the United States prior to the German invasion of Poland.

"House Joint Resolution Two Thirty-Seven has been introduced in the House," began Hopkins. "The Judiciary Committee should pass the resolution to the full House this week or next. The Speaker and the Majority Whip believe they have sufficient numbers for passage. The Senate is a little more problematic, at the moment. The Majority Leader has a few hard resisters, who may filibuster the bill."

"Do we know who they are and what can I do to help?"

"The House is fairly sound, so no action there. I did ask Senator Barkley, yesterday, if he needed our assistance. He declined and asked us to stand back and to standby. Based on what we know today, Congress is on track for passage and presentation to you for affirmation next month."

"OK, Harry. Stay on this one. We need this done as soon as possible. I do not want Congress getting distracted."

"Yes sir."

"What is next?"

"Van Bush should be waiting for your scheduled 11 o'clock meeting. He wanted to brief you on their current status with the uranium project."

"Very well. I'm afraid this one will have to be between Van and me, Harry."

"As you wish, Mister President."

"Please show him in and tell Grace no interruptions, except for an emergency."

"Yes sir."

Doctor Vannevar 'Van' Bush held a doctorate in electrical engineering jointly from the Massachusetts Institute of Technology and Harvard University, a bit of a rarity by itself. He had been the Chairman of the National Advisory Committee for Aeronautics (NACA) since mid-1938. He became the Chairman of the National Defense Research Committee (NDRC) in June 1940. On the 28th of June 1941, President Roosevelt signed Executive Order 8807 that established the Office of Scientific Research and Development (OSRD) with Van Bush as its inaugural director. OSRD assumed overall supervision for all defense related research and development, subsumed several other research groups including NACA and NDRC, and more importantly, took over control of the secret S-1 Uranium Committee from Lyman Briggs.

"Good morning, Van," the President said in a robust voice upon seeing Bush enter the Oval Office. "Great to see you, again." Hopkins had already pushed his wheelchair to the usual position between the two long couches.

"Thank you, Mister President, and good morning to you."

The President gestured for Bush to take a seat on his right and closest to him. "Have you settled in?"

Bush chuckled. "This is a very big job, and I am not certain I will ever be able to settle in. There is so much going on, but everyone has been exceedingly helpful. I cannot thank you enough for your support."

"You are most welcome. The importance of your assignment cannot be overstated, Van. War is coming whether we like it or not, and as we have seen in England, so much will depend upon what Churchill calls the 'Wizard War.' We are counting on you."

"I shall strive to meet your expectations, Mister President."

"I'm sure you will."

"You had indicated at my appointment that you had a keen interest in the uranium project. OSRD has assumed direct control of the project, and we relieved Mister Briggs, with your approval." Roosevelt nodded his agreement. "As this is our first meeting since the restructuring under OSRD and I do not know how familiar you are with the state of our research, I shall touch upon the key elements and please stop me, if you are familiar with any of this." Again, Roosevelt nodded his agreement. "The Germans were the first to accomplish fission – the process of splitting a uranium atom in December 1938, although it was a German refugee of Jewish descent, Otto Frisch, working at a physics laboratory in England, who actually confirmed and recognized what the German physicists had accomplished. By the spring of 1940, Frisch and another German refugee physicist of Jewish heritage, Rudi Peierls, had taken on and produced a scientific paper, referred to as the Frisch–Peierls memorandum, which clearly established the theoretical potential of a fission device with enormous explosive potential." Roosevelt sat quietly listening, although he had heard earlier everything said so far; however, he wanted to hear Bush's take on the research. "The British were convinced, and as a consequence of the memorandum, the government formed a specific committee of the top nuclear physicists in Great Britain, well actually, in all of the British Empire. That group of scientists was referred to as the MAUD Committee. They issued their final report in July of this year." Roosevelt nodded his head, again. "We began scientific collaboration with the British shortly after your letter from Albert Einstein and at your direction. The Tizard Mission, begun at the height of the Battle of Britain, brought everything into focus. Any questions so far, Mister President?"

"None. I must say, you have done an exceptional job boiling down and articulating this vital project," the President said.

"Thank you, Mister President. "I should take a slight, short detour from the primary topic to acknowledge the extraordinary and incalculable benefit given to use by Sir Henry Tizard and his team. Further, one of the incomplete tasks of his mission initiated last year was finally fulfilled a week ago. A fully functional Whittle W-1 turbine engine was dismantled, loaded in the bomb bay of a B-24 heavy bomber and delivered to the GE Aeroengines facility in Lynn, Massachusetts. GE management indicated yesterday that they have secured a manufacturing license and expect to have a locally manufactured prototype unit available in six months."

"Excellent. Thank you for the update. Please continue."

"Yes sir," responded Bush. "One of the exchange sub-teams has been and continues to be devoted to the development of a nuclear explosive device. To give you an idea of the potential, using the Frisch–Peierls calculation and MAUD refinements, an aircraft-deliverable size unit might yield in the tens of thousands of tons of TNT explosive equivalency—a really big and powerful bomb in a comparatively small container."

"So, Einstein was correct."

"Yes sir, he definitely was . . . at least theoretically. As I am sure you can imagine, there is a monumental difference between a laboratory experiment and a practical operational device. Have you seen or read any of these documents?"

"No, and I trust you and the scientists to have done so."

"Yes, we have . . . many times over. If you should ever feel the urge to read them, we retain high quality copies, and they are available to you within hours."

"Thank you, Van, but I only need to be convinced you and your team have a handle on things and what support you need from the government."

"Yes sir."

"How far along are the Germans?"

"That is an essential question, and the short answer is, we do not know. When Hahn and Strassmann experimentally demonstrated fission, they did not recognize the implications of what they had accomplished. Since then, another renowned German physicist, Walter Heisenberg, has been appointed chief of their nuclear development program. We know with certainty the Germans have a parallel development effort underway. They may well be ahead of us, and that is precisely the British fear. I have met with Colonel Donovan several times already on this topic . . . and others."

"Excellent," Roosevelt interjected. "I was going to ask that."

"He is keenly aware of our intelligence needs as well as the significance and sensitivity. Everything associated with the uranium project, including the COI collection effort, has been classified top secret and further compartmented to protect our work."

"Excellent."

"I must also tell you, Mister President, the British are frustrated with us. They feel we do not appreciate the urgency or magnitude of this development effort. They truly and genuinely fear German achievement of a practical device before us."

"That is quite understandable, since they would likely be the primary target for such a device."

"Quite so," responded Bush. "They need our industrial support. Development and production would exceed their time-capacity – the ability to produce the necessary material in a reasonable time."

"Yes," Roosevelt said. "You are not alone in being a focus of British urgency in this matter. Prime Minister Churchill gave me quite the arm-twisting two months ago. Since this raises the political dimension, please allow me to be direct and frank. I have explained this to Churchill, but I do not want it to go farther than you, for now. Am I clear?"

"Yes sir."

"I want to provide you all the resources you need and more. I share Churchill's since of urgency. The prospect of Hitler gaining use of such a powerful bomb is beyond our imagination and worst nightmares. However, I cannot go to Congress for proper funding. I just do not trust them with such a sensitive subject . . . at least at our current political state. You do have some appropriations under the rearmament laws, but I have been funding the uranium project from presidential discretionary funds to keep it hidden, for now. Very few people inside the government know of this work, and even fewer still outside the government. We are doing the best we can at this state of affairs. When war comes, as I most assuredly believe it will, my pool of discretionary funds will expand dramatically. I am afraid you must suffer the complaints of our cousins for the interim. I know this temporary funding constraint makes your task more difficult, but that is a cross we must bear for now."

"I understand, sir. I shall do my best to keep things moving and assuage the British fears. We are all convinced of the 'what,' through the background work. We know the theoretical 'how,' but we do not yet have a clear path to the practical 'how.' The engineering task is daunting to say the least; yet, on the positive side, we have framed the experiments necessary to clarify the path. This is a really big project . . . perhaps one to two billion dollars over the next three to four years. The 'why' belongs to you, Mister President. The potential of a weapon such as this is so far beyond the realm of contemporary explosive materials. Once the 'what' was determined, more than a few of the scientists began to raise the question of why? We think we appreciate the destructive potential, but none of us believes we know what that potential may ultimately be. We may spend a lot of Treasury funds to build a device that might never be used."

"Thank you for your assessment, Van. Yes, the political aspects of employment are staggering to anyone with a conscience. However, we must balance the destructive potential against the loss of life and damage done without it. The 'why,' as you call it, does not need to be answered today, but I certainly acknowledge the question and my responsibility to answer that particular question. Is there anything else we need to discuss today?"

"No sir. We have covered the broad strokes. We can bore into anything you wish. If you need to go very much deeper, I may need some assistance from the physicists."

"That shouldn't be necessary. Thank you for your time, Van, and your extraordinary efforts at OSRD. Please keep me informed of your progress, of obstacles that need my assistance, and especially of significant changes in the position of our British colleagues."

"Yes sir, will do. Thank you for your time and support, Mister President."

"You are most welcome," Roosevelt said and extended his right hand to Bush. They shook hands. "Keep up the great work." The President turned his chair and wheeled himself back to his desk. "Please ask Mister Hopkins to come back in."

"Yes sir. Good afternoon, Mister President."

"Thank you, Van, and the same to you."

—

Friday, 10.October.1941
Office of the Coordinator of Information
National Institutes of Health Building
2430 E Street Northwest
Washington, District of Columbia
United States of America

The three-story, ordinary looking building on the grounds of the U.S. Navy Bureau of Medicine and Surgery became the headquarters of the Coordinator of Information shortly after the President's order forming the organization last July. The offices and furniture were not new except for the security doors, gates, checkpoints and strong rooms installed in the building to control access to the people and information that would be the lifeblood of COI. Although no signs identified the organization inside or even the name, those working in or with COI referred to the building and designated area as the E Street Complex.

Secretary of War Henry Stimson and Secretary of the Navy Frank Knox agreed to meet with the COI, Colonel 'Wild Bill' Donovan, in the latter's office at the E Street Complex. They had been negotiating responsibilities even before the President's COI order was issued.

"Welcome to our humble offices," Donovan said, greeting his guests.

"This is certainly out of the way," commented Stimson.

"Better for COI, it seems to me." Donovan motioned to the small table. They took their seats. "As I understand our discussions of last week and your concurrence yesterday, we are agreed that COI will take up clandestine operations in the manner of the British SOE."

"Agreed," answered Stimson.

"Yes, agreed," Knox added. "We do not see any conflict in that arena. We still have serious concerns about overlap and conflicts within ONI and G-2 over intelligence collection and analysis. How are you organizing your intelligence department?"

Donovan smiled. "I hadn't considered that two such powerful cabinet secretaries were so concerned about the organization of this little COI group."

"No need to play coy with us, Bill," Stimson said. "You are not so naïve. Frank and I, our chiefs, and our directors of intelligence have voiced our concerns since before you went to England last year. You know perfectly well why Frank asked the question."

"Bill," interjected Frank Knox, "I recommended you to the President as the best man for the task he needed, we needed, to be done. You have done your job in spades. Both Henry and I have endorsed your COI proposal. Neither of us has been assuaged. Our concerns remain. I do believe all three of us want to fulfill the President's charge to us . . . work together, avoid conflict, and provide the President, the government and our military organizations with accurate intelligence. Do I need to repeat my question?"

"No," Donovan responded. "We are structuring the intelligence branch of COI around regional desk groups. The first was the German desk that includes the occupied territories. We have a Russia desk focused upon the Soviet Union and its possessions. We have just recently formed the UK, Africa and Middle East groups. Frankly, we are struggling to gain sufficient expertise to get a hold on the Japanese and Pacific situation."

". . . like MI-6," Knox observed.

"Yes Frank, that is the model that seems the most relevant given our charge and circumstances. If there is one thing I have learned from our British cousins, it is good intelligence evolves from not just passive listening and watching, but also from offensive collection operations, making contact with dissidents, developing networks, and aggressive probing for the information we seek. I have also proposed a joint intelligence committee to share intelligence, resolve conflicts, establish joint intelligence objectives and the general plans to achieve those objectives."

"We like that proposal," Stimson said, "and we hope it achieves the results you suggest."

"We shall do our best," Donovan answered. "Now, I would like to get back to the purpose of this meeting." Stimson and Knox nodded in agreement. "I am also using the British SOE as a model for our clandestine service. I have recruited a number of connected individuals with combat experience during the Great War to become field agents."

"We've heard," Stimson commented. "We are not exactly sure how your rich friends are going to be valuable field agents in the service of strategic intelligence."

"Before I respond to that, Henry, please allow me to inform you both that I have been working with Bill Stephenson and his British Security Coordination group in New York City, and through him the Canadian and British intelligence services, to form a unique offensive field agent training facility. The Canadians have offered a very rural site near Whitby, Ontario, which should be ready to begin operations within two months. We will begin clandestine operations four months after that. The mission of our clandestine service will be intelligence collection and offensive activities to disrupt enemy operations against our allies and us. We intend to operate where our traditional intelligence services do not and cannot. It is our intent to complement the activities of G-2, ONI and State . . . not to interfere with established activities."

"The proof is in the pudding, Bill," observed Stimson.

"None of us can guarantee the future, Henry. The President . . . the country . . . is depending upon us to make this work. I will do my part, and I trust you will do yours. We all acknowledge the risks in this endeavor, but failure is not an option."

Silence filled the room for a handful of seconds. Stimson broke the quiet. "You have avoided my concern for your choice in so-called 'agents.'"

"Not by intention," Donovan responded. "I chose men whom I trust, who have extensive contacts with foreign nationals in their assigned areas, and are wealthy enough to minimize their susceptibility to bribery or compromise."

Both Knox and Stimson stared at Donovan, as if to judge the veracity of his statement. "There is logic to your criteria," pronounced Secretary of the Navy Knox. "Although I must admit, there is risk to your scheme."

"We need a secret offensive capability behind enemy lines," added Stimson. "We want your COI organization to be successful."

"Thank you, Henry. We are off to a good start."

Knox cleared his throat. "Let us plan to reconvene in two weeks' time to discuss progress."

"Certainly."

Stimson stood, followed by Knox. Donovan was the last to rise. The three men shook hands. Knox and Stimson departed. *Well, that went better than I expected.*

———

Saturday, 11.October.1941
Chequers Court
Ellesborough, Buckinghamshire, England
United Kingdom
18:15 hours

"Thank you for coming, Stewart," Churchill greeted the Director-General of the Secret Intelligence Service."

"My pleasure, Prime Minister. Thank you for the invitation."

"It will be just us tonight for dinner. Mrs. Churchill is away visiting her sister."

Well, I am honored and sorry to miss Mrs. Churchill. She is always such a warm and lively hostess."

"Indeed, she is. So, to business, I understand you have a few news items from the Eastern Front."

"Yes, I do. I did not bring the Buff Box, so if you would like to read the deciphered messages, I need to make telephonic arrangements for a courier at your convenience."

"Not necessary. Your memory is sufficient."

Menzies nodded his head in acknowledgement. "Hitler issued his Order 42, which orders the immediate summary execution of suspected commandoes and other clandestine agents."

"Oh my!"

"While the order was directed at the three army groups operating in Russia, the order could be interpreted by zealous SS and SD officers to apply to all German forces and operating areas. Before you ask, yes, I met with Hugh Dalton personally at Baker Street to discuss the essence of this order without disclosing the source. He indicated their training of clandestine field agents has included that assumption, so he believes his agents are fully aware of the risks they face within enemy territory."

"Very well. Thank you. If you have the opportunity, soon, I would suggest the same conversation with Bill Donovan, and short of that with Bill Stephenson, who can ensure Donovan gets the information."

"Excellent point. I will handle it as soon as I return to London."

"In a back-handed sort of way, I suppose such an order suggests there are various partisan and resistance movements forming within the occupied territory in the Soviet Union."

"That is a fair statement and our assessment as well. We are in the process of instructing our field agents in the region to carefully, as best they can, make contact with the resistance and provide what assistance they can. Hugh Dalton is doing the same. I imagine Bill Donovan will so as well."

"By the way, as it is on my mind of late, I presume you are aware that Donovan initiated a joint training facility with the Canadians in rural Ontario, near the village of Whitby I do believe. Dalton and Brigadier Gubbins have met with Donovan and the Canadians, and they have agreed to provide associated curricula and instructors."

"Hugh did not mention all that, but it is logical and a progressive move. I look forward to learning more and seeing how we might collaborate."

"I also believe they plan on an exchange program, once the Canadian-American facility is up and running, and producing field operatives."

"Excellent. The Canadians are behind even the Americans in forming a strategic intelligence apparatus and more so a clandestine service, so the Ontario facility should aid that process."

"What other little tidbits do you have?"

"The Joint Mission informs us that Stalin issued orders for the formation of at least three all female specializing squadrons in night time operations—a night fighter squadron and a night attack squadron. We are not sure what the mission of the third squadron will be, just yet."

"Are they that short of men?"

"No, we do not think so. We believe it is Stalin's recognition that they have a substantial number of female pilots who are clamoring to add their skills to the national defense."

Churchill smiled broadly. "Will wonders ever cease?" he asked rhetorically.

"It was an interesting development. We shall remain as attentive as we are able to see how these new squadrons do in combat. It is also interesting that they are focusing the women on nighttime air operations, particularly difficult tasks from what I understand."

"Indeed it is. Do we know if the Russians have solved the night intercept problem?"

"No. We have no evidence on that point. We have had multiple meetings with the Air Ministry's Technical Section of their Intelligence Branch to appreciate the technology as well as the problem . . . to educate ourselves. We have seen no evidence, as yet, of any comparable technical development program, so we are not sure how they might be operating."

"That is worth adding to the collection list," added Churchill. "I would hate to think those women might be groping around in the dark, hoping to run into something."

"We'll see what we can find out. Is there consideration by the Government to share our research work with the Russians?"

"The War Cabinet has discussed that possibility. To date, we have less confidence in the Russians than we did with the Americans a year ago. Our collective opinion, and here I mean with the Americans, is to wait for two primary reasons: one, to allow our technology to mature and to be far enough into operational deployment, and two, to make sure the Russians survive the winter as an adversary of the Germans."

"Very well. We shall keep that in mind. My last news item comes from the west."

"Do tell."

"Earlier this week, two of our MAUD Committee scientists met with the Americans. I'm not sure how current you are, so allow me to say, the new Office of Scientific Research and Development, or OSRD as the Americans like to say, has subsumed the Briggs Committee. Doctor Cockcroft met with Doctor Bush, the President's choice to lead the research office earlier this week. He acknowledges the awareness of the Americans to the threat and time risk; however, he remains seriously concerned over the level of support to this particular development project."

"I have raised this matter with President Roosevelt in personal, private communications. I believe your assessment is correct, and the President acknowledges the perceived lack of adequate support. I am constrained in what I am able to share. Let it suffice to say, the President faces a daunting challenge of walking a very fine line in supporting the fission development project and resistance in . . ." A knock on the study door broke Churchill's thought.

The Prime Minister's duty assistant private secretary, this weekend John Martin, opened the door. "Excuse me, Prime Minister. Mrs. Landemare has indicated supper is ready."

"Thank you, John. Give us a few minutes. I must finish my thought."

Martin nodded his acknowledgment and closed the door.

Churchill turned back to Menzies. "As I was saying, the President is dealing with an isolationist prone Congress. He has a pronounced concern regarding calcifying the resistance in Congress and jeopardizing the project. I am becoming somewhat apprehensive about the continued pressure on the Americans. We all want an all-out commitment by the Americans for this development; however, I'm afraid we must trust the judgment of President Roosevelt. I will ask Professor Lindemann privately to talk to Doctor Cockcroft and the other liaison scientists working with the Americans to tone down their pressure on the Americans. We cannot risk alienating them. We must allow them time and space to find their footing. We are agreed that American industrial capacity is critical to achieving a functional device ahead of the Germans. Unless someone can present more specific information, I remain convinced the Americans are doing the best they can in the circumstances in which they find themselves."

"I trust your judgement, Prime Minister. Likewise, we will convey your guidance."

"Excellent. Now, I'm afraid we've kept Mrs. Landemare waiting long enough. Let's enjoy a sumptuous meal. We will have to constrain our discussions to the secret level until we can return here for brandy and cigars."

The two men rose and walked together to the dining room. Their conversation would continue through dinner and into the evening.

—

Thursday, 16. October. 1941
Oval Office
The White House
Washington, District of Columbia
United States of America
10:50 hours

The double knock on the door preceded Harry Hopkins' entry. The President's special assistant stepped inside and closed the door behind him. "Mister President, Foreign Minister Eden and Colonel Donovan are here for your scheduled meeting," announced Hopkins.

"Very well. I'd like you to sit in on this one, Harry." Hopkins nodded his acknowledgment. "Let's get this started."

Harry pushed Roosevelt's wheelchair to the usual position between the two couches, and then opened the door and invited their visitors to enter.

Eden entered first. "Good morning, Mister President."

"A pleasure to see you, again, Anthony," Roosevelt said and extended his right hand to the British Foreign Minister. The two men shook hands warmly. "Welcome back to Washington." He gestured for everyone to be seated.

"The pleasure is mine, Mister President."

"Thank you for joining us, Bill. I presume you and the Foreign Minister are well acquainted."

"Yes sir. We were introduced quite a number of years ago, now, and we've crossed paths many times since."

"Excellent. Shall we get started?" the President said, looking to Eden.

"First, the Prime Minister sends his warmest regards and best wishes for your continued health."

"Thank you. He is a good friend."

"Second, the situation continues to darken in Russia. The Red Army resistance has stiffened, but they have yet to stop the German advance. They have evacuated Odessa, and worse, the Germans are less than one hundred miles from Moscow. The Russian capital is in turmoil. Stalin is wavering between evacuating Moscow and defending the city. The prospects are not good. The Prime Minister asked me to discuss the 'what if's with you during my visit."

"Should we have Hull and Stimson, and perhaps Knox, here as part of this discussion?"

"Yes, certainly," Eden responded. "This was not intended to be the definitive or exclusive exchange, only the preface."

"Very well, continue."

"The Prime Minister and the War Cabinet advocate for mutual contingency plans should the Soviet Union collapse. The strategic calculation would likely change dramatically."

"Winston and I discussed the potential at Placentia Bay in August."

"Yes, so he informed the War Cabinet. The situation in Russia has worsened since then, and the end for the Soviets is certainly closer than it was last August. Winter is approaching, but not fast enough. Their rate of progress since Field Marshal von Bock initiated the current offensive two weeks ago has been fast . . . faster than expected . . . and the Meteorology Office indicates the first snowfall for the Moscow area has only marginally slowed the German advance. The War Cabinet assessment is that Stalin is likely to lose the race for winter, thus Winston's charge to me for raising the contingency plans."

"Do you have anything to add, Bill?" asked the President.

"No sir. We concur completely, as does the G-2."

Roosevelt thought for a few moments, and then looked to Eden. "What do you, the Prime Minister and the War Cabinet have in mind?"

"If the Germans are successful, they will have access to the Caspian and Romanian oil fields and untold resources. Based on what we have seen so far in the Baltic States and the Ukraine, I dare say the Germans will gain substantial manpower for their army, once the Red Army, NKGB and NKVD are eliminated as threats." Eden paused to give the President an opportunity to ask any questions. Roosevelt gestured somewhat impatiently for him to continue. "At the bottom line, we see a renewed invasion threat, and the Germans will have learned the lessons of SEALION from last year. We are stronger, but if they can redirect their resources, amplified by the additional personnel they are able to mobilize from the newly occupied territories . . . well, we may not be strong enough the next time."

Roosevelt waited several seconds. "So, Prime Minister Churchill wants to know what it will take for the United States to join in the defense of Great Britain?"

"Not to put too fine a point on it," Eden said, "yes, that is the question."

Roosevelt looked to Donovan, and then Hopkins. "What is said from this point is classified TOP SECRET – PRIME, and not to be discussed with anyone outside this room and this group. Is that understood?"

"Yes, Mister President," answered Donovan.

"Yes sir," Hopkins added.

Roosevelt glanced to Hopkins for a confirmatory head nod, and then turned back to Eden. "The direct and frank answer . . . I do not know. What I do know is, we are not to that threshold, as yet. I know Winston is keenly aware of the isolationist movement in this country. We have made inroads to mobilize the country for what I believe is the inevitable conflict. Hitler cannot tolerate a free nation on his flank. Winston and I agreed a year ago that we would supply Great Britain and do our utmost under the law to sustain the British people in this fight. We further agreed at Placentia Bay to supply the Soviet Union. I stretched Congress to the point of resistance. The last thing any of us needs is Congress rejecting our efforts, or the Press and the people to stiffen in their resistance to the inevitable war. We were less prepared than you were two years ago. We are still not prepared for war. The answer to your question today is, I do not know. I cannot see the path to that answer. I understand the essence and necessity of the question. I wish I had the answer. I think the best I can offer at this moment is the plan Winston and I set in motion two months ago. We must pray for time."

"The Prime Minister does not expect an answer, Mister President. He only seeks contemplation . . . to prepare for what might come, should things in Russia continue to worsen."

"We must consider our situation and our response to your question," Roosevelt said. Eden nodded his head in concurrence. "What I can assure you and the Prime Minister, whatever we decide, we will have a joint plan. I think the appropriate British saying is, in for a penny, in for a pound."

Eden chuckled. "Yes, Mister President . . . quite right. I shall communicate our conversation to the Prime Minister and the War Cabinet." The President nodded his head. "Now, if I may, we are disturbed by events in Japan."

"As are we," the President added and looked to Bill Donovan.

"Yes sir. Prime Minister Fumimaro Konoye resigned under pressure yesterday. We received information earlier this morning that General Hideki Tojo has replaced the civilian premier. Tojo apparently heads a military junta that has replaced the civilian leadership."

"That is our intelligence as well," added Eden.

Donovan continued, "The events in Tokyo over the last 48 hours point to a more aggressive military stance by the Japanese . . . a further militarization of their society. We have not been able to establish a reliable forecast or prognosis for what they intend to do."

"Our focus has been Germany."

"Understandably so," commented the President.

"We have less insight into Japanese intentions. Our Australian colleagues have been most helpful in that regard, but even they acknowledge their sun blindness."

"Interesting analogy, Anthony," Roosevelt said, "I'll have to remember that one. Bill has been working to improve our strategic intelligence capability."

"Not fast enough," interjected Donovan, "for events unfolding before us."

"You've only been at this business for three month, Bill. No need to be so harsh on yourself."

"I'm struggling to find field agent recruits who fit the criteria I'm looking for," Donovan confessed. "I've got Europe reasonably covered for starters, but Japan, China and Asia in general have been a little more problematic.

"I do have field reports," he continued, "from several reliable sources in Shanghai that naval and transport ships began gathering in port about a month ago, and several divisions of infantry and at least one armor division have bivouacked on the north side of the city. They appear to be staging forces for something. Tojo's ascension coupled with other tidbits we have been picking up suggests they are preparing to strike south to the resource rich Dutch East Indies. If that is their intention or objective, it is our opinion they cannot leave Hong Kong, Singapore or the Philippines as staging areas on the flanks of their supply lines. We have alerted the Philippines of our concerns."

"We share the same concern," Eden said. "Yet, the Admiralty maintains that Singapore is well-armed and virtually impregnable. Our forces in Hong Kong are probably insufficient to defend the territory, but they can certainly make the Japanese work for their conquest, if that is their intention. The Prime Minister has proposed to the War Cabinet that we move two battleships, the new Prince of Wales and the older Repulse, to Singapore and the South China Sea, as a deterrent to any Japanese designs on the Dutch East Indies."

"The same ship that transported Winston to Placentia Bay," observed Roosevelt.

"Yes, one and the same."

"Magnificent ship, I must say. The Japanese do not impress me as being susceptible to being affected by such demonstrations," the President said. "We have nearly a dozen battleships and two aircraft carriers deployed to the Pacific, and none of our efforts at strength projection have appeared to alter their course toward general war. With the military apparently taking over the government, I suspect they will be less impressed. With the deposing of Konoye, I suspect diplomacy has failed, although we shall pursue every possibility to the end." Roosevelt paused and looked to Donovan. "Our task now is to anticipate their next few moves before any blood is spilled. We are still not ready for war." Donovan nodded his head in clear recognition of his charge. Roosevelt looked back to Eden. "We need more time. To be candid, our material support for your country, and now the Russians as well, is sapping our capacity to rearm and prepare under the current political constraints."

Foreign Minister Anthony Eden smiled and nodded his head. "We are most grateful for your support, Mister President. It has been an essential ingredient to our recovery. Now, I am quite mindful of your precious time. I'm afraid I have already overstayed my allotted time. Should we adjourn to tomorrow's planned meeting?"

Roosevelt glanced to Harry Hopkins behind his left shoulder. Harry nodded his head once discreetly. "Very well. Tomorrow it is."

"If you have no objection, Mister President, I would appreciate the attendance of Lord Halifax and Bill Stephenson."

"By all means. We will ensure they are on the access list. If you think of anyone else before the meeting, please let Harry know. He will see to it." Roosevelt looked to Donovan. "I presume you will attend, Bill."

"It is on my calendar, Mister President. I shall be there."

"Excellent. Then, we are adjourned." Eden, Donovan and Hopkins stood. "Thank you for the preface visit, Anthony. Always a pleasure to see you."

"Thank you, Mister President. Likewise, I am certain." The two men shook hands. Eden and Donovan departed. Hopkins pushed the President's wheelchair back to his desk. His next appointment was undoubtedly waiting with Grace Tully.

—

Chapter 15

We have to distrust each other.
It's our only defense against betrayal.
-- Thomas Lanier Williams
AKA Tennessee Williams

Monday, 20.October.1941
No.10 Downing Street
Whitehall, London, England
United Kingdom
16:20 hours

"**W**elcome back, Anthony," Churchill greeted the Foreign Minister in his office. "I understand you arrived yesterday afternoon."

"Thank you, Winston. Yes, rather long flight, I must say, with two intermediate stops for fuel and provisions. It is good to be home, although I must say it is distinctly refreshing to be in a place where they do not have to extinguish the exterior lights at night."

"Quite so, I should think. I wanted a quick word before we assemble the War Cabinet for your report." Eden nodded his agreement. "Pray tell me of your conversations with Franklin and the Americans."

"At the bottom line, I believe the Americans are more apprehensive about the situation in Russia than we are."

"We are closer to the tip of the spear."

"Yes, well, they perceive they have much more at stake to lose."

"What of our proposal?"

"The President seemed to be a little apprehensive when I first presented the question in the Oval Office. However, the next day with State, the service secretaries and the military chiefs, he was more supportive. We are all agreed that our first line of defense is sustaining the Russians to winter, and then resupply them as best we can before renewed field operations in the springtime. We are also agreed that aerial lift on such a scale is out of the question, which leaves us the arctic sea-lanes until the water freezes over."

"That does not give us much time," observed Churchill. "And, that will not help resupply the Red Army during the winter slowdown. On the positive side, we received word from the Moscow embassy that Stalin has directed the NKGB and NKVD to quash any signs of panic, evacuation or retreat, and they are doing so with ruthless and brutal efficiency. Apparently, he has decided they will stand at Moscow. The Germans are a scant 65 miles from the outskirts of Moscow, as of this morning. We must find another way to supply them during the winter months. The Bosporus, perhaps."

"Turkey will be a problem to that end."

"How about Persia and through the Caucuses?"

"That might be more feasible diplomatically, but what about the Germans?"

"Yes, well, that is indeed a complication. The German Army Group South is making the best progress of the three groups. It might become a race against time. Could we build a connecting rail line from Basra before the Germans reach Baku and cut the line?"

A knock at the door diverted Churchill's attention. Principal Private Secretary John Martin poked his head in past the door. "The War Cabinet is waiting, sir."

"Thank you, John."

Martin closed the door, again.

"One more item before we attend to our colleagues," said Eden. Churchill locked eye contact with his foreign minister and nodded his head slightly. "President Roosevelt raised the situation in Japan on the first day. My impression was he was more concerned about the mounting list of negative signs emanating from Tokyo and their holdings in Asia than he was about the situation in Russia. Bill Donovan reported what certainly appears to be a build-up of military forces and shipping transport in the vicinity of Shanghai and probably in their Home Islands. They sense general war rumblings. I did not, I could not add to their concerns other than we are picking up chatter."

"Well, it seems we are both a little thin on intelligence from within the sphere of Imperial Japan. The replacement of Konoye with General Tojo is the most ominous to me. We cannot and must not neglect our responsibilities to our colonies is South Asia and Asia proper, but frankly that is a side show for us with Germany on the cusp of subduing the Soviet Union. No matter what happens in the Pacific, we must find the means to keep the Americans focused on defeating Germany." Churchill paused to think. "Germany first, and then we will sort out the Japanese. Italy will fall by its own weight. I cannot see Mussolini sustaining his ill-gotten gains without Germany."

"The Americans apparently feel Hong Kong and Singapore are at risk."

"They may well be, but we are stretched dreadfully thin. General Percival maintains that he has sufficient forces and coastal defenses to withstand any attack on Singapore or Malaya. Did you inform the President of our plans to deploy *Prince of Wales* and *Repulse*?"

Lieutenant General Arthur Ernest Percival, CB, DSO & bar, OBE, MC, had been General Officer Commanding Malaya Command since the previous April, with his headquarters at Singapore.

"Yes. Their chiefs voiced considerable apprehension that battleships might be deployed without an aircraft carrier for air defense."

Churchill grunted. "Perhaps they should get busy building us a few more aircraft carriers. The Admiralty contends there are none for reassignment. We have insufficient carriers for the Home waters and the Mediterranean Sea."

"I am just the messenger here, Winston."

"Thank you for being a faithful messenger, Anthony. Now, we are late. Let us be off to the War Cabinet. You will undoubtedly need to rehash these items with the full cabinet, but let us confine the discussion to Germany. We must not be distracted from the immediate threat."

The two men stood and departed the Prime Minister's office for the short walk to the Cabinet Room for their scheduled meeting.

—

Tuesday, 21.October.1941
Standing Oak Farm
Winchester, Hampshire, England
United Kingdom

Flying Officer Brian Drummond arrived at Charlotte's farm later than usual due to some unspecified railway delay en route. He paid the driver with a generous gratuity and thanked him for his perfect service. As he stepped out of the old 1936 Austin Low-Loading Taxi and closed the door, Brian turned to see Charlotte running toward him from the barn. He spread his arms and braced for impact. Charlotte literally leapt into his waiting embrace. They wrapped around each other, and kissed in a prolonged and passionate manner. The crunch of the taxi's tires on the gravel diminished to silence, and yielded the modest clapping and cheers of Lionel, Horace and young Jacob brought the two lovers back to reality.

Charlotte pulled her head back just enough to speak. "You missed the afternoon milking."

"I guess so. Something happened on the railway that stopped us for over an hour between Farnborough and Basingstoke."

"Not to worry, my love. You are here now.

Brian lowered his wife to the ground. They remained arm in arm and led the way into the main house. Edith was inside the main house with Ian, and she had done most of the evening meal preparation. Brian took Ian from Edith to have some personal time with his son, and to allow Edith to assist

Charlotte in completing and serving the evening meal for the crew. Brian was truly fascinated just watching Ian's eyes swallow everything around him, and the infant boy intermittently smiled and cooed for his father.

The evening meal was simple but complete and filling. Charlotte fed Ian while she ate. The conversation was sparse and lighthearted with no war talk whatsoever . . . not even about checkpoints. When the men finished their meal, they excused themselves, and headed back to town and their homes. Brian helped Edith clear the table and wash the dishes, while Charlotte finished feeding Ian. Eventually their son was sated, fell sound asleep and detached from his mother's breast. Brian kissed his son's forehead, and then kissed Charlotte for the first time since arriving at the farm. Edith carefully took Ian from Charlotte and walked upstairs to change the infant's diaper and put him in his crib.

"Let's get some fresh air," suggested Charlotte.

Brian nodded his head, donned his overcoat without his uniform tunic, and then helped Charlotte into her overcoat. Charlotte told Edith they were going for a walk. The sun had set, and it was mid-twilight. The sky was clear. There was no detectable breeze, and the temperature was chilly but not cold. They walked hand-in-hand at a rather leisurely pace. There was no need to hurry.

Brian spoke first. "I don't recall mentioning to you my admiration for your work on this path and the sitting area under the oak tree by the lake."

"No. You didn't."

"Most impressive, I must say.

"Well, thank you."

"Jacob and I came out here for his question time last May. I'm sorry, I forgot to mention your excellent work."

"He helped put it together. We spread the rock because I just did not want to deal with the mud on rainy days anymore. Jacob was most appreciative of your time and candor. He really idolizes you."

"It was a pleasure." They reached the sitting area and sat next to each other on the long bench, still holding hands.

"I've found myself coming to the lake more these days for contemplation."

"That sounds serious."

"Not really . . . just reality."

Should I ask? Brian hesitated. *Silence is golden . . . until she is ready to talk.*

"The birth of Ian has changed my life," Charlotte said.

"And mine."

"Really?" challenged Charlotte. "You are doing what you have always done. With Ian in our lives, I worry even more about your well being and being part of our lives. But, I did not bring you out here to chastise you for your love of flight and sense of duty. I have accepted you as you have been and as you are. I wanted the peace and serenity of this place to discuss our future."

"I'm sorry, Charlotte. I'm confused. Has something happened? Are you or Ian in danger?"

Charlotte shook her head in the negative, and then smiled. "No, my darling. My apologies for being so cryptic. I know but you cannot see what is in my mind." Brian simply nodded his head in agreement with confusion registered by his expression. "I have two topics. One, I cherish Ian and the joy he has brought to my life, but I simply must insist we take precautions until this war is done and I know you are safe."

"We agreed to that months ago . . . before Ian was born. What has changed?"

"Perhaps nothing. Perhaps in your absence, I suppose I feel more vulnerable. I do not want my temporary . . . momentary fear of procreation to interfere with our intimacy."

"It hasn't, yet. Has it?"

"No, I am concerned about the thoughts I have rumbling through my mind. The responsibility largely, if not solely, rests upon me, but I feel compelled . . . the urge . . . to share my thoughts with you. I am seriously concerned."

"I think I have understood that since our first discussion. You do not want to raise children alone and you want your husband to share in that duty."

"I would not say it quite like that, but yes that is the basis."

"I will do my part . . . what has to be done," Brian said.

"Thank you. The second item is land." Again, a puzzled expression bloomed across Brian's face. "Just in the last two months, since your last visit, two adjacent landowners have approached me about buying their land."

"Is it good land?"

"As good as ours."

"Then, let's buy them out."

"I . . . we own this land outright, Brian. I do not have that kind of cash and I'm quite reluctant to take out a mortgage these days to raise the necessary funds."

"How much?"

"Harris is asking £3200 for his 85 acres."

"So, roughly £40 per acre. That's $160 per acre . . . rather pricey, it seems to me. As I learned last February, farmland back in Kansas is roughly $70 per acre."

"Less available land in Hampshire, I dare say."

"Perhaps so. Is £40 per acre a fair price in this area?"

"Yes. It is not the rate but the total I am concern with here."

"OK. Then, what would we use that land for?"

"For now, the same as Mister Harris, wheat and barley."

"What about the other parcel?"

"That is the Brownfield farm. He is growing oats with several score of mature pigs. He is asking £4500 for his 110 acres."

"Basically, the same rate."

"Yes."

They discussed the value of each farm in terms of revenue and expenses, the management of those farms, if they acquired both properties, as well as future development potential. Charlotte clearly felt good about the acquisitions, but worried about the money. Brian had the money and talked enough to satisfy himself of the worthiness of the purchase.

"So, both properties are £7700, or US$30,000. I'll contact Jonas Braddock to transfer US$50,000 to your bank account to purchase the properties and have a little extra, just in case."

"Are you sure?" asked Charlotte.

"As sure as I am of you . . . of course. In fact, now that I think of it, you should have contact information for Braddock as well as Travis Atherton, our foreman in Kansas, and Bobby Joe Sales, who is running our aviation company. You already have Gerty Bainbridge's address."

"You are most generous, my darling. Now, it is getting chilly and twilight is nearly gone. The house should be quiet. Let us go to bed, so you can show me how much you love me."

"Yeah, baby!" Brian exclaimed with enthusiasm.

He pulled her tightly to him, again, kissed her passionately, and then released her, took her hand, and led her back to the house. They made quick work of getting to the point and delighted in rejoining their union.

—

Thursday, 23.October.1941
Headquarters, Special Operations Executive
No.64 Baker Street
Marylebone, London, England
United Kingdom

"We have been eagerly anticipating your return, Diamond," Gubbins announced, as Trevor Andersen entered the small conference room.

"Welcome back," added Hugh Dalton.

They shook hands and took seats at one end of the rectangular table with Dalton in the middle.

"Congratulations on your promotion, sir" Andersen said, upon noticing Gubbins' three pips under crown on his epaulettes. The senior officer was now a brigadier.

Brigadier Colin McVean Gubbins, DSO, MC was a well-known special operations officer, who lead the Independent Companies during the Norway Campaign in the spring of 1940; he was awarded the Distinguished Service Order for that work. Gubbins took up his assignment as director of Operations & Training for the Special Operations Executive (SOE) a year ago.

"Thank you, Trevor. Happened last spring, thanks to Minister Dalton.

"Justly deserved," Dalton added.

"So, what did you learn?" asked Gubbins.

"The alias and cover clearly worked, as I am here. It took a week to sort out things. As best I can determine given the constraints of circumstances, the White Rose Society is a rather loose student resistance movement at the University of Munich. I was on campus ostensibly for metallurgical research. I could not ask a lot of questions, but the group appears to have at its intellectual core Kurt Huber, a professor of psychology and music. He seems to be the anchor. However many members, they may have is beyond my reach at the moment. The driving force, such as it is, appears to be a number of medical students at the university, who attended Huber's lectures and served a stint in the medical corps on the Eastern Front. Hans Scholl and Alexander Schmorell are two names that popped up more than a few times, along with Hans' younger sister Sophie. I happened to be listening when another student claimed Hans had become disillusioned with National Socialism during his tenure in the Hitler Youth in the 30's."

"That last bit makes me very nervous," interjected Gubbins. "Too many fanatics have sprouted from that indoctrination group."

"So, it is true. Our information was accurate," Dalton observed.

"I would not go that far, sir. The indications appear to suggest accuracy, in that a rather loose group of student dissidents does exist. However, to the best of my knowledge, they have taken no steps toward action. It is intellectual machinations at this point. Also, I have purposefully not made contact out of concern for risk to them and to myself. More than a few medical students are upset at being pulled from their studies for medical service on the Eastern Front. I could not establish precisely what is going on there, but what I have seen and heard suggests medical students are required to serve for several months at a time in field hospitals in the occupied territories of Russia. Once their service is completed, they are allowed to return to their studies. The government wants them to complete and gain certification as medical doctors, but this seems to be the way they have chosen to augment their field medical staff without fully conscripting these students short of certification."

"Interesting twist," said Dalton.

"I thought so," Andersen responded. "It seems to be a fertile ground to cultivate whatever dissent we can."

"So, you would advocate periodically returning to Munich, to nurture that dissent," Gubbins said.

"Yes."

"Very well, then, let us plan on a follow-up visit next spring. Winter adds too many complications, especially in Bavaria."

"Yes sir. That was my thought as well."

"Anything else, Trevor?" asked Dalton.

"Not at the moment, sir."

Gubbins raised his hand slightly. "You said your cover worked. Was that cover story tested?" asked the Director of Operations and Training.

"Yes, a couple times – once at a roadway checkpoint inside Germany and another by a local constabulary officer in Munich, not far from the university. They seemed to be routine checks, almost pro forma from what I absorbed."

"So, you would recommend holding that cover story?"

"Yes. I was not the only industrialist travelling back and forth to Switzerland."

"Excellent," said Gubbins. "By the way, I have not had the opportunity to chat with you since your mission to re-establish contact with Major Merton. He is back in battery, as they say, and doing quite well. We have given him the code-name 'Rouge'."

Andersen laughed. "For his alias, Moulin?"

"Exactly. I thought you should know, in case you run across the code-name in your travels. Anyway, well done on that mission to recover Merton, and what appears to be a very successful mission to Munich. We shall keep this information very close-hold."

"Thank you, sir. Will there be anything else?"

"No, Trevor. Well done."

Andersen stood. He left the two men seated at the table, and departed the conference room and the building. He would take an extended weekend to visit his parents and relax a bit before returning to the saddle.

—

Friday, 24.October.1941
No.10 Downing Street
Whitehall, London, England
United Kingdom
11:10 hours

Colonel Menzies arrived without the Buff Box or any other carrying case. The door closed behind him.

"Good morning, Prime Minister," 'C' offered.

"And, good morning to you, 'C.'" Churchill gestured to the comfortable chairs and joined the SIS Director-General. "To what do I owe the pleasure of this unusual early morning visit?"

"First, a duty to perform . . . ," he said, reaching into his suit jacket inside pocket. "I was just conferring with the Foreign Minister and indicated I was due here next. He asked me to carry this for you." Menzies handed the single piece of letter paper to Winston.

SECRET

```
Washington (via U.S. Embassy London)
October 24, 1941
Dear Winston:
     I have established the position of
Coordinator of Information, with authority to
collect and analyze all information and data
which may bear upon our national security;
to correlate such information and data, and
to make such information and data available
to me and to such departments and officials
of the Government as I may determine; and to
carry out, when requested, such supplementary
activities as may facilitate the securing of
information important for national security
not now available to our Government.  I have
designated William J. Donovan, whom you know,
as Coordinator.
     Colonel Donovan tells me that he has had
most helpful cooperation from the officers of
```

```
His Majesty's Government who are charged with
direct responsibility for your war effort.
     In order to facilitate the carrying out
of the work of the Coordinator with respect
to Europe and the occupied countries, I have
authorized Colonel Donovan to send a small
staff to London.  I trust this is acceptable to
you.
     All's well; best of luck.
As ever FDR
```

"The letter was delivered by special courier from the ambassador this morning, and apparently transmitted via the embassy without classification, even though the letter was clearly meant for you, as a personal communication."

"Have you read it?"

"Yes. Minister Eden shared it with me."

"None of it is news, except for the last paragraph. What do you think?"

"First, such a staffing move was entirely expected. I would have been surprised if they did not have direct representatives here. Second, I welcome the direct contract with Donovan via his direct representative. Third, this should improve the timeliness of our communications. Lastly, as I informed Mister Eden, I would strongly recommend enthusiastic acceptance and support."

"I am happy to hear your assessment and share it completely. Please draft an appropriate reply for Eden's review and my signature. I must say I look forward to meeting Donovan's London field chief."

"Yes sir . . . as do I. I'll get the draft done straight away." Menzies took the letter back and returned it to his jacket pocket. "I also have other news from the Eastern Front and multiple dissimilar sources including Boniface." Churchill nodded. "The German Army Group South has taken Kharkov and cutoff Crimea, which cannot hold out long without resupply and replacements."

"Can we supply them via the Black Sea?"

"Where there is a will there is a way. However, given the situation as we know it, a convoy of supply ships, even from Alexandria, would take longer to organize and transit than the time remaining to the Red Army in Crimea." Again, Churchill nodded his head. "Army Group Center is now a mere 50 miles from the outskirts of Moscow. The mud has slowed them down somewhat, but they are still progressing. Temperatures are dropping, but the hard freeze is probably still several weeks away, and there has been snow but no accumulation, yet."

"Then, there is still hope."

Menzies smiled. "Always the optimist, Prime Minister. We are keeping a very close eye on developments in that theater of operations."

"Good. Things are getting dreadfully close to the precipice. Anything else?"

"No sir. That was it for me."

"I have one items I need to make you aware of," Churchill said and went to his desk. He leafed through a small stack, but did not find the item he was looking for. He depressed the intercom lever on his desk telephone box. "Jock, would you be so kind to retrieve President Roosevelt's October 11th letter."

"Yes sir," came a scratchy reply. Within a few seconds, Colville entered with a single piece of paper. Churchill pointed to Menzies, and Jock handed the letter to the Director-General.

"Thank you, Jock."

"Yes sir," he answered and departed closing the door behind him.

"I received this letter last week. I want you to be aware of it, in case of you come across associated information or communications."

TOP SECRET - MAYSON

The White House

Washington, DC

October 11, 1941

My Dear ~~Mr. Churchill~~ *Winston*:

It appears desirable that we should correspond or converse concerning the subject which is under study by your MAUD committee, and by Dr. Bush's organization in this country, in order that any extended efforts may be coordinated or even jointly conducted. I suggest, for identification, that we refer to this subject as ~~MAUDSON~~ *MAYSON*.

FDR

TOP SECRET - MAYSON

Churchill chuckled softly. "I think he was trying for Son of MAUD, but rightly adjusted. For now, we have accepted and will comply with his designation, although we will retain TUBE ALLOYS for our research and related work."

"Very well. So noted," 'C' answered and handed the letter back to Churchill. "If there is nothing else, Prime Minister, I shall be on my way."

Menzies departed and left the door open. Churchill returned to his desk and paperwork.

—

Friday, 31.October.1941
No.10 Downing Street
Whitehall, London, England
United Kingdom
21:45 hours

"The First Lord and First Sea Lord have arrived to see you, sir," announced John Martin.

Prime Minister Churchill had been working alone in his office, reading various documents, not requiring a response or action. The Admiralty had called over with breaking news.

"Show them in, please, John."

"Good evening, Prime Minister," said A.V. Alexander, as he led Admiral Pound into the office. Martin closed the door behind the Navy minister and naval chief. "Thank you for delaying your journey to Chequers."

"Good evening to both of you. So, pray tell, what have you?"

Admiral Pound spoke first. "We have been monitoring a wolf pack attack on fast convoy H X One Five Six, west of Iceland. The urgent news from the convoy commodore is the American destroyer *Reuben James* has been sunk, while on screen service for that convoy."

Churchill smacked his hands together one time. "I'll be damned. The Germans finally took the bait." The confused expressions on the faces of both naval men demanded explanation. "Last August, at Placentia Bay as you know, Roosevelt and I sanctioned the Americans extending their convoy coverage to the Icelandic longitude. President Roosevelt needed a *raison d'être* for escalating U.S. involvement in the war. What are the casualties?"

"So far, just the *Reuben James*," answered Sir Dudley. "The attack came shortly after dawn as the Americans were transferring escort responsibility to us. The report we received indicated the *Reuben James* had begun maneuvering to investigate a 'Huff-Duff' contact when a single torpedo struck her and apparently penetrated the forward magazine."

'Huff-Duff' was the military slang reference to High Frequency Direction Finding equipment and procedures, used by both British and American navies, especially on escort duty. The German U-boats communicated with their naval headquarters using high frequency radio signals. Such contacts were most often

U-boats, as they were the only unaccounted for sources of high frequency radio transmissions in the middle of the Atlantic Ocean.

Admiral Pound continued, "The ship went down in a matter of minutes. We were waiting for the casualty report and subsequent action. We received the report when we called your office. The convoy commodore radioed they had to abandon further search operations. They rescued 44 sailors . . . that means 115 are presumed dead."

"Terrible loss of life. Any other losses?"

"Not yet," Pound responded. "They are just entering the primary engagement zone. The next three days will be difficult. 'Boniface' has been good for the last few days. The 'Huff-Duff' contact the *Reuben James* was investigating was in fact the German attack message from U-552, commanded by none other than Lieutenant Erich Topp, one of their more successful submarine captains."

"Did 'Boniface' indicate Topp knew who he was attacking?" asked Churchill.

"Not directly, as yet," the First Sea Lord answered.

Alexander added, "We think there is little doubt, however."

"The *Reuben James* was an older, four-stack, flush deck destroyer of the same class and vintage as the destroyers the Americans lent to us last year," Pound interjected. "So, configuration alone would not have been sufficient to identify their target."

"They know the Americans have been covering the western half of the transit. We have seen that in message traffic. The attack came at a point west of the transition longitude. The location has been clearly established."

"Very well," said Churchill. "I will send a simple condolence message to Roosevelt and will not embellish the incident. That said, from what you have reported, this incident will not sit well with the President. Whether it is sufficient to stimulate the Americans is yet to be determined. We do know a legislative bill repealing their Neutrality Act has passed the House, and by last report from Lord Halifax, appears to have the votes in the Senate. The Americans are inching ever so much closer to joining us. This incident may be the straw that broke the camel's back." Alexander and Pound nodded their agreement. "We shall see. Please keep me posted with any additional news on this incident and this convoy."

"Yes sir," they responded in unison, and both stood.

"Before I let you go . . . ," the Prime Minister said and reached for a simple folder on the table between them. "Do you recall this message?" he said and handed the single page contents of the folder to the First Lord.

Alexander read it, and then Sir Dudley.

PERSONAL and MOST SECRET

```
London [via U.S. Embassy]
Oct. 9, 1941, 2 A.M.
For the President from the Former Naval Person.
Fully understand situation which can quite well
be coped with here.  We definitely prefer your
second alternative of sending our troops via
Halifax for transhipment and onward passage to
Near East in United States escorts so far as
needful.  This plan lessens greatly dislocation
of complex escort programmes and delay in
subsequent convoys.  Furthermore, your valuable
fast ships would not run any appreciable risk
from U-boat attack.  If you agree our experts
can make a firm programme whereby nine British
liners arrive at Halifax with 20,800 men
comprising the eighteenth division and start
transhipment to your transports on November 7.
```

W

PERSONAL and MOST SECRET

"Are we ready to execute this transport arrangement?" asked Churchill.

"Yes. We are ready. In fact, as I recall," Alexander began and paused to look for a confirmatory head nod from Sir Dudley, "the 18th Division has staged near Halifax Harbor since our discussion last week. We are awaiting ship assignments from the Americans, which 'Intrepid' is facilitating as well."

"Very good. This is one more opportunity to get the Americans pregnant." Both men laughed. Churchill stood as well, thanked the men for coming to inform him at the late hour, and then called for his private secretary.

"John, please retrieve one of the duty stenographers. I must send an immediate message to the President of the United States."

John Martin performed his task with efficiency. The briefing by the First Sea Lord instigated a cascade of follow-on actions that would keep the Prime Minister and secretariat staff working until nearly dawn. Churchill possessed an effervescence the staff had not seen in a while. They liked the change.

Chapter 16

The shrewd guess, the fertile hypothesis,
the courageous leap to a tentative conclusion –
these are the most valuable coin of the thinker at work.
-- Jerome Seymour Bruner

Sunday, 9.November.1941
Chequers Court
Ellesborough, Buckinghamshire, England
United Kingdom

The day had begun peacefully by the standards of the day. Prime Minister Churchill had awaken mid-morning, taken his breakfast in bed, and dispatched the daily paperwork in his morning Dispatch Box before lunch. It had begun in a more routine manner.

After a quiet, simple lunch alone, Churchill bundled up for an afternoon stroll in the immaculate gardens of the country estate. He had a heavy, long overcoat over his usual siren suit overalls with a nice cashmere scarf around his neck, fur-lined gloves on his hands and a bearskin cap given to him by the Russian ambassador on his head. He needed the solitude and quiet of the gardens, even the chilly and foggy air of a very English autumn afternoon. The chill eventually returned him to the manor house.

The Prime Minister's duty assistant private secretary Anthony Bevir waited for him at the rear door. "Sir, Colonel Menzies is waiting for you in your study."

"Thank you, Anthony. Did we expect him?"

"No sir. He arrived unannounced."

As Churchill walked through the passageways toward the large study, he shed his gloves, cap, scarf and eventually his overcoat, handing the items to Bevir.

Menzies stood as the Prime Minister entered the room. Bevir closed the door behind the Prime Minister.

"Did you get lunch?" asked Churchill.

"Yes sir. I did, actually. Thank you, sir."

"Well, pray tell what have you?" Churchill said, as he sat in the adjacent, overstuffed, leather-upholstered chair. "It must be important for 'C' to make the journey in person on Saturday."

"First and foremost, good news . . . Force 'K' under the command of Captain William Gladstone Agnew aboard *Aurora* intercepted an Italian-German supply convoy bound for Libya. They engaged early this morning, just after midnight, southwest of Calabria. Reports received by the Admiralty indicate they sank all seven of the transports, including two ammunition ships that went up in spectacular fashion, along with at least one of the escort destroyers and perhaps two or three others seriously damaged. We had only one destroyer slightly damaged by shrapnel. I note this encounter not in place of the Admiralty, but only to illuminate that ULTRA played a vital role in staging the battle. We happened to be on a high point with their 'Blue' codes and were able to inform the Admiralty of this convoy's content, escort order of battle, sailing date and route."

"Not much more you could have given them," Churchill commented.

"No sir. It was the perfect amalgamation of intelligence and operational forces. We were also able to provide sufficient lead time that our forces out of Malta orchestrated a cover operation to mask any suspicion of our foreknowledge. The results were impressive. It was a night action with a waning three-quarter moon, a couple of hours short of zenith. The combination of the advance warning and especially the radar-directed main battery guns of the cruisers *Aurora* and *Penelope* were merciless."

"That will hurt Rommel."

"Without question. Our analysis estimates that convoy alone was roughly half of his monthly operational supply needs. We are watching carefully to see how he reacts to this loss, which will tell us the real impact."

"Excellent. We need more of these successes. I am certain A.V. or Sir Dudley will brief me on the full details later. Congratulations to you and Station 'X' for your exceptional work. Please pass along our most grateful appreciation for what Bletchley was able to accomplish."

"Gladly, sir. On the less but still positive side of the ledger, our sources in Moscow reported more Siberian divisions passed through the city in apparent high morale and deployed immediately in the cities defenses. We count more than a dozen in the last month, which implies Stalin is no longer concerned about the Japanese on his eastern border, or at least does not perceive them as a sufficient threat. Unfortunately, as of this morning, the Germans of von Bock's Army Group Center were within 40 kilometers of the city."

"Twenty-five miles."

"Yes sir."

"Nearly artillery range."

"Yes sir. There are also reports of shelling on the outskirts of the city already."

"Are they going to stop them?"

"Our current estimate is yes, we think they will. First persistent snow came last week. Temperatures have been at or below freezing. If winter has not arrived, it is very close. Now, it is only a matter of how the Germans handle and adapt to the cold. They are not used to fighting in such conditions. The prisoners of war recently taken outside Moscow reflect a serious lack of winter clothing and equipment. If the army as a whole is represented by these prisoners, winter should burden their forces swiftly and ruthlessly."

"They have come dreadfully close to their prize."

"But, close does not count in such affairs. Winter in the open of a battlefield will sap their strength, while it will bring a modicum of respite for the Red Army to reinforce. The Reds are far more accustomed and prepared to carry out combat operations in winter conditions, so we suspect we shall see some Soviet successes during the winter months."

"The next few weeks should tell the tale."

"Quite so. I can assure you, we are all quite attentive to what is happening outside Moscow."

"We must ask for the utmost attention. The situation in Russia is critical, and the outcome is vital to our war effort."

"We all recognize that reality, sir."

"Anything else?"

"Yes sir. The Germans entered and secured Kursk on the 3rd. In an interesting twist, Hitler himself ordered an entire air corps to redeploy immediately from Army Group Center to Sicily and Cyrenaica on the 5th. Lead elements of that transfer are already en route."

"They are so close to victory. The evil corporal is either extraordinarily over-confident in taking Moscow, or he is comparably out of his bloody mind . . . to jeopardize his ultimate achievement at such a critical time."

"Or, perhaps he is more concerned about the situation in North Africa."

"That is even more strange, since Rommel is rolling up the Army of the Nile. Wavell has not found the foothold to stop their advance. The need for an entire air corps to be moved from Russia to the desert is not obvious . . . 25 miles, I'm sure he can practically taste victory."

"We will keep watch, as we always do, Prime Minister."

"Good! Now, I'd like to return to the naval situation." Menzies nodded his head. "The Admiralty briefed me a week ago on the wolf pack attack on Convoy 'H' 'X' One Five Six. How did ULTRA play in that battle?"

"Short answer: not as much as we wanted. For reasons we know not, the communications traffic for U-boat skippers was uncharacteristically sparse. We were not able to help anywhere near as closely as we did this morning."

"What I want to know specifically is, did the U-boat captain that attacked the American destroyer *Reuben James* know he was attacking an American warship?"

"The First Lord asked me the same question a week ago. We have deciphered all of the naval message traffic for the week surrounding that incident. We have seen no indication either way. We received and passed on their attack message, but it said nothing of specific targets. They reported one escort sunk, but nothing about flag or configuration."

Churchill lapsed into contemplation for a minute. "That convoy made Liverpool a few days ago without further loss . . . well other than the Reuben James and a Sea Hurricane they catapulted to fend off a Condor, and then had to ditch, as they had no means of recovery."

"I shall instruct Denniston to pay close attention."

"Please be careful. I do not want that question widely known. I have traded general messages with President Roosevelt, and his government is considering their response. If the German captain knew, then it was an intentional attack and provocation, which in turn might color the President's response."

"Yes sir. I understand. I will make sure Denniston is aware of the sensitive nature of that question."

"Thank you, Stewart. Anything else?"

"No sir. That is it . . . at least for this morning."

"Would you care to stay for dinner?"

"Thank you, sir, but if I may, I should respectfully decline. I would like to stop at Station 'X' before heading back to London."

"Very well. Thank you for bringing the information directly. Safe journey."

"Thank you, sir. Good day."

Menzies departed. Prime Minister Churchill sat motionless for several minutes lost in thought. So many moving parts, this war was a long way from being concluded. Churchill felt the urge to press Roosevelt. In the end, he placed himself in the President's position and acknowledged he would appreciate the pressure. He would have to give Franklin the space he needed to get where he needed to be and to pray that the President's place coincided with their common purpose. Time would tell the tale. He had to wait.

—

Monday, 10.November.1941
Headquarters, Special Operations Executive
No.64 Baker Street
Marylebone, London, England
United Kingdom

Brigadier Colin Gubbins welcomed Trevor Andersen back to the headquarters of SOE and the home of what some were beginning to call the Baker Street Irregulars. Minister Dalton was tending to politics somewhere in

Whitehall, so this particular meeting would only be Gubbins and Andersen, and this time in Gubbins' modest office.

"Thank you for coming in, Mister Andersen."

"My pleasure, sir."

"Please take a seat. This will take a few minutes." They sat in opposite chairs across a small low table with nothing on it. "I have another potential mission we have not discussed before that I wanted to present to you." Trevor nodded his head. "We are in the final stages of planning and preparation for an important operation in Libya, North Africa. Are you familiar with the Special Air Service – SAS?"

"Only that such a service exists. I am not familiar with any details."

"Very well. Please allow me to raise your knowledge." Andersen listened and did not respond. "Last July, we formed a unique group we called the Special Air Service Brigade. It was hardly a brigade size . . . but, we have aspirations and we wanted to imply a more pervasive organization should the enemy become aware of the unit's existence. In this instance, 'L' Detachment, again to imply a much larger group, will deploy shortly to Libya. Their mission will be to capture *General der Panzertruppe* Erwin Rommel, *Befehlsgeneral Panzergruppe Afrika*," he said in high German, "and failing that to assassinate him."

"A rather ambitious objective, don't you think?"

"Yes, quite so. We received intelligence from a reliable source . . . ." *That has to be Enigma*, Trevor thought. ". . . that leads us to believe he will arrive at his forward headquarters in Apollonia, near the village of Susah, east of Benghazi, in a few days. The team feels they need a fluent German speaker, who looks German . . . ."

"Aryan."

"Yes, to be blunt . . . to enable their penetration of Rommel's field headquarters. Your name immediately came to mind."

"*Prima facie*, that sounds like a rather sporty operation."

"That is the nature of our business, Mister Andersen."

"How large is this detachment, and how do they intend to get close enough to snatch him and make a successful escape?"

"Six men. You would make the seventh." Andersen exhaled audibly and shook his head in disbelief. Gubbins ignored Trevor's reaction. "The primary means is by sea and submarine, since his headquarters area is closest to the coastline. The alternate is by the desert. An extraction team will be simultaneously staged at a concealment site roughly 20 miles inland should the submarine extraction fail."

"When do they deploy?"

"In two days." Gubbins paused, as if anticipating a reaction from Trevor, which did not come. "They will fly to Alexandria via Gibraltar, board a submarine, and if all goes to plan, they will land by rubber boat on the 14th. Our intelligence indicates his headquarters is in field tents located in the ruins of the old Roman city of Apollonia, which is on the coast. The plan calls for a day, perhaps two, of surveillance to establish the presence of the intended target. That is where you come in. You will be dressed in a rather dirty, dusty field uniform of an armor company commander to allow you limited access. Once he is localized, the team will enter three hours before dawn, sedate the target and haul him to the beach. The critical phase will be entry to at least three miles to sea. You will rendezvous with the submarine ten miles off the coast to make good your escape."

"What is my intended role exactly?"

"The short version . . . to confirm the general's presence."

Andersen laughed robustly. Gubbins did not react and allowed the laughter to subside. Trevor proclaimed, "I dare say, sir, I think you have an overly-inflated view of my skills."

Gubbins stared quite intently at Trevor without blinking or diverting his gaze. "You do not need to under-sell your skills either, Mister Andersen. The task is not to walk into his command tent and shake hands with the man. You are a tank company commander fresh from combat and on a mission for your company, like other combat commanders in the area. The most recent aerial photographs of the camp will be provided before boarding the submarine. We expect MI6 to have the photographs properly annotated so that you understand the orientation and your visit points. You cannot be expected to be fully familiar with the site, so you can ask questions of other officers. This is a combat, field headquarters site, so there is inherently confusion. Your only task will be to act like you belong there and gain any confirmation that Rommel is there. The team will handle the rest. You may need to communicate with the general once he regains consciousness. To our knowledge, he does not speak English, or any other language other than German." Trevor smiled. Brigadier Colin Gubbins did not smile. "Well?"

"As I said earlier, you have outlined a very ambitious plan."

"Are you in?"

"What about White Rose?"

"As we discussed, we will not attempt another contact until spring. This mission will not take more than a few weeks to complete *in toto*."

"What will you do with the general, if we are successful?"

"If nothing intervenes, he will be initially interrogated at Broadway House, and then interned at Trent Park, where he will be monitored until the end of the war, or some of other purpose is created for him."

"What happens at Trent Park?"

This time Gubbins smiled. "Let us just say, various intelligence agencies pay attention to officers detained at Trent Park. The intelligence value remains incalculable."

"Then, I surmise this mission is a unique opportunity and worth the risk."

"Yes, that is an accurate assessment."

"Very well, then. Where do I meet the team?"

Gubbins answered, "They have a house at Sennybridge in Brecon Beacons. You will join them this evening. You will fly out of RAF Hereford on Wednesday."

"Then, I should be off."

"There is a Hudson at RAF Northolt waiting for you. An SAS officer will meet you at RAF Hereford and transport you to the team."

"Very well, then. Off we go."

"Good luck, Trevor."

"Thank you, sir. I just hope this is worth it."

"We all do, Trevor."

Andersen stood, shook hands with Brigadier Gubbins, turned and departed the office and the building.

—

Monday, 12.November.1941
RAF North Weald
North Weald, Essex, England
United Kingdom
10:30 hours

The second request for Press interviews of the No.71 Squadron pilots in the matter of an hour proved to be the straw that broke the camel's back. Squadron Leader Tug Meares commanded Corporal Harris to report the squadron as Unavailable and stormed out of the Dispersal Hut an hour ago.

"What is the skipper pissed about?" Salt Morton asked.

"My guess," responded Whitey Whittington, "these Press requests are distracting from our mission."

"We are not here," Pete Peterson said, "for the entertainment of anyone else. We are here only to kill Germans."

It was cold outside. The pilots remained inside the Dispersal Hut and uncharacteristically quiet. They all had to wonder what was next.

Meares returned. Corporal Harris opened his mouth to speak, but stopped when their commander raised his right hand. "Gather up lads," he said. "Unless otherwise directed by me, personally, none of us will grant any further interviews with journalists of any nationality at any time." The pilots unanimously clapped and cheered. "I have requested and the base commander has agreed to seal the base to anyone other than authorized Royal Air Force personnel."

Wing Commander Jason Bradley Treadle had taken command of the airfield and its operations last summer. He had seen the nearly incessant parade of journalists from a dozen countries, predominately from the United States of America, so he was predisposed to Meares' request. Treadle had immediately instructed his base security department accordingly.

Meares continued, "This is not to say that you should avoid any conversations with the Press, when you are accosted outside the base. Whether you talk to the Press is your choice and decision. However, I would caution each of you to be courteous and respectful regardless of your choice, and above all mind your operational security protection. If you do talk to the Press, I would encourage you to inform me. I will in turn inform Wing Commander Treadle, who will inform the Air Ministry. We must remain focused on our flying and killing Germans. Are there any questions?"

"What about telephone calls?" asked Sweet Sweeny.

"If the telephone contact comes from the Press or anyone who is not personal to you, then the order stands. We are not to be disturbed while we are on duty."

"Does this include the Mess after we are released?" Salt Morton asked.

"Is the Officer's Mess on base?"

"Yes sir."

"Then, you have your answer. Yes, it applies. As I said and want to emphasize, henceforth, we must have a sanctuary of sorts to eliminate as many distractions as we can." Meares waited to see if there were any other questions from his pilots. "Very well, then, let us get back to business. Corporal Harris," he said, turning to look at his operations clerk, "kindly report the squadron at Available Status."

"Yes sir." Meares went to his office. Harris lifted the blue phone and completed his task.

Four of the new pilots – Sloppy, Jimmy, Hick, and Buddy – returned to their card game.

Dusty shifted seats to sit next to Brian. "You are usually the point of that Press attention, Hunter." Dusty said. "What do you think of the Skipper's order?"

"I think it's an order."

"Yeah, but do you agree with it?"

"We don't have to agree with it. An order is an order."

"Come on, give me a break."

Brian smiled. He had given his right wingman enough guff. "Yes, I say, it's about time. I have never been a fan of these . . . these . . . inquiries. I just want to fly. I do not want to be famous or see my name in print."

Frog joined them and sat in an open chair on the other side of Brian.

"Whether you like it or not," Dusty said, "you are the most famous pilot in this squadron."

"I don't care."

Dusty thought for a few seconds. "But, apparently a lot of other people do care."

"I do not fault the Press for reporting on things. That is their job. It is just that we don't need the distractions. There will be time to talk about what happened. I just don't think now is that time."

Frog leaned forward and looked to the Green Section Leader. "You know, Brian, a year ago September, the beginning of my senior year at university, a bunch were listening to the radio one evening and heard Edward R. Murrow's London After Dark broadcast about you." Forcier stared at Brian, glanced at Dusty and back to Hunter. "So, I ask you, how many pilots have had their flying exploits broadcast to the world?"

"I did not ask for any of that. I was just flying as we all do every day."

"No," protested Langford quietly. "None of us have done what you have done, Brian. Humility is admirable, but let's not lose sight of reality."

Brian ignored Dusty's comment, stood and walked to the window. "Looks like we're not going to be flying today . . . well, at least not this morning."

Langford followed his section leader and softly said, "Smooth subject change." Brian nodded his head but did not speak. "You don't like people talking about you, do you?"

"No. I don't," Brian answered. "I'm just a guy who enjoys flying – nothing more."

The two men looked out the window at the moderate but steady rain that would make the ground too soggy to move their three-ton aircraft. Dusty added, "Whether you like it or not, Brian, you are the best recruiting tool Colonel

Sweeny has in getting more Americans over here to help out." Brian shrugged his shoulder and again did not speak. Langford left Brian at the window.

A few minutes later, Squadron Leader Meares appeared in the Ready Room. "It's lunch time, lads. Corporal Harris, kindly report us Available-in-30-Minutes."

"Aye sir."

A covered truck waited behind the Dispersal Hut to take the pilots to the Officer's Mess for their mid-day meal. *If we're lucky, we'll get released after lunch. I need a beer*, Brian thought.

As they entered the Mess chattering about various insignificant topics, the duty desk clerk called out, "Mister Drummond, you have a message."

Brian went to the desk and took the small, folded paper from the aircraftman. Charlotte had called two hours ago and asked for him to call home. The note also said it was not an emergency. "Thank you." Brian gestured to Dusty and Frog to go on in to lunch.

All of the six telephone booths were opened. Brian picked one of the middle units, closed the folding doors and sat down on the small wooden seat. He lifted the handset and tapped the saddle three times to signal the operator. "Winchester Four Three Seven Nine, please," he said when the operator came on line. It took a little over a minute and he could hear the various connections being made.

"Standing Oak Farm," came Charlotte delicate, melodious voice.

"It's me. What's up?"

"Just wanted to tell you that we are now the proud owners of more than double our land holdings in Hampshire," she said with clear excitement."

"Congratulations, sweetheart."

"Thank you so much for all your help."

"I'm glad to be of service. How's Ian?"

"He is perfect and sleeping at the moment. I just fed him and he ate well. The doctor stopped by a couple of days ago to check on us. He declared all is well."

"Excellent. We've a short time for lunch, sweetheart. Is there anything else you need to tell me?"

"Yes," she paused and Brian waited. "I love you."

"As I love you. Again, congratulations, Charlotte."

They concluded their telephone call and Brian joined his mates at the Mess table in the dining room.

—

Thursday, 13.November.1941
Cabinet Room
No.10 Downing Street
Whitehall, London, England
United Kingdom
16:00 hours

"Gentlemen," Cabinet Secretary Sir Edward Bridges said strongly, "let us come to order." He waited for the last few men to be seated and for quiet to come to the refurbished Cabinet Room. "The Prime Minister has asked for a combined War Cabinet and Defense Committee meeting to discuss news recently received. First Lord, I believe you have the floor."

"Thank you, Sir Edward." First Lord of the Admiralty A.V. Alexander cleared his throat. "We just received notification from Commander Force 'H' and from Gibraltar that our aircraft carrier *Ark Royal* was struck by a single torpedo 25 miles east of Gibraltar. Her escorts are still prosecuting her attacker. As of a few minutes before we convened, there is no word, yet, regarding the destruction of the *Ark Royal*'s attacker. The ship reports the strike inflicted a grievous wound. Battle damage control measures are underway, but the captain has issued an alert that those measures within their capacity may not be sufficient to save the ship."

"Tragic," commented Clement Attlee. "Such a glorious history, including her part in finishing off the *Bismarck*."

"Quite so," Churchill injected. "How did the submarine get past the carrier's escort screen?"

Alexander looked to First Sea Lord Admiral Sir Dudley Pound to answer.

"We do not know, as yet. Admiral Somerville's flag is aboard *Nelson*. He is still gathering information."

Vice Admiral Sir James Fownes Somerville, KCB, KBE, DSO, had served under Vice Admiral Sir Bertram Home Ramsay, KCB, MVO, during Operation DYNAMO – the BEF evacuation from Dunkirk. He had assumed command of the newly formed Force 'H' just prior to being the central element for the execution of Operation CATAPULT – the neutralization of the French fleet at Mers-el-Kébir and Oran, Algeria, in July 1940. His Force 'H' had played an important role in forcing the battlecruisers *Scharnhorst* and *Gneisenau* to port and sinking of the battleship *Bismarck*. Since assuming command, the primary mission of Sir James and Force 'H' had been protecting resupply convoys to Malta.

Sir Dudley continued, "He sent a preliminary report that the torpedo was fired from an extremely long range, which if validated, made that a very lucky shot."

"Lucky or not, that shot still struck home."

"Yes sir, it did. Admiral Somerville knows we will all want to know how and why it happened, so that we can amend our tactics."

"Have the Germans developed some new torpedo?" asked Churchill.

"Not that we are aware of," Pound responded, "but that must be part of the investigation. I asked Admiral Pike that precise question"

"What were they doing and what are they doing now to save the carrier?" asked Greenwood.

"They were returning to Gibraltar to resupply and pick up the next convoy planned for early next week. *Ark Royal* is still making way, although at a seriously diminished speed. She is listing badly. As the First Lord indicated earlier, some of the destroyers were dispatched to get the submarine before she can take another shot. They've dropped several score of depth charges already with the effort still on going."

"Is there anything we can do to help?" Greenwood pressed.

"No sir, not that I can think of. Admiral Somerville believes they are doing everything that can be done to ensure the returning merchant ships make the safety of Gibraltar Port, help the *Ark Royal*, as they are able, and destroy that submarine."

Sir Edward sensed the end of the War Cabinet's inquiry. "Shall we move on?"

"Before we do," Churchill responded, "please convey to Admiral Somerville that he must do his utmost to save *Ark Royal*."

Sir Dudley said, "I know he will, Prime Minister."

"Then, tell him also are prayers are with them."

"Yes sir."

"Now," Churchill said, to change the topic, "I thought it appropriate to report to the combined War Cabinet and Defense Committee on my recent communications with President Roosevelt. There are several elements to our exchange. First, he, his service chiefs and his administration are watching the situation in the Pacific Ocean quite intently. I might even say they are preoccupied with the rapidly mounting signs the Japanese are preparing for a major operation. The intelligence chiefs have not been able to determine the object or mission of whatever it is they are preparing for, but let it suffice to say, it is something serious."

"So," interjected Greenwood, "he is not going to take action after the sinking of the *Reuben James*? He is not going to do one of his famous fireside chats with the American people and the world?"

"We discussed that incident across several messages. I suppose the short, direct answer based on my understanding of his words is no, probably not."

"I thought that was the anticipated incident from the Placentia Bay conference," Attlee said.

"Yes, it was. However, as I said, he is looking for the situation in Asia to clarify."

"Then, the usefulness of the *Reuben James* incident may pass," added Attlee.

"He is aware of that reality, I can assure you. An incident like the *Reuben James* was precisely what we discussed at Placentia Bay. However, his apprehension with respect to the rapidly evolving situation in Asia has captured his attention, and I am reticent to attempt diversion at the moment. I believe we need to allow things to play out for the time being. The best information we have from MI6, the Admiralty or any other intelligence sources available to us remains quite foggy at best. We do not know what the Japanese intend to do with their military forces. Further, we do not want to risk wearing out our welcome in Washington. To take Arthur's point, I have no indication from Franklin whether he will take action or not, and I do agree with Clement, the window of usefulness of the *Reuben James* incident is closing. The best I can counsel today is we wait and see what turns up . . . at least for the time being."

"I cannot disagree with that assessment, Winston," Attlee said.

"Me, either," added Greenwood.

The other members of the War Cabinet – Sir John Andersen, Sir Kingsley Wood and Anthony Eden – gestured their concurrence. With those items discussed, the War Cabinet moved onto the routine matters of logistics, rationing, housing, reconstruction and actions to be taken to improve the quality of life of their citizens during the struggle of wartime.

—

Thursday, 13.November.1941
Oval Office
The White House
Washington, District of Columbia
United States of America
15:15 hours

"Thank you for stopping by on such short notice, Bill," the President said, as his Coordinator of Information 'Wild Bill' Donovan entered the room and shook the President's proffered hand.

"My pleasure, Mister President." Donovan sat on the long couch to the President's right, when Roosevelt gestured for him to sit. Just the two of them occupied the Oval Office.

"My apologies for not providing the topic, so you could prepare." Donovan nodded his acknowledgement. "What do we know about this so-called Japanese Indo-China treaty?"

"First, it is hardly a treaty. The Japanese have used their alliance with Germany to their advantage. The French had a veritable pistol to their heads, granting the Japanese rights of occupation to all of Indo-China."

"For what purpose?"

"While we know some of the details of the French concessions by MAGIC intercepts, we do not know exactly what the Japanese intend to do with their ill-gotten gains and the situation remains unclear."

"What is your best guess?"

"The intelligence agencies continue to assess that specific object. We are all watching as closely as we are able. As of yesterday afternoon, we believe the Japanese are preparing for a major operation against the Malay Peninsula and specifically Singapore, and likely Burma and beyond."

"India?"

"That objective is certainly quite plausible. However, it is more likely they are looking to the Dutch East Indies . . . far more raw resources they need to sustain their war machine. Australia cannot be discounted. The Japanese with the support of the Germans tried to intimidate the Dutch, as they did the French, without success. The British are apprehensive about French Indo-China, but they are far more concerned about the Dutch East Indies. Beyond the raw material, those islands would be the ideal staging area for an invasion of Australia."

"Churchill seems assured that Singapore is impregnable, and as long as Singapore stands, Australia will be beyond their reach."

"There is credibility to the British position, but we know from Admiral Pike and the Admiralty intelligence office that the Royal Navy is more apprehensive than the government apparently is at the moment. They have been ordered by the War Cabinet to move the front line battleships *Prince of Wales* and *Repulse* to Singapore as a show of force. However, the *Ark Royal* is the carrier assigned to the task force deployment and she was hit by a torpedo earlier this morning. She is reportedly listing badly. While she has not sunk, yet, the damage will undoubtedly leave the task force without an aircraft carrier and without air cover beyond the range of their land-based fighters. The government is certainly considering its options."

"The *Prince of Wales* is a magnificent ship. I've been aboard her, you know." Donovan nodded his head. "We need to be careful with that information,

Bill. The last thing we need at this time is to expose any growing rift within the British government."

"Yes sir. Only my British desk chief and me, and now you, are aware of my communications with Admiral Pike. I will keep it that way."

"Very good. So, if I understand your current analysis, the Japanese military preparations are likely pointed south and there is no risk to our forces?"

"Oh no, Mister President, I have erred if I left you with that impression. The Philippines could easily be their next target, but that would only be to eliminate the immediate threat of our forward-based naval, land and air forces. The Philippines do not have the resources of the Dutch East Indies. We have not seen the scale of forces necessary for an invasion attempt on the North American continent. However, the Territory of Hawaii is not beyond their reach."

"Do you really think they would attack Pearl Harbor?"

"They could, not that they would, sir."

"Should we warn them?"

"Of what, Mister President? We have only information they are preparing for something, but we truly have no idea what. They might be massing forces for a flanking maneuver within China. The Navy seems to shuck off suggestions they might attack us in the Philippines or elsewhere. Our MAGIC material is scarce and tenuous at best. If we react prematurely, we might raise suspicions within the Japanese military establishment that we are reading their mail."

Unbeknownst to the President or his Coordinator of Information, the Navy had suppressed the MAGIC intercept of September 24th and transferred the director of naval intelligence to a fleet command in October. The leadership of the Navy could not fathom the Japanese being so audacious to attack the powerful U.S. Pacific Fleet.

"We need to know what the Japanese intend to do, Bill."

"I am well aware of that information requirement, Mister President. We are working the sources we have and continuously trying to expand our sources. MAGIC remains our best source, but decryptions are far too sporadic. The cryptographers have as their number one priority improvement of MAGIC production, but we are just not there, yet."

"Is there anything I can do to help?"

Donovan considered the question and his reply for a few moments. "To be blunt, sir, probably not, without doing more damage than good. I can assure you they are doing their utmost. From what I know of the British successes and processes with Enigma and ULTRA, a closer collaboration with the British Station 'X' would benefit both cryptologic services."

"That will have to wait, I'm afraid. We have not reacted to the sinking of the *Reuben James* and my window of credibility is closing. Escalating our efforts against the Germans with the potential of a major Japanese attack would appear foolish and juvenile. I need clarity to the Asia situation, soon, Bill."

"Yes sir.

"So, to summarize your response to my question, you do not place much weight on the Japanese-French Indo-China treaty."

"No sir. There are far too many potential purposes from routine administrative to more sinister. We have compared notes with MI6 twice, so far, and we are all still in agreement, in and of itself, we cannot attach more significance to the agreement other than the Japanese with the Germans successfully strong-armed the Vichy French."

"Very well, then. Please keep me informed, as you are able and I know you will."

"Absolutely, Mister President." Donovan stood.

"Now, if you would be so kind, please ask Harry to join me. Good luck, Bill. We are all counting on you to make sense of this mess."

"Thank you, Mister President. We shall do our best." Roosevelt nodded his concurrence and wheeled himself back to his desk. Colonel Donovan departed. Harry Hopkins appeared shortly, thereafter.

—

Chapter 17

Danger close
-- friendly forces are within
close proximity of the target

Monday, 17.November.1941
Headquarters, Secret Intelligence Service
No.54 Broadway
Westminster, London, England
United Kingdom

"Good afternoon, 'C,'" Head of SIS Operations Carl Ambrose Acton said upon entering the office of the Director-General, Colonel Stewart Menzies.

"Is it really good, Carl?"

"You shall be the judge, sir."

"What have you?"

"We just received coded confirmation from a well-placed asset in the Berlin city police that validates a clear text broadcast this morning that *Generaloberst* Ernst Udet has died. Our source indicated he shot himself in the head and left a suicide note railing against Göring's incompetence and mismanagement of the Air Force. He blames Göring directly for losing the Battle of Britain and he apparently believed Operation BARBAROSSA was a disastrous decision that will be the ruin of Germany. Udet reportedly had been talking to his girlfriend on the telephone when he did the deed."

"A blaze of glory, I suppose. Well, I wonder how the Nazis are going to conflate this one?"

"Hard to say. They will suppress the suicide note for sure. Goebbels goes into convulsions with such negative news. They might use the usual, he died heroically on the battlefield, or they could make him the scapegoat for their air force losses to the Russians and us. We have known for some time now, from various unrelated sources, that Udet was not pleased with Göring or Milch, the SD, SS, or the Nazis in general. Udet was surprisingly vocal with notable foreigners about his displeasure."

"Isn't Milch a *mischlinger*, as the Nazis like to say?"

"Yes sir, a half-Jew. *Generalfeldmarschall* Erhard Milch was Aryan-ized by the stroke of a pen before the war began. When he was promoted late last year, he was given the title of air inspector general and reportedly given responsibility for all aircraft design, development and production, reporting only to Göring himself. I imagine that the reality of his mixed blood heritage remains like Damocles' sword over his head."

"It never ceases to amaze me how atypically German the Nazis are with the application of their so-called hereditary standards."

"Indeed."

"I recall indications of Udet's dissatisfaction, but I had no idea he was or even could be that despondent. Could this have been an execution with the subterfuge of a suicide note?"

"That potential can never be discounted with the SS involved. However, the sense from the confirmatory message tends to substantiate the authenticity."

"Do you see any reason to take this to the Prime Minister or the War Cabinet?"

"That is your call, sir."

"We have just another dead German."

"Yes sir. Although . . . it was Udet himself who confirmed the execution day for Operation BARBAROSSA to the Swiss ambassador. More than a few believe Udet was the most likely source of the Oslo Report."

" . . . as was Canaris."

"Yes, quite so. Some on our German desk believed Udet was a prime candidate to be turned."

Menzies was surprised. "I had not heard that before."

"We did not talk about it . . . like souring the deal before the chickens have hatched."

Menzies considered Acton's statement for a few moments. "Are you aware of the pending mission by SOE into Bavaria with the objective of engaging the White Rose student group?"

"Yes sir. We have been supporting Brigadier Gubbins and his assigned agent Trevor Andersen. We participated in several training sessions, before Andersen deployed to North Africa."

"Which reminds me," interjected 'C,' "what do you know about FLIPPER?"

"That is the operation code name for the mission I just mentioned . . . to capture or kill Rommel."

"Where are we on the mission plan?"

"We have one specialist assigned to the team and onboard HMS *Torbay*—the submarine taking them in and bringing them out. As I recall, they inserted by rubber boat day before yesterday. They are going to try to snatch Rommel from his field headquarters in Apollonia, Cyrenaica. Their intended execution time is tonight, shortly after midnight local time."

"Andersen is part of the SAS team, is he not?"

"Correct sir. He was assigned by Gubbins himself and is serving as the team's language and disguise expert."

"I sure hope nothing happens to him. FLIPPER is a very risky mission."

"We expect to resume our support of Andersen's training upon his return. It is a good plan with a reasonable probability of success."

"If the Gestapo does not get to them first."

"Yes sir, that is always a threat with such infiltration missions. Andersen is one of our best when it comes to language and blending skills."

"Yes, he is . . . and far more skilled and accomplished than most folks will ever know."

"I have never met the man."

"I have . . . many years ago . . . before the war started, actually. He attended a meeting we had with Churchill before the old man returned to the government to consider a very sensitive mission. I've met him several times since, and I have been impressed every time. Good man . . . goes by the codename 'Diamond,' if you ever see the reference in traffic."

"He has quite the reputation."

"Deservedly so, I must add. Anything else?"

"No sir. I just wanted to make you aware of circumstances of Udet's death, in case it should arise in your conversations."

"Thank you, Carl. Keep up the great work."

"Yes sir. Thank you, sir," Acton replied, before standing and departing the director's office, and then closing the door behind him.

—

Tuesday, 18.November.1941
No.10 Downing Street
Whitehall, London, England
United Kingdom
15:15 hours

Minister of Economic Warfare and chief of the Special Operations Executive Hugh Dalton arrived unannounced for an impromptu meeting with the Prime Minister with what had to be an important matter of some urgency, before the Prime Minister's scheduled 15:30 meeting. "Pardon my intrusion, Prime Minister, and thank you for seeing me on such short notice. I just received this message and felt you should see it immediately, so I decided to bring it myself."

Churchill extended his left hand from behind his desk and grasped the single piece of paper from Dalton. He read the message several times, intently.

SECRET

```
ZZZZ/1194DCA3589/AD-S0/SSRTABEG/836/ZZZZ
DATE: 18.11.41 0147[Z] HOURS
TO: ADMIRALTY SOE
FROM: TORBAY
SUBJECT: OPERATION FLIPPER
BREAK
FLIPPER TEAM APPARENTLY DISCOVERED AND LOST
BREAK TWO MANAGED TO ESCAPE MAKE RENDEVOUZ AND
ARE ON BOARD BREAK VULTURE NOT PRESENT BREAK
INTEND TO REMAIN AS PLANNED FOR TWO REMAINING
RENDEVOUZ WINDOWS PER OP PLAN BREAK DEBRIEFING
TWO RECOVERED TO LEARN MORE
END
ZZZZ/1194DCA3589/AD-S0/SSRTABEG/836/ZZZZ
```

SECRET

"I presume you do not know any more than this?" asked Churchill.

"No sir. I checked with A.V. before I left Baker Street. He had just received the same message. The Torbay is still at sea, loitering for one or two more recovery windows to service. I suppose there is always hope as long as she is in the area. The preliminary report was to alert us that bad news will likely follow."

"We took a serious risk and we apparently lost. It was still a worthy mission," Winston commented.

"Yes, and although we do not know much, it was like the enemy knew the team was coming and they were waiting for them."

"It was a small team, as I recall, Hugh. It could have also been a fateful chance encounter that compromised the team. Those things happen in war."

"Certainly."

"I am most concerned for the welfare of the team," Churchill added. "They will not likely communicate further until she has safely made Alexandria."

"That is our expectation."

"Please keep me informed as you learn more. Thank you for coming over straight away. Can you leave this message with me?"

"Certainly. Again, thank you for seeing me. Now, I must be on my way, as I am on the verge of encroaching upon your scheduled 15:30 appointment."

Churchill nodded his head and returned to his paperwork on the desk before him. Dalton departed. Winston had lapsed into contemplation about what might have happened. The thought of their communications being compromised as they were reading the German message traffic had to always be considered. It was the nature of the beast; it was not a pleasant thought.

The duty private secretary, John Martin, announced, "Colonel Menzies is here for your scheduled meeting."

"Thank you, John. Please send him in," Churchill said and repositioned from this desk to the informal area of his office.

"I saw Hugh Dalton on the way out," Menzies offered, as Martin closed the door. "He told me about FLIPPER."

"Yes, tragic. What have you?"

'C' placed the Buff Box on the short table between them. Churchill retrieved the only key and unlocked the case.

"This is not a good message."

Churchill read the only message.

MOST SECRET - ULTRA

```
SECRET SENSITIVE
DATE: 1 NOVEMBER 1941
TO: ARMED FORCES HEADQUARTERS
FROM: HEADQUARTERS AFRICA CORPS
BREAK
STAR SENDS BREAK STAR DEPARTED BY AIR FOR
EMERGENCY MEETING IN ROME THIS MORNING AT DAWN
BREAK ANTICIPATE RETURN THIS HQ ABOUT 18 NOV
END
SECRET SENSITIVE
```

MOST SECRET - ULTRA

"I presume 'Star' is Rommel."

"Yes. That is our assessment."

"A day late and a quid short, I'm afraid," mumbled Churchill.

"The bloody Germans altered their encryption process to and from Africa on the first of the month. It took us nearly three weeks to break it. I brought this one, as I knew it reflected upon FLIPPER, although I did not know how much until just now."

"Yes, nearly the whole team was lost, probably killed or captured. Two managed to escape and were recovered, but we do not know much more. Hugh just informed me with this message. Churchill handed the message from HMS *Torbay* to Menzies.

"Dear God. Of all the times to have a slow decryption. Hopefully they were captured rather than killed. It is most unfortunate, but these things will happen. With your permission, Prime Minister, I would like to confer with Denniston to sensitize the Station 'X' personnel even more that timely decipherment is critical to the war effort. Just one or two days might have made the difference to save the team."

"That is the point. We have had successes. We will have failures. The thought of capturing and interrogating Rommel was just too bloody tempting."

"On the positive side, we did get independent confirmation from a reliable source that Udet definitely committed suicide rather than being executed by some SS thugs. He was literally talking to his girlfriend when he pulled the trigger."

"Good to know."

"If I understand correctly, Wavell kicked off his counter-offensive," said Menzies.

"Yes... Operation CRUSADER. That was another part of the temptation in the FLIPPER mission. The stars seemed to line up. In addition to the prospects of capturing Rommel, we had hoped to disrupt the German command in Africa to amplify the potential for success of CRUSADER."

"At least the Germans know we made an attempt. That should rattle their cage a little."

"There is always hope."

'C' retrieved the sensitive message, returned it to the proper folder, and locked the Buff Box. Menzies left without another word between them. Once again, Churchill was left with his thoughts of what might have been.

—

Wednesday, 19.November.1941
RAF North Weald
North Weald, Essex, England
United Kingdom
15:30 hours

They had flown three sorties on the day's mission – a transit to and from RAF Manston for refueling and rearming, and a cross-Channel CIRCUS raid. The object of the day's mission was strafing a railroad-marshaling yard and nearby logistics staging area used by the Germans. No.71 Squadron was the sole squadron assigned this particular CIRCUS mission. While they were overwater, they had throttled back to conserve fuel, and opened their coolant and oil heat exchangers to push their operating temperatures to the low end of the safe band, to store up thermal capacity for the high speed, low level run into their target. The tactic had worked. The anti-aircraft fire was above them. The fighters sent to intercept them wound up chasing them without success.

All of the North Weald squadrons had been committed to cross-Channel missions of one form or another, and the Eagle squadron was the first to return to home base. They landed with all their pilots and aircraft, although a handful of the latter did have some non-serious new holes; fortunately, none of the former experienced any wounds.

The intelligence debriefings were still on going and had taken longer than usual, as their debriefers probed for details and annotated common sketches of their target area.

Brian had been one of the first few pilots to complete his debriefing. It was a sunny but chilly day common to mid-autumn at this latitude, yet Brian found it refreshing to sit outside against the south-facing, downwind wall of the Dispersal Hut, around the corner from the main entrance. He kept his fleece-lined flight jacket and trousers on, and closed his eyes to doze for however long he would be allowed. Brian was distantly aware that two people had joined him for enjoyment of the afternoon sun. Thankfully, no one spoke for an indeterminate, welcome while.

Frog Forcier was the first to disturb the peace. "Wow!" he proclaimed loudly. "That was sumpin', wasn't it."

Brian did not twitch or respond. Dusty Langford took the bait. "Yeah, Frog, it was a good day."

"There were so many secondary explosions . . . black smoke from fuel and brilliant explosions from ammunition, even . . . did you see that steam plume from the locomotive."

"Yeah, we did," came Dusty's calm, cold, unemotional response. "We were on the same mission."

Brian kept is eyes closed and smiled inside, not wanting to offer any response. *Dusty is becoming quite the veteran.*

"We were bobbing and weaving all the time we were down there to avoid those clouds."

"Except the skipper."

"Yeah, he was first, so he didn't have to dodge nuthin'. What did you think, Hunter?"

Brian wanted to ignore his left wingman, but he knew he should not do so. He answered simply, without opening his eyes. "Dusty covered it rather well, I should think."

"Yeah, but wasn't that an exciting mission?"

"It might have turned out differently, if those One Oh Nine's had caught up to us," Brian answered calmly and rather matter-of-factly.

"But, they didn't."

"Nope. Fortunately for us, they did not show up until after we made our last turn for the run home."

"Damn it, Frog, can't you see the man was trying to rest?"

Brian opened his eyes to see Dusty sitting next to him with his wooden chair leaned back against the Dispersal Hut wall, and Frog, Hick and Sloppy standing in front of them. "It's OK, Dusty. This is part of the learning process. If they had caught us on our eastbound strafing leg, we would have given them a pursuit turn to close the gap, and with our temperatures already on the high side, and low on fuel and ammunition, we would have had a very different problem. As exciting as it was to shoot things up, don't ever forget, those guys are out there to kill you. They are really good at what they do. We were very lucky today."

"Better lucky than good," Sloppy chimed in.

"Yes," responded Brian, "that is what we say, isn't it."

They watched as one Spitfire and two Hurricane squadrons returned for landing within minutes of each other. Brian wondered, *were they all on the same mission?* No matter how many times he heard the melodious tones of Merlin powerplants, Brian truly marveled at the sound, even the spit and sputter of idling engines. They counted 12 for each squadron. *Yes, it had been a good day.*

Sweet came around the corner of the building and announced, "Skipper said we are released for the day."

"I guess everyone completed their debriefing," Dusty said.

"Guess so," Brian answered, as he stood up. He was among those who needed to return his flying kit to his assigned peg before securing for the evening.

The pilots walked in small groups from the Dispersal Hut to the Officer's Mess. They had much to celebrate. Tonight would be a good night.

—

Monday, 24.November.1941
Cabinet Room
The White House
Washington, District of Columbia
United States of America

Harry Hopkins pushed the President's wheelchair into the Cabinet Room. Everyone stood and stopped talking.

"Please be seated, gentlemen," President Roosevelt said, as Harry rolled him to the center of the long, rectangular, highly polished, mahogany, conference table. The men in attendance took their seats as directed. As soon as the President was appropriately situated, Hopkins took the seat behind the President against the outside wall.

In the Cabinet Room with the President and Hopkins were:

-- Secretary of State Cordell Hull;

-- Secretary of War Henry Stimson;

-- Secretary of the Navy Frank Knox;

-- Chief of Staff General George Marshall;

-- Chief of Naval Operations Admiral 'Betty' Stark;

-- Deputy Chief of Staff for Intelligence (G-2) Brigadier General Sherman Miles, USA [USMA 1905];

-- new Director of Naval Intelligence (N-2, ONI) Rear Admiral Theodore Stark 'Ping' Wilkinson, USN [USNA 1909], who had replaced Captain Alan Kirk the previous month;

-- Director of War Plans (N-3) Rear Admiral Kelly Turner, and

-- Coordinator of Information Colonel 'Wild Bill' Donovan.

"Well, Frank, you called this meeting. What do we know?"

Knox cleared his throat and began, "We intercepted a message two days ago that took us that amount of time to decipher from the Japanese high command ordering their combined fleet to sea. We received confirmation," he paused to check his wrist watch, "90 minutes ago that all warships and transport ships assembled at Shanghai sailed just after dawn this morning, which was 18:30 our time last night, 15 hours ago. We have not been able to confirm the sortie of the main fleet from the Home Islands, but it is our considered opinion," again, he paused to look to Admiral Stark and Admiral Wilkinson for confirmatory head nods, "a large portion of the home fleet sortied as well."

"So, whatever they have planned, they are now in the pre-execution phase," Roosevelt said.

"Yes, Mister President," answered Knox. "That is our assessment."

"Do we know any more?"

"No, Mister President," came the reply from the ministers in unison.

Roosevelt looked to Donovan. "Bill, do you have any additional information?"

"No, Mister President."

Unfortunately, the MAGIC intercept of two months prior had not been widely disseminated. None of the men in the Cabinet Room, except Turner, had seen or were aware of the request for information from the military liaison in the Japanese Foreign Office to the consulate in Honolulu, Hawaii. If they had, the latest news might have appeared more ominous than it did. Such was the course of intelligence and military operations in 1941.

"Well, something is about to happen. No one moves that many ships for no reason. So, what are we to make of the movement?" asked the President.

"The signs point to Southeast Asia, Mister President," answered General Marshall. "They could be executing a flanking operation in Southern China, or reinforcing units already in French Indochina, and we cannot discount Thailand or the Malay Peninsula."

"What about our forces in the Philippines?" Roosevelt asked.

Everyone except Roosevelt, Hopkins and Donovan looked to General Miles, who responded, "We think the Japanese are unlikely to take on the United States directly, or the United Kingdom and its Commonwealth for that matter. Thus, the Philippines, Hong Kong, or Singapore are unlikely targets."

"How confident are you in that assessment, General?"

"We can never be 100 percent assured, Mister President. I would say we are as confident as any intelligence agency can be."

"I am somewhat reticent to be confrontational, but I fear I must . . . do you agree with that assessment, Colonel Donovan?"

"We have no indications to validate or offer to the contrary here, Mister President. The COI cannot argue with the G-2's analysis. I must say, the Dutch East Indies would have to be included in their potential target list. Operations into the South China Sea and beyond cannot ignore our military and naval forces stationed in the Philippines, or the Royal Navy at Singapore."

"The size of our forces," interjected Admiral Stark, "are hardly sufficient for offensive operations. The British have fewer than we do. We must not read too much into this Japanese movement . . . to over-exaggerate their intentions."

"They may well know that, don't you think, Admiral Stark? They may believe they have free reign to do as they please."

"We have nine battleships and three aircraft carriers in the Pacific Fleet. We intentionally stacked the Pacific Fleet as a deterrent after the oil embargo in July."

"Yes, we did. What if they are not deterred?"

General Marshall jumped in, "The most likely mission of the Japanese is Hong Kong, or at least to isolate Hong Kong, and consolidation of their holdings in China."

"Where are we on moving the China Marines?" the President asked.

"The lead elements departed two days ago," answered Admiral Stark. "The bulk of the 4th Marines will depart on the 27th and the remainder of the regiment on the 28th. Two American President Line ships were contracted to move the regiment out of China, to execute the order we gave them earlier this month. They should be in Subic Bay, the Philippines, by the end of the month. They will join General MacArthur's Army Forces Far East for the time being."

General Marshall added, "MacArthur has a hodge-podge of American and Philippine units . . . barely two divisions worth at best. The Marines will be a welcome addition, but hardly sufficient for a proper defense of the territory. The current war plan calls for an aggressive defense of the Bataan Peninsula to buy time for reinforcements to arrive. A brigade was sent in July to bolster his forces, but much of the equipment he needs remains at West Coast depots awaiting shipping."

"We have insufficient lift . . . ," responded Admiral Stark, stopping when the President raised his right hand, palm out.

"This is not an order of battle debate, gentlemen," the President interjected. "I need to understand the threat . . . the risk to our interests and forces. The Japanese are not moving their forces of this magnitude for practice. Something is about to happen."

"Yes sir," answered Marshall. "I believe we are agreed on that point. My point earlier was we are not likely the object of their operation."

"Under-estimating the Japanese can be dangerous," Donovan added.

The President did not wait for the lance to be picked up. "What is State doing?" the President asked Secretary Hull.

"Dialogue between us and the Japanese Foreign Office continues. They claim to seek a diplomatic solution to the current situation and want the sanctions against them removed."

"Our position has not changed."

"No, it hasn't, and we have clearly communicated that reality to the Japanese government. I must say that since Konoye's resignation and General Tojo's ascendency last month, the Foreign Office seems to be going through the motions rather than conducting a purposeful or productive negotiation. It is like they are playing for time. They want to keep us focused on diplomacy."

"A distraction?" asked the President. "Subterfuge?"

"Given our discussion this morning," said Hull, "we must assume that is the case."

"Then, it seems to me," the President paused to consider his words, "we must issue a war warning to our forces in the Pacific region, to prepare them for what may be coming their way. I will send a personal note to Prime Minister Churchill, and you should be sending similar communications to your subordinate commands and your British counterparts," he said, looking directly at Hull, Stimson and Knox, "to inform them of our assessment. They should expect aggressive action by Japan in the next few days to a week or two at the outside."

"How far do we extend this war warning?" asked the Chief of Naval Operations.

"I thought I was fairly clear on that point," answered Roosevelt rather tersely.

"Then, you wish us to extend the war warning to the West Coast, Alaska and Hawaii, and perhaps if so, we should warn the Canadians," replied Stark.

Roosevelt stared at Admiral Stark, glanced at Frank Knox, and then back to Stark. "I take your point, Admiral. Unless our intelligence professionals inform us otherwise, I would suggest the warning should be focused on the Western Pacific . . . perhaps as far east as Guam. I have heard no threat to Midway, Wake or the Hawaiian Islands, or the West Coast of the United States. So, we must be careful not to overreact and cause undo concern, especially to the western states."

"Very well, Mister President," Secretary Knox replied. Stimson and Hull nodded their agreement.

"Then we are agreed. Please keep a close eye on things in the Pacific. We must know of any changes in the information we gain or any indications of Japanese intentions. We have entered a critical period. Let us remain vigilant and prepared. Is there anything else we need to discuss?"

"No sir," answered Stimson.

President Roosevelt glanced over his left shoulder to Hopkins and nodded his head toward the door. Hopkins wheeled the President out of the Cabinet Room and back to the Oval Office.

—

Wednesday, 26.November.1941
Headquarters, Secret Intelligence Service
No.54 Broadway
Westminster, London, England
United Kingdom

Carl Acton entered the Director-General's office and closed the door. Menzies motioned for them to set at the small conference table. "I want to brief you on a clarifying picture we have of the FLIPPER disaster, but first, I need to make you aware of two news items." Menzies nodded his agreement. "We have confirmation from Donovan that at least one and probably two Japanese fleets have sortied from Shanghai and probably the Home Islands. The combined fleets are believed to number in the hundreds of warships and support vessels as well as three or four full divisions. Donovan indicated the service chiefs were directed to issue war warnings to their Far East commands. We have not seen the messages, yet, but we expect them any day now."

"I will need to discuss this with the Prime Minister and the War Cabinet as soon as we have confirmation of those alerts. Also, if you and Admiral Pike are satisfied with their confirmatory evidence, then we must notify our forces and the Commonwealth countries in the region that something serious is afoot. On second thought, I'd better talk to the Prime Minister straight away."

"Yes sir. I'll see to the notifications as soon as I return to my desk."

"What else?"

"As you know, Udet's suicide was confirmed as well. The Germans have decided to declare his death an heroic combat fatality and gave him a resplendent state funeral attended by Hitler, Göring, and other Nazi notables on the 22nd. Of interest, Udet's funeral Guard of Honor was comprised of the German Aces of Aces including Galland, Gollob, Lutzow, Oesau, and Peltz among others. Captain Werner Mölders was returning from Crimea to be in that Guard of Honor when he was killed while travelling as a passenger in a Heinkel 111 that crashed while attempting to land in a snowstorm at Breslau, Poland. He was credited with 24 aerial victories and believed to have had at least 30, as Hitler banned him from risking any more in combat."

"An odd twist . . . that an accomplished fighter pilot should die in a simple transport accident."

"Indeed. It also appears his was not the only fatality while attempting to attend Udet's extravagant funeral. Lieutenant General Helmuth Wilberg also perished in a transport aircraft crash near Dresden in the same snowstorm enroute to Udet's funeral. By the way, I might add as a footnote, Wilberg was another one of Göring's favored air force officers to be Aryan-ized by the stroke of a pen; his mother was Jewish."

"Oh my . . . how things twist."

"Anyway, that is the news I have at the moment. Now . . . to the FLIPPER disaster. HMS *Torbay* has not made Alexandria harbor, as yet. She completed her recovery operations per the plan. No additional personnel were recovered, and she cleared the area. She is expected to dock the day after tomorrow. We should know more when we can interview the two survivors. Both men have now been identified: SAS Sergeant Jeremy Helper and SOE Field Agent Trevor Andersen. The latter apparently completed his task and established that Rommel was not present. He and Helper made the first extraction window. As they were paddling out for the rendezvous with the submarine, heavy gunfire was seen and heard in the vicinity of the team's staging area. No one made the second or subsequent extraction windows. Security forces in the vicinity of Rommel's forward headquarters could have easily outgunned the team once they had been discovered. The end of the fight came when the two survivors were about halfway out to the rendezvous. That is all we know so far. As I said, we may be able to get more in a few days when we can talk to Andersen and Helper. Both SIS and SOE field supervisors will be waiting to debrief them."

"From the sounds of it, we may not know more until we get the POW listings from the Germans, and even then, we may not know the rest of the story until the end of the war, and perhaps ever."

"We will keep looking. That's all I have 'C.' Acton concluded.

"Very well. I think it prudent to brief the Prime Minister and Foreign Minister straight away."

"Yes sir." Acton departed and Menzies raised his telephone to take the next step.

—

Chapter 18

And that's the way it is.
-- Walter Cronkite

Friday, 28.November.1941
Office of Naval Intelligence
The Admiralty
Whitehall, London, England
United Kingdom

His visitor had requested an urgent private meeting and Admiral Pike readily consented. The First Sea Lord had indicated just yesterday that changes might be happening at a very fast pace.

Special Naval Observer Rear Admiral Bob Ghormley, USN, entered Jumper Pike's office. Pike's aide closed the door behind Ghormley. "Good afternoon, sir," greeted Ghormley.

"I hope it is a good afternoon, Bob. To what do I owe the pleasure?"

"As you will note, I have been instructed to bring three messages regarding actions taken by the United States Government, to inform you, the Admiralty and His Majesty's Government."

"Do tell?"

Ghormley opened a small leather, brief case and removed a red-striped, bordered folder. He then removed the first single page message from the folder and handed it to Vice Admiral Pike.

TOP SECRET

```
NO
PA KTA NR 483
TS 271809Z NOV 41
FM CNO
TO CINCPACFLT CINCAF
INFO CINCLANTFLT SPENAVO LONDON
T O P   S E C R E T
BT
THIS DISPATCH IS TO BE CONSIDERED A WAR WARNING
BREAK NEGOTIATIONS WITH JAPAN LOOKING TOWARD
STABILIZATION OF CONDITIONS IN THE PACIFIC
HAVE CEASED AND AN AGGRESSIVE MOVE BY JAPAN IS
```

```
EXPECTED WITHIN THE NEXT FEW DAYS BREAK THE
NUMBER AND EQUIPMENT OF JAPANESE TROOPS AND
THE ORGANIZATION OF NAVAL TASK FORCES INDICATES
AN AMPHIBIOUS EXPEDITION AGAINST EITHER THE
PHILIPPINES THAI OR KRA PENINSULA OR POSSIBLY
BORNEO BREAK EXECUTE AN APPROPRIATE DEFENSIVE
DEPLOYMENT PREPARATORY TO CARRYING OUT THE
TASKS ASSIGNED IN WPL 46 BREAK INFORM DISTRICT
AND ARMY AUTHORITIES BREAK A SIMILAR WARNING IS
BEING SENT BY WAR DEPARTMENT BREAK BREAK
SPENAVO LONDON INFORM BRITISH BREAK CONTINENTAL
DISTRICTS GUAM SAMOA DIRECTED TAKE APPROPRIATE
MEASURES AGAINST SABOTAGE END
BT
COPY TO WPD WAR DEPT
NNNN
```

TOP SECRET

"War warning, ay?"

"Yes, so I am informed."

"Your government has included the Kra Peninsula and Thailand, which would point toward Singapore, and potentially Burma and India."

"Exactly why I am here, I do believe."

"I am not familiar with War Plan 46. Can you give me a summary outline?"

"Sure, in summary, given Japanese hostilities in the Western Pacific region, it directs Pacific Fleet to initiate carrier-based air raids on Japanese holdings in the Marshall Islands in an attempt to relieve pressure on the Philippines and Allied territory."

"Is Admiral Kimmel prepared to do so?" asked Pike.

"He had better be. War Plan 46 is part of the Rainbow operations plans derived from the Joint Staff Conversations we held early this year in Washington, so they are fairly fresh. We are not quite sure what appropriate defensive deployment preparatory to War Plan 46 directs or offensive operations means in this instance. It is my understanding the War Plans Department is attempting to clarify the instruction."

"Very well."

"I felt an earlier message sent four days ago was germane to this situation, so I included it as well." Ghormley retrieved the second message and handed it to Pike.

TOP SECRET

```
NO
PA KTA NR 466
TS 242301Z NOV 41
FM CNO
TO CINCPAC CINCAF COM11 COM12 COM13 COM15
INFO SPENAVO LONDON CINCLANT
T O P   S E C R E T
BT
CHANCES OF FAVORABLE OUTCOME OF NEGOTIATIONS
WITH JAPAN VERY DOUBTFUL BREAK THIS SITUATION
COUPLED WITH STATEMENTS OF JAPANESE GOVERNMENT
AND MOVEMENTS OF THEIR NAVAL AND MILITARY
FORCES INDICATE IN OUR OPINION THAT A SURPRISE
AGGRESSIVE MOVEMENT IN ANY DIRECTION INCLUDING
ATTACK ON PHILIPPINES OR GUAM IS A POSSIBILITY
BREAK CHIEF OF STAFF HAS SEEN THIS DISPATCH
CONCURS AND REQUESTS ACTION ADEES TO INFORM
SENIOR ARMY OFFICERS THEIR AREAS BREAK UTMOST
SECRECY NECESSARY IN ORDER NOT TO COMPLICATE AN
ALREADY TENSE SITUATION OR PRECIPITATE JAPANESE
ACTION BREAK GUAM WILL BE INFORMED SEPARATELY
END
BT
COPY TO WPD WAR DEPT OP12
NNNN
```

TOP SECRET

"This was a precursor message to the war warning?" asked Pike.

"Yes."

"The combination of the two messages leave me with the impression the U.S. Government is convinced war will break out with Japan very soon."

"That is my impression as well, although I must caution that there are no offensive orders, as yet. These are all preparatory or defensive messages."

"Understood."

"Just this morning, the Chief of Staff, General Marshall, issued a parallel message to Army forces Far East, which is the third message." Ghormley handed Pike the third paper.

TOP SECRET

```
NO
PA KTA NR 492
TS 2809239Z NOV 41
FM CNO
TO COMNPCF COMSPCF CINCPAC CINCAF
INFO CINCLANT SPENAVO LONDON
T O P   S E C R E T
BT
SPENAVO LONDON INFORM BRITISH
RESEND
FM COS
TO CMDR WEST DEF COM CINCAFPI SIXTH ARMY
NEGOTIATIONS WITH JAPAN APPEAR TO BE TERMINATED
FOR ALL PRACTICAL PURPOSES WITH ONLY THE BAREST
POSSIBILITIES THAT THE JAPANESE GOVERNMENT
MIGHT COME BACK AND OFFER TO CONTINUE BREAK
JAPANESE FUTURE ACTION UNPREDICTABLE BUT
HOSTILE ACTION POSSIBLE AT ANY MOMENT BREAK
IF HOSTILITIES CANNOT REPEAT NOT BE AVOIDED
THE UNITED STATES DESIRES THAT JAPAN COMMIT
THE FIRST OVERT ACT BREAK THIS POLICY SHOULD
NOT REPEAT NOT BE CONSTRUED AS RESTRICTING YOU
TO A COURSE OF ACTION THAT MIGHT JEOPARDIZE
YOUR DEFENSE BREAK PRIOR TO HOSTILE JAPANESE
ACTION YOU ARE DIRECTED TO UNDERTAKE SUCH
RECONNAISSANCE AND OTHER MEASURES AS YOU
DEEM NECESSARY BUT THESE MEASURES SHOULD BE
CARRIED OUT SO AS NOT REPEAT NOT TO ALARM CIVIL
POPULATION OR DISCLOSE INTENT BREAK REPORT
MEASURES TAKEN BREAK A SEPARATE MESSAGE IN
BEING SENT TO G2 NINTH CORPS AREA RE SUBVERSIVE
ACTIVITIES IN THE UNITED STATES BREAK SHOULD
HOSTILITIES OCCUR YOU WILL CARRY OUT THE
```

```
TASKS ASSIGNED IN RAINBOW FIVE SO FAR AS THEY
PERTAIN TO JAPAN BREAK LIMIT DISSEMINATION
OF THIS HIGHLY SECRET INFORMATION TO MINIMUM
ESSENTIAL OFFICERS BREAK UNQUOTE. WPL 52 IS
NOT APPLICABLE TO PACIFIC AREA AND WILL NOT BE
PLACED IN EFFECT IN THAT AREA EXCEPT AS NOW IN
FORCE IN SOUTHEAST PACIFIC SUB AREA AND PANAMA
NAVAL COASTAL FRONTIER BREAK UNDERTAKE NO
OFFENSIVE ACTION UNTIL JAPAN HAS COMMITTED AN
OVERT ACT BREAK BE PREPARED TO CARRY OUT TASKS
ASSIGNED IN WPL 46 SO FAR AS THEY APPLY TO
JAPAN IN CASE HOSTILITIES OCCUR END
BT
NNNN
```

TOP SECRET

"Hostile action at any moment is rather direct wording," Pike observed.

"Yes, it is."

"And, the United States desires that Japan commit the first overt act seems rather passive given the tone of these messages . . . that offensive, hostile, military action seems to be imminent."

"General Marshall goes on to direct commands to take appropriate active measures to ensure proper defense within their areas of responsibility."

"May I retain these messages, so the First Sea Lord and other officials can read the words themselves?"

"There are no instructions beyond inform, which means verbally rather than physically. Therefore, I have not been authorized to disseminate these documents, as they are classified top secret. I chose to share the messages with you because of our relationship and to reinforce the seriousness of the situation."

"What am I permitted to share?"

"You have read the messages. You can share the information you have seen with any appropriate individuals within the government. Given what these messages represent, I suspect there will be direct, official communications within days on this matter and situation. I believe I was asked to share them with you as a heads up—an alert, so to speak."

"Very well, so it shall be. The First Sea Lord told me just yesterday that events may be moving very fast, and now I know why. I shall notify the First Lord and First Sea Lord as soon as we are done. I am certain they will want to notify the Prime Minister and War Cabinet quickly as well. Thank you for the information, Bob."

"Thank you for seeing me on such short notice, Sir Geoffrey. These are troubled times."

"Yes, they are. Godspeed and following winds to you and your countrymen. Our prayers are with you."

"Thank you. Good day, Bob."

Both men stood shook hands and departed Pike's office. Ghormley turned left to leave the building, while Pike turned right, heading directly to the First Sea Lord's office.

—

Monday, 1.December.1940
Station HYPO
Administration Building, Naval Station Pearl Harbor
Russell Avenue and Port Royal Street
Pearl City, Honolulu County, Oahu, Hawaiian Islands
Territory of the United States of America
13:10 hours

The Fleet Radio Unit Pacific, commonly referred to as FRUPAC or Station HYPO, occupied a windowless, reinforced, concrete room in the basement of the Administration Building. The code name came from the phonetic word for 'H' used by the military at the time. No one could remember precisely whether the 'H' stood for Hawaii, their general location, or for the radio antennae array at He'eia, on the west side of the island, which they used for their radio intercept and communications work. It was a fine point of detail that no one really thought much about. There was only one entrance into or out of the secure room, closed off by a large, thick, heavy, locked, solid steel door and guarded around the clock by a series of two, heavily armed, very serious Marines. The men who worked inside affectionately referred to the secure workspace as the 'dungeon.' The large room was crammed with tables full of boxes containing cards and printouts from tabulating machines that continually spit out messages intercepted from the various receiving stations around the Pacific and sent to HYPO. The dungeon housed the cryptanalysis unit of the Pacific Fleet and was manned 24 hours a day by a handpicked group of Navy code breakers and Japanese language specialists.

Lieutenant Commander Joseph John 'Joe' Rochefort, USN, served as officer-in-charge of the naval signals intelligence center at Station HYPO and occupied a common desk on the floor with the rest of his staff.

The duty watch officer Lieutenant Junior Grade Jason Holiday, USNR, and Senior Chief Petty Officer Bradley Robertson approached Rochefort's desk together. Robertson was the duty Japanese analyst. The Senior Chief had spent considerable time in Japan after the Great War and in China supporting the 4th Marine Regiment. He spoke the Japanese language fluently with a keen appreciation of most idioms used by the Imperial Japanese Navy. Like Rochefort, Robertson had met *Kaigun-Taisho* Isoroku Yamamoto, the commander-in-chief of the Combined Fleet, several times both professionally and socially, during their prior duty assignments in Japan.

Holiday handed the single sheet of paper to Rochefort. "We just deciphered this message, sir." Rochefort took the proffered paper, but listened to Holiday. "It was received at noon and took us an hour to decode. Chief Robertson thinks it's pretty important."

TOP SECRET - MAGIC

```
TOP SECRET
DATE: 19411202 1730 HOME
FROM: SUPREME COMMANDER TOKYO
TO: COMBINED FLEET
CLIMB MOUNT NIITAKA 1208 REPEAT 1208
NAVY SERIAL 10
TOP SECRET
DECIPHERED: HYPO 020937Z DEC 41
```

TOP SECRET - MAGIC

"What is your assessment?" Rochefort asked Robertson.

"Hard to say, sir," Chief Robertson replied. "It would appear to be something fairly important. This is from Admiral Yamamoto himself to the Combined Fleet, like an ALNAV. However, in this instance, we have assumed the reference to be the units that sortied last week. It is clearly a fully coded message containing pre-arranged code words, which would leave one to believe they are seriously concerned about operational security, thus pretty damn important."

"Agreed, but what about the text?" Rochefort pressed.

"Taken on its face, Mount Niitaka is the highest mountain in Japanese territory – 13,000 feet on Taiwan and not a particularly easy ascent from what I'm told. The implication could be their operation is a challenge of the highest

366   Cap Parlier

order with considerable risk. What that is exactly we have no clue that I am aware of, yet. The 1208 is important in that it is the only element repeated. It could be an execution date . . . in this instance, the eighth of December, given their common date notation format."

"So, the seventh here."

"Yes sir. But, it could be anything . . . a location, a target number, even a time, anything."

"Could this be an attack command?"

"No way to know, sir. It could mean anything. I suspect only a handful of senior Japanese admirals would know the exact meaning."

"Not much I can take to Intel or the CinC."

"No sir. But, they do not send this kind of message for grins."

"Point taken, Brad. OK, let's forward it to Op-20G. Maybe they have more relevant data than we do.

"Yes sir," answered Holiday.

They returned to their work, looking for a needle in a field of haystacks. It was these disassociated hints that haunted Rochefort.

—

Sunday, 7.December.1941
Chequers Court
Ellesborough, Buckinghamshire, England
United Kingdom
10:15 hours

Winston Churchill was finishing his breakfast in bed, still dressed in his morning robe, his daily, morning, Despatch Box of communications from last evening, David Smithfield, the Churchill's butler, knocked and opened the door. "Mister Peck, a moment please," Smithfield said.

Duty Assistant Private Secretary John Peck looked first to Churchill, who nodded his consent. He stepped out of Churchill's bedroom. "Yes, Smithfield."

"Sir, Mister Menzies has just arrived unannounced."

"Please show him up. I know the Prime Minister will need to hear what he has brought."

"As you wish, sir."

Smithfield walked down the hallway, leading to the stairs. Peck returned to Churchill.

"Sir, Smithfield has gone downstairs to retrieve Colonel Menzies."

"Thank you, John. We will need to clear the room."

"Yes sir," Peck responded and ushered the two duty stenographers out of the room.

Churchill finished the last few bites of his eggs and potatoes, placed the lap tray on the other side of the bed, and moved the Despatch Box to the night table. He instinctively checked the presence of the key at his chest, straightened his robe and smoothed the *duvet* at his waist. The double knock on the door announced his guest's arrival.

"Sir, Colonel Menzies is here to see you," said Peck.

"Please show the Colonel in."

Menzies entered straight away, dressed in a dark business suit with the Buff Box manacled to his left wrist, and said in a calm but upbeat voice, "Top of the morning to you, Prime Minister."

"And, the rest of the day to you, Stewart. What have you for me in the box this dreary, near winter morning?"

"Just one message, sir." 'C' placed the box on the bed, next to Churchill's left hip. Winston leaned slightly forward, unlocked the case cover, and extracted a single, pink sheet of paper.

MOST SECRET - ULTRA

```
SECRET SENSITIVE
DATE: 7 DECEMBER 1941
TO: ARMED FORCES HEADQUARTERS
FROM: FOREIGN OFFICE
RETRAN: IMPERIAL JAPAN EMBASSY BERLIN
BREAK
EFFECTIVE 0001 1941 DEC 7 HOME ALL JAPANESE
DIPLOMATIC STATIONS SHALL DESTROY ALL CODE
BOOKS AND REVERT TO BRIDGE
END
SECRET SENSITIVE
```

MOST SECRET - ULTRA

"If I read this correctly, the Japanese Foreign Office has directed all diplomatic code books to be destroyed."

"Correct. It is highly doubtful the German Foreign Office would issue such a message to the military without direct communications from Japan to this effect."

"Which in turn is a common prelude to war."

"That is our interpretation."

"What is bridge?"

"We do not know for certain. However, our educated assessment is a code word for their new, probably contingency codes to be used during the initial stages of whatever is going to happen, until new, operational, code books can be issued."

"Do the Americans know this? Do we have anything from them?"

"We have not shared this information, as yet. We only just deciphered this message," Menzies glanced at his wristwatch, "not quite two hours ago."

"Please transmit it as soon as possible to the President, Secretary Hull and Colonel Donovan."

"Yes sir. I will go directly to Station 'X' and transmit from there."

"Make sure to ask them if they have confirmation by their sources and if they have any additional information."

"Yes sir."

"So, whatever they have been up to is going to happen soon . . . in the next few days, I presume."

"That is our assessment."

"The Americans issued their war warning on the 27th. What do they know that we do not?" Churchill returned the message to the Buff Box and locked the cover.

Menzies checked the cover to make sure it was secure and stepped back. "Based on my discussions with Donovan's man in London, they are just more jittery with the developing picture. Much of their information has come from us. Hong Kong is prepared as best they can be. Singapore and India are alerted, but they are unlikely targets. Perhaps we should issue our war warning for Commonwealth and territorial holdings in Asia."

"I tend to agree with you, Stewart. However, for an action like this, I need to give the War Cabinet the opportunity to weigh in. I'll call a special session for this afternoon. We will most likely meet in the Cabinet War Rooms. I will ensure we get notice to you for your attendance. Do you think you can tend to the communications at Bletchley, and then make it back to London, by say, four this afternoon?"

"I do not anticipate a problem. If I may, I would like to telephone Denniston from here to meet me at Station 'X', to ensure he knows precisely what is going on and he can increase their sensitivity for intercepts."

"By all means. You are also welcome to bring him along for this afternoon's Cabinet meeting."

"If you don't mind, sir, please allow me to assess the situation at Station 'X' and decide on his attendance."

"Very well. Agreed."

"Thank you, sir. Now, if you will excuse me, I shall be off. I will see you later today."

Churchill waved his hand, as if shooing a fly. "If you would be so kind, would you ask Peck and Sawyers to see to me?"

"Yes sir," Menzies answered, bowed his head slightly and departed.

The first to arrive was Sawyers. "I need to get dressed immediately, Sawyers. I must be off to London in short order." Churchill stood, doffed his robe, and with Sawyers expert assistance, he began to dress. Before his valet could retrieve the components of a vested business suit, Winston decided on his medium blue, siren suit – the overalls designed for him to don quickly in an air raid. The one-piece outfits resembled flight suits worn by some pilots. Peck arrived before Churchill had completed dressing. "There you are, John. Please contact Sir Edward immediately. I need an urgent meeting of the War Cabinet this afternoon. Now that I think of it, I will need the chiefs of staff in attendance as well. Let us shoot for sixteen hours. Ensure Sir Edward invites Colonel Menzies to attend. Have Mrs. Landemare make up a quick picnic basket of sandwiches and drinks. We shall have to lunch on the way into the city. Sergeant Carrick will drive with Thompson, Ismay, you, one of the stenographers, and me of course." Churchill chuckled softly, as it he had made a joke. "We shall have to finish the Despatch Box on the road. It will be a little cozy, but we can snuggle in for the two hour drive."

"Anything else, sir?"

"Yes, please convey to our guests my apologies for the regrettable recall to London. I have sudden, most urgent business I must tend to this afternoon."

"Certainly sir," Peck responded. Churchill nodded his head for Peck to get on with his tasks. "By your leave, sir." Churchill nodded again.

Once Peck had departed, Sawyers asked, "Will you be returning to Chequers after your Cabinet meeting, sir?"

"Most likely not, Sawyers. Please see to it that you, Smithfield and Landemare have transport to the Annexe. I suspect we shall have a very busy few days."

"Yes sir. I will see to it."

Churchill checked his appearance in the full length, standing mirror. Satisfied, Winston left the bedroom for the driveway. He would work in the rather luxurious, Rolls-Royce Phantom III limousine, while the final arrangements and collection of travel-mates were completed.

—

Sunday, 7.December.1941
Cabinet War Rooms
New Public Offices
Whitehall, London, England
United Kingdom
16:15 hours

"**W**e are all present," announced Cabinet Secretary Sir Edward Bridges. "The floor is yours, Prime Minister."

Churchill scanned the underground conference room one more time. The entire War Cabinet and the Chiefs of Staff Committee, with General Ismay, and both Menzies and Denniston in attendance as well. Everyone was cleared for ULTRA information. None of their usual staff support was in the room, and the door was closed. He also knew there were two, armed, Home Forces guards posted outside the door to ensure there was no disturbance of the meeting. "Thank you for amending your Sunday afternoon." Colonel Menzies had already gone through the Buff Box opening ritual and carried the ULTRA message into the War Room. "MI6 and Station 'X' retrieved this message," he held up the pink paper, "this morning." He passed it to Clement Attlee, to his right. The ULTRA message moved quickly to each attendee. "I believe we are in agreement," he said, looking to and receiving a confirmatory head nod from the Director-General of the Secret Intelligence Service (MI6), "that message is a certain sign offensive operations by the armed forces of Imperial Japan are likely to strike in the next few days. 'C', if you will, do we have any new information on their objective since this morning?"

"No sir."

"We have discussed the Japanese intentions numerous times. I see no need to rehash what we know and don't know. As 'C' informed us, we have no new information regarding their intentions or objectives with this latest deployment, just that it is very large and nearly total. This morning's ULTRA message," Winston said, pointing to the pink paper two-thirds through the initial reading by those present, "makes it very clear that whatever they are up to, it is a rather serious operation and it is happening soon. The question before the chiefs and the War Cabinet is, should we strike first, before the Japanese have entered the execution phase of their plan?"

"What exactly are you suggesting?" asked Arthur Greenwood.

"The prevailing opinion of both our intelligence services and the Americans is the Japanese objective is most likely in the vicinity of the South China Sea . . . most probable objective being the Dutch East Indies and their bountiful resources. I am proposing we order Force 'Z' to sortie immediately from Singapore and interdict the invasion fleet, to disrupt their plans."

"Prime Minister," interjected First Sea Lord Admiral Sir Dudley Pound. "I am compelled to remind you and the War Cabinet that Force 'Z' is without air cover, after the loss of *Ark Royal*, and then the grounding of *Indomitable* last month. While *Prince of Wales* and *Repulse* are exceptional ships of the line under competent leadership, they are vulnerable to air strikes without air cover. The Japanese will undoubtedly have a carrier or two with their invasion armada. Sacrificing Force 'Z' as a gesture hardly seems appropriate."

"Well, that is quite an objection, First Sea Lord. Apparently, I have more confidence in Admiral Phillips than you do."

"I must strenuously object, again, Prime Minister. My counsel has absolutely nothing to do with the commander's competency; it has everything to do with the reality of modern naval warfare. While capital ships were formidable entities in the last war, the aircraft carrier and modern aircraft have altered that balance. Two battleships and their escorts without proper air cover are simply too vulnerable to air attack."

"Are you suggesting, Sir Dudley, our first line warships are simply show pieces for the harbor at Singapore?" Churchill said with a rather irritated tone.

"No, Prime Minister, I am not. To be frank, I feel myself in a comparable state as Sir Hugh was in May of last year," Pound said, referring to the calmly defiant resistance of Air Chief Marshal Sir Hugh Dowding, during the tragic days of the Battle of France. "The Royal Navy is stretched very thin. We have not been able to replace *Ark Royal* for Force 'Z'. I am urging you and the War Cabinet to not sacrifice Force 'Z' for political reasons."

Churchill stared at Admiral Pound, clearly not pleased with the First Sea Lord's counsel and words of response. "Very well, then, First Sea Lord, what do you propose we do?"

"We need either reliable, accurate intelligence of their position and destination, or some other specificity to their actions. Admiral Phillips is well aware of the situation and the Japanese deployment, as well as our assessment of the information we do have from Admiral Pike's Intelligence Branch and MI6. Force 'Z' is simply too small and vulnerable to air attack to be left wandering around the South China Sea searching for a likely, vastly superior, Imperial Japanese Navy fleet. The Japanese have shown no caution in their naval actions to date. They are not likely to be intimidated or dissuaded by Force 'Z', to be frank."

"All's well and good, Admiral; however, if I can interpret your words, you want Force 'Z' to wait until we have clear indications regarding the enemy's objectives."

"Yes sir, that is correct. We need to know where to point Force 'Z' before the trigger is depressed, so that Admiral Phillips can properly evaluate his air cover situation. Several squadrons of Brewster Buffalos are hardly capable of a concerted defense, set aside Force 'Z'."

"There is a New Zealand Hurricane squadron at Singapore," Air Chief Marshal Sir Charles Portal added.

Pound looked askance at Sir Charles, as if to say really, you think a dozen Hurricanes are going to make a difference. "Yes, well, my opinion remains the same. Admiral Phillips must have a clear target to avoid squandering scarce and valuable assets."

Churchill lowered his head like a ram about to charge his adversary. "Are there other contributions to this particular topic?"

Clement Attlee leaned forward. "I have listened intently to this discussion. I am inclined to agree with you, Winston; however, I am unable to ignore or discount the First Sea Lord's informed counsel. We must husband our resources and use them wisely."

"I agree with Clement," added Greenwood.

Sir John Andersen, Sir Kingsley Wood and Ernest Bevin simultaneously said, "Concur."

Churchill looked to Anthony Eden, who responded, "An essential principle of warfare remains . . . the best defense is a good offense. Any action is better than inaction."

"With all due respect, Foreign Minister," interjected Sir Dudley, "I am not suggesting inaction. I am only urging prudence in the actions we are about to take." Eden did not reply.

"Very well, then. The informed wisdom of this august body advocates caution. So be it! Please ensure Force 'Z' has clear instructions. I dare say given this," Churchill said, holding up the single pink page that had been returned to him, "we shan't have long to wait for the requisite information. The Admiralty must keep the War Cabinet immediately informed of Admiral Phillips' decisions and actions, at least until this situation is clarified."

"Very well," answered A. V. Alexander, First Lord, speaking for the first time.

"Then, we are adjourned for now. I urge everyone to remain close to your communications for the next few days. Something dreadful is afoot, and we must be prepared to act promptly."

No one spoke as they filed out of the War Cabinet conference room and turned left to ascend the stairs out of the Cabinet War Rooms bunker. Churchill did not move. He simply turned his swivel chair and stared at the map on the far wall depicting Southeast Asia. After several minutes of contemplation, he too ascended a different set of stairs to the No.10 Annexe on the ground floor of the New Public Offices building.

—

Sunday, 7.December.1941
U.S. Army Early Warning Radar Site No.3
Opana Point, Oahu, Hawaiian Islands
Territory of the United States of America
07:02 hours [12:32, Washington, DC; 20:32, London]

The antenna for the Signal Corps Radio Set SCR-270 radar unit dominated the clearing at the crest of the coastal hill. The long axis of the rectangular dipole array stood vertically to a height of 55 feet on a rotating base, mounted on a mobile trailer with four stabilization and leveling legs. A separate gasoline-powered generator truck provided the necessary electrical power for the unit. The operations truck housed the electronics to operate the radar unit on the back half of the truck bed, metallic compartment. The front half contained the actual operating station for two operators with its large, etched, display scope that allowed the operators to determine the direction, range and altitude of aircraft within range of the antenna's radio signals. The assembly of equipment at the Opana Point site on the northern tip of Oahu was actually serial number 12 – the 12th operational unit built by Westinghouse Electric Company – and had only become operational the previous July. The Opana Point unit was one of four SCR-270 radar units deployed around Oahu.

"It's time to secure the unit," announced Private Joseph 'Joe' Lockard, the supervising operator on this particular shift.

By direct order of the Commanding General, Hawaii Department, General Short, the SCR-270 units were only operated for four hours from 03:00 to 07:00, each day. None of the operators understood why they were restricted to only four hours per day, but they liked to joke that the Army brass wanted to make the complicated and undoubtedly expensive equipment last longer.

"Wait, Joe, do you see this?" said Private George Elliot – the duty scope operator this particular morning – and pointed to the unusually large, green luminescent return painted on the radarscope. Lockhard looked over Elliot's shoulder.

"Remember, George, the lieutenant told us a flight of B-17's is inbound and supposed to arrive from California this morning."

"This return is a lot bigger than six B-17's, Joe, and whatever this is, they are coming out of the north . . . too far north for a flight from California."

"Perhaps the lead navigator got off track and is just recovering."

"They would have to be pretty far off track for this direction . . . and look at it. The return is getting bigger. We'd better call this in."

"OK. You're the operator. Call it in."

Elliot lifted the telephone handset off its cradle and called the Pearl Harbor Intercept Center at Fort Shafter, on the north side of Naval Station Pearl Harbor. The Intercept Center served as the common point for the four radar sites and the duty controller would order any action to be taken on detections by any one of the radar sites. The duty controller for this shift was First Lieutenant Kermit Arthur Tyler, USAAF. Tyler's day job was the Executive Officer of the 78th Pursuit Squadron, based at Wheeler Army Airfield, on the south side of Schofield Barracks in the middle of Oahu. Pilots from the various Oahu squadrons rotated through a stint as intercept controller, ostensibly to give the pilots a better appreciation of the detection and engagement process. Two months ago, his squadron had completed their transition to newly delivered P-40 Warhawk fighters.

"Sir, this is Private Elliot at Opana Point. There's a large formation of planes coming in from the north." George paused for a reply. "Yes sir. I picked them up on the scope at 140 miles, zero zero three degrees from our site at 8,000 feet on up. They're now at 130 miles, still inbound at the same altitude." Again, Elliot paused for Lieutenant Tyler to respond. "With all due respect, sir, this return is far larger than six B-17 heavy bombers." George shook his head and looked over his shoulder at Joe. He looked back at the scope. "Based on what I see right now, I would say several hundred aircraft at the same altitude." Elliot continued to shake his head as he listened. "Very well, sir. Thank you, sir." Elliot returned the handset to its cradle.

"What did he say?"

"Not to worry about it."

"OK. Let's shut down the unit and get outta'here."

The two men worked together, precisely following the well-rehearsed procedures for switching off the scope, the radar, and the power supply. As they opened the control compartment door, they squinted their eyes against the bright sun in the eastern sky. The skies were crystal clear, not a cloud could be seen. They closed all the access and cooling doors, and secured the power truck and operations van. They rechecked everything to make sure all the necessary locks were secure. After months of working the radar unit, they thought nothing of the procedures they had been ordered and trained to follow. The site would remain unmanned and unfenced until they returned at 02:30 hours the next day to prepare for their operational shift – just another day in the life.

"OK," said Lockard. "Let's go get some breakfast and get back to post. It's another beautiful day in paradise. We might actually get lucky and get some beach time."

"Sure."

The two soldiers jumped into their assigned, new, U.S. Army Truck, General Purpose, 1/4 ton, 4x4, or Jeep, the phonetic contraction of General Purpose – GP thus Jeep, as they commonly called the vehicle, and headed down the hill. They would stop at a local café at the bottom of the hill for breakfast, before returning to their barracks at Schofield.

—

Sunday, 7.December.1940
Station HYPO
Administration Building, Naval Station Pearl Harbor
Russell Avenue and Port Royal Street
Pearl City, Honolulu County, Oahu, Hawaiian Islands
Territory of the United States of America
07:27 hours [12:57, Washington, DC; 20:57, London]

Commander Joe Rochefort nervously scanned both routine and deciphered communications traffic. He instinctively felt the nearly imperceptible electricity of impending events. He was not often in the office on a Sunday morning, but his building anxiety rendered him sleepless. Troubling information had trickled into the confines of the dungeon in drips and drabs, and seemed to be picking up its pace. Radio traffic began to mount at increasing frequency, some of it disturbing, open, friendly radio communications received an hour ago from the USS *Ward*, attacking a submerged submarine detected south of the entrance to Pearl Harbor.

Petty Officer 3rd Class Johnny Jacobs jumped up from his receiver station four rows left from Rochefort, knocking over his chair with a loud bang and starting to run to the chief, ripping his headphones from his head. "Sir, I just heard in Japanese a clear voice transmission. I could only get a cut on the source . . . north-northwest of the antenna array." Rochefort shrugged his shoulders, as if to say 'and.' "The message was one word repeated three times. *Tora, tora, tora.*"

"Are you sure that is what you heard?"

"Yes sir . . . plain as if the sender was sitting right here with us."

"In Japanese?"

"Clear, distinct, educated."

"Tiger, tiger, tiger?"

"Yes sir."

Rochefort thought for a moment, remembering his Japanese language training in Tokyo before the war. "Tora could be a code word of significance to those who know the meaning; but, it could also be a contraction I remember hearing years ago at several naval parties in Japan – *totsugeki raigeki*, literally meaning 'lightning strike.' Several high-ranking Japanese naval aviation officers liked that term. Either way, combined with your azimuth cut, it cannot be good. Get back on your set, Johnny. See if you can pick up anything else. I need to get on the horn."

"Yes sir," answered Jacobs, and then he swiftly walked back to his station.

Rochefort lifted the telephone handset and dialed the number he knew by heart – Commander Edwin Thomas 'Ed' Layton, USN [USNA 1924], Naval Intelligence, Pacific Fleet. "Commander Layton, please." He listened. "OK. Thank you." Rochefort dialed another number he knew from memory – Layton's residence. He waited for the answer he expected. "Commander, this is Joe. We have a problem . . . I suspect a big problem. We just picked up a clear broadcast in Japanese from north of the island that I suspect is an attack signal. Yes, Ed." He paused for a reply. "Tora, tora, tora. It is either the word for tiger, or it could be a contraction meaning lightning strike or attack, repeated three times." Another pause. "Exactly. I think we'd better call the CinC. I think an attack is imminent, if not underway," Rochefort said, referring to the Commander-in-Chief Pacific Fleet Admiral Husband Edward Kimmel, USN [USNA 1904]. "OK. Thank you," he said and hung up the telephone handset.

The duty officer Ensign Lawrence 'Larry' Stern, USNR, stood in front of Rochefort's desk. "What did Commander Layton say, if you don't mind me asking, Joe?"

"He would run it up the chain."

"That's all?"

"Yep. We have done our part. Let's focus on our mission, Larry. Let's make sure all of the operators are not distracted. Lives may depend upon what we can hear."

"Yes sir," Stern relied and walked toward the row of receiver stations along the wall.

What Rochefort and HYPO did not know at the time was the meaning of the broadcast they heard. The words had been radioed in the clear from *Kaigun-chusa* Mitsuo Fuchida, the leader of the first wave of Japanese attack aircraft, to *Kaigun-chujo* Chuichi Nagumo, Carrier Battle Group commander aboard his flag ship, the aircraft carrier *Akagi*, one of six carriers in the task force. The signal was a code word meaning complete surprise had been achieved and the second wave should proceed according to plan.

—

Chapter 19

In bello parvis momentis magni causus intercedunt.
(In war trivial causes produce momentous events.)

-- Julius Caesar

Sunday, 7.December.1941
Oahu, Hawaiian Islands
Territory of the United States of America
07:55 hours [14:25, Washington, DC; 20:25, London; 03:25 (8.Dec.), Tokyo]

The Japanese naval aviation strike force had indeed achieved complete surprise. The first wave had split into several attack groups that began their planned simultaneous attacks from the north and west, on the U.S. fleet anchorage at Pearl Harbor, and the Army and Navy air bases at Wheeler, Hickam, Bellows, Ewa, Halewa and Ford Island. The Japanese strike force was composed of 180 fighters, dive-bombers, torpedo planes and high-level bombers. They went about their work with well-rehearsed precision and skill.

08:45 hours

The first wave attack ended and the Japanese attackers withdrew to their carriers. Commander Fuchida remained high overhead to assess the inflicted damage and observe the second wave attack beginning.

08:50 hours

The second wave of 170 aircraft with no torpedo planes had launched at 07:15. En route, the second wave leader, *Kaigun-shosa* Shigekazu Shimazaki, received the expected signal for them to press their planned attack. The second wave began their attack from the east, adding the Marine Corps Air Base Kaneohe to their targets that also included repeat attacks on virtually all of the first wave targets as well.

―

Sunday, 7.December.1941
Polo Grounds
West 155th Street and Frederick Douglass Boulevard
Manhattan, New York City, New York
United States of America
15:25 hours

The National Football League game featured the traditional cross-town rivalry between the Brooklyn Dodgers and the New York Giants. The last regular season game was also celebrated as 'Tuffy' Leemans Day, to honor the Giant's star fullback Alphonse 'Tuffy' Leemans. A full stadium of 55,000 fans, all bundled up against the 32-degree chill, watched the game with enthusiasm, as the Dodgers appeared to dominate the game.

As the third quarter approached its conclusion, the public address system blared out. "Attention, please. All active duty military personnel are to report to their posts immediately. I say again, all military personnel are to report to their posts immediately."

"That cannot be good," Bill Donovan said softly to his friend and law partner Carl Elbridge Newton.

"What do you think it means?" asked Newton.

"I have no idea, but I would bet good money that whatever is happening, it is serious and not good."

"Do you need to go?"

"I am not in the military . . . at least not at the moment."

"Didn't you say the President intends to reactivate you?"

"That is what he indicated, but I am not pressing for it."

"I know you cannot tell me what you are up to, but it must be important for you to leave the law firm you founded."

"Leave of absence . . . let's not forget, I've not sold my share of the partnership."

Newton laughed. "*Touché.*"

"I'm not sure how long this assignment is going to last . . ." His sentence was interrupted.

"Attention, please," the PA speaker blared, again. "Here is an urgent message: Colonel William Donovan, come to the box office at once. There is an important phone message. I say again, Colonel William Donovan, come to the box office at once. There is an important phone message."

Bill looked directly at Carl as he stood. "Serious, indeed. I may not be back, Carl. The Giants had better get things sorted out, if they are going to get back in this game." Donovan reached the aisle, but did not lose Newton's eyes. "Anyway, enjoy the rest of the game. If I do not return in a few minutes, I may have been called to Washington. Would you mind stopping by the apartment to tell Ruth what happened?"

"No problem, my friend."

"Tell her I will call as soon as I am able."

"Sure thing, Bill. Good luck."

Donovan ascended the stairway and was grateful the spectator traffic was light. He found the west ticket office, talked to the manager, and was let in to use the office telephone. He lifted the handset off the table. "This is Bill Donovan."

"Mister Donovan, this is Manhattan Operator 73. I have an urgent call from Operator 19 in Washington, DC."

"Thank you." He heard the clicking and crackling on the telephone line, as the New York operator worked to reestablish the connection. Bill knew Operator 19 was actually and solely devoted to White House telephonic communications. He heard Operator 19 come on the line and his operator telling him he was connected and to go ahead.

"Colonel Donovan?" asked Operator 19.

"This is Colonel Donovan."

"You have an urgent call from the White House. I must remind you, this is not a secure line. I'll connect you now."

Donovan did not respond and waited the few seconds for the connection to be made.

"Colonel, this is Jimmy," meaning James Roosevelt II, the President's second child. Jimmy did not wait for a reply. "The Japanese have attacked Pearl Harbor, Hawaii. The President needs you back here as soon as you can get here."

"I'll try . . . ," Donovan began but was interrupted.

"No sweat. There is a plane waiting for you at the military terminal of Municipal Airport. Get there as quick as you can. Get a police escort if you need it."

"Please inform the office I am on my way," Bill replied and placed the telephone handset in its cradle.

Donovan turned to the box office manager. "I need to get to New York Municipal Airport immediately," Donovan said to the manager. "Do you have a fast car available and a good driver?"

"Yes sir. Mister Mara called down right after the announcement. He expected you might need a good car. Mister Mara offered his Cadillac and driver. They are standing by right outside the gate. I will take you to them."

"Please convey my sincerest gratitude to Mister Mara for his understanding and graciousness."

"Yes sir. Gladly," the manager answered, as they walked swiftly toward the nearest exit gate.

"Also, is there a police officer close by?"

"Yes sir . . . by the gate and he has a call box next to him."

"Excellent," Donovan said and turned to leave.

"If I may ask, sir, is it true?"

"What?"

"Have the Japs attacked Pearl Harbor?"

"Apparently so, but that is all I know so far."

"Damn, chicken ass bastards." He stopped himself. "My apologies for the profanity, sir."

Donovan did not respond. There was not much he could say, anyway. 'Tim' Mara's driver and excellent car were indeed waiting for him. Donovan looked for the police officer, but the manager had already gestured to the officer, who was approaching.

"Is there a problem?" asked the officer.

"Yes," the manager said, before any sound exited Donovan's open mouth. "This is Colonel Donovan." The officer saluted and Bill returned the gesture of respect. "He has to get to LaGuardia Field immediately."

"You need a police escort?"

"That would be most helpful, officer. Time is of the essence."

"So, it's true?" the officer asked.

"Yes, so I am told. I know little more than you, at this stage."

"Bastards." The police officer turned and stepped off to his call box. He shouted without looking back, "Get going, Colonel. I'll have an escort for you before you hit the street."

Bill thanked the manager for his assistance and reminded him to thank Mister Mara. He did not have the opportunity to thank the police officer.

True to his word, two police motorcycle officers were waiting at the street gate, complete with red lights flashing, fore and aft. They did not wait for instructions. Donovan heard their sirens start up, even inside the closed limousine.

Blessedly, the Sunday afternoon traffic was comparatively light and unusually cooperative as well as compliant. The limousine driver and police motorcyclists only slowed less than a handful of times during the journey from the Polo Grounds, across the Triborough Bridge, and to LaGuardia Field airport. The motorcycle police officers somehow knew precisely where Donovan needed to go.

They pulled up to the substantial aircraft hangar structure with an attached, one-story, office building of similar construction. The sign simply said in bold red letters, government property, keep out.

—

Sunday, 7.December.1941
Oahu, Hawaiian Islands
Territory of the United States of America
09:45 hours [16:15, Washington, DC; 22:15, London]

The Japanese air attack was finally over. A third wave had been planned to attack the logistics support facilities including oil/fuel storage, ammunition dumps, supply depots, communications facilities and dockyards. Based on the limited damage assessment information available at the time, Admiral Nagumo felt a third wave would be stretching their good fortune too far. He knew from Fuchida's debriefing reports that the American carriers were not in port as they had understood and expected. Those aircraft carriers could be anywhere, even within striking distance, and that was simply too much risk for what might be gained.

The Japanese naval strike force had inflicted devastating losses on the armed forces of the United States in the Pacific region, including 2,403 Americans killed and 1,178 wounded; nine battleships sunk or seriously damaged plus nine other cruisers, destroyers and auxiliary ships sunk, destroyed or damaged; and, 188 aircraft destroyed and 159 damaged, 155 of them on the ground, of the 402 total American aircraft of all services in Hawaii. By contrast, the Imperial Japanese Navy lost 55 airmen and nine midget submariners, with one pilot captured; and, of Japan's 414 available carrier aircraft, 29 were lost during the attack (nine in the first attack wave, 20 in the second), with another 74 damaged by antiaircraft fire. Adding insult to injury, some of the American losses were due to friendly fire in the chaotic confusion of the attack.

—

Sunday, 7.December.1941
Military Terminal
New York Municipal Airport
East Elmhurst, Queens, New York City, New York
United States of America
16:47 hours

An Army first lieutenant in full service uniform and wearing the shiny silver wings of a newly designated aviator above the left breast pocket of his tunic was standing at curbside. The lieutenant opened the passenger door and saluted. As Donovan stepped out, the lieutenant said, "The airplane is waiting for you, Colonel. Everyone else is on board."

"Thank you, Lieutenant." Bill Donovan did not follow the lieutenant. He went to both police officers, extended his right hand, shook hands with each of them, and shouted over the sound of the engine, "Thank you very much." They each saluted him. Bill turned back to the driver and did the same. "Thank you so much for your skills. Please convey my gratitude to Mister Mara."

"I will, sir. Good luck to you, Colonel. You're going to need it," the driver responded.

Donovan nodded his head and turned to the lieutenant. "Lead on, Lieutenant."

The smart looking Army officer asked, "Do you have any baggage, Colonel?"

Bill laughed. "I was at the football game when the call came. No, I don't have any baggage. Just get me on that airplane."

"Yes sir." The young man stepped out confidently and led Bill Donovan through the office portion of the building, out onto the tarmac, and toward a C-39 twin-engine transport aircraft in Army Air Forces, olive-drab livery.

An enlisted man, presumably a junior crewmember, who was initially dancing back and forth, trying to generate some body heat in the cold air, sprang into action when he saw the two men approaching. The man ascended the short ladder and opened the hatchway. He called to someone in the interior. Back on the tarmac, beside the ladder, the man came to a position of attention and saluted. Dressed in his business suit and overcoat, Donovan saluted, climbed up the ladder and into the interior. The aircraft must have been a hastily dispatched airplane, since it was not configured with an executive interior. One row of seven, standard, Army, canvas, forward facing seats on each side of the fuselage. There were only three passengers already on board. Presidential adviser Judge Samuel Irving 'Sam' Rosenman stood from the left seat, as Donovan moved up the aisle way, and sat next to Postmaster General Frank Comerford Walker, vacating the seat next to Vice President Henry Agard Wallace. The No. 2 engine started before the door closed behind him and locked.

"We've been waiting for you, Mister Donovan," said the Vice President with discernible irritation. As a very liberal, verging upon socialist Democrat, Wallace had never taken kindly to Roosevelt's penchant for selecting staunch Republicans for key national security positions in the administration. The Vice President seldom lost an opportunity to convey his displeasure to Donovan, whom he considered to be a minor, peripheral player.

"Thank you for that, Mister Vice President," Bill responded with calm demeanor, as he sat in the stiff seat next to Wallace. *He probably does not want to hear that I was at the National Football League game in the city.*

The No. 1 engine was started and the aircraft began to taxi out in surprisingly swift sequence. Donovan buckled his seat belt.

"What do you know?" asked Wallace.

"Not much. Rumors and conjecture for the most part. I got the call and talked briefly to the President's son Jimmy, who confirmed the attack at Pearl Harbor."

"We know that much," Henry said with more irritation.

"I've got to get back to the office to see what my field officers have learned."

The Vice President waved his hand dismissively and looked out the window to his right.

The aircraft did not stop or slow once it had begun moving, which meant the pilot had received priority clearance to take off ahead of all other aircraft. Within minutes, they were airborne and headed south for the hour and a half flight to the nation's capital.

—

Monday, 8.December.1940
Cabinet War Rooms
New Public Offices
Whitehall, London, England
United Kingdom
00:30 hours

"Gentlemen, please," Prime Minister Churchill said with impatience, not waiting for Cabinet Secretary Sir Edward Bridges, who nodded his acceptance. Only Denniston was missing from the previous afternoon's meeting. "We have urgent business to tend to tonight." He waited, oddly shifting his weight from one leg to the other, adding to the image of impatience. Once the assembled War Cabinet and Chiefs of Staff Committee had taken their seats and quieted, he began, "I believe everyone has heard the news of the Japanese surprise attack on the American naval base at Pearl Harbor, Hawaii. We did not have to wait long from this afternoon's assessment until the Japanese executed their evil plan . . . at least in part. In addition to the Americans, we have confirmation from General Percival that Japanese air and ground forces have landed in Thailand and the Northern Malay Peninsula. Their attack began an hour and a half ago. I do believe Admiral Phillips has his target, as we agreed this afternoon."

The First Sea Lord responded, "Admiral Phillips has the intelligence we have plus the field intelligence from the Malay landing at Kota Bharu. He expects to sortie Force 'Z' this afternoon with the specific objective of engaging the enemy invasion fleet. He intends to operate within range of RAF fighters in Malaya."

"Godspeed and following winds, Admiral," Churchill commented in an uncharacteristically muffled tone. "Very well. There are other developments coming in rapid-fire succession. 'C', if you would be so kind, please inform the War Cabinet of what you know as of this moment."

"By all mean, Prime Minister," answered Menzies. "The Japanese have also landed an invasion force at Samut Prakan, Thailand, at dawn. We have uncorroborated information that the Japanese Army with Air Force support have crossed the frontier from French Indo-China into Thailand, but we have not been able to substantiate the information, as yet. Our analysts believe the enemy's immediate objective is land and air isolation of Singapore, with an ultimate objective of taking Singapore. Commonwealth forces under General Percival engaged the Malay invasion force at the beach. We have coordinated with the Thai government to aid in their defense. Also, Japanese military activity clearly indicates their intentions to invade the American territory of Guam and the Philippines as well as our colony at Hong Kong. The execution of those plans has already begun and the physical landing of ground forces at each site is expected later this morning. We have been in near constant coordination with our American counterparts."

"If I may interject," Churchill said, "to that end, we have received quite a few open channel communications from virtually all of the American sites. I have singled out two in particular as relevant to our task at hand." Churchill passed the two sheets of paper to Clement Attlee on his right. "This first is the initial message from the Commander U.S. Pacific Fleet at Pearl Harbor, Hawaii, received at 19:25 hours this evening."

```
NPEK
AN RTT NR 275
OI 071725Z DEC 41
FM CINCPACFLT
TO ALL CMDS
BT
UNCLAS
AIR RAID ON PEARL HARBOR BREAK THIS IS NOT A
DRILL
BT
NNNN
```

"The second message was received 40 minutes ago from the American Chief of Naval Operations Admiral Stark to the Pacific region Navy forces under his purview. Both were un-coded, open channel messages."

```
NO3A
AA RAA NR 507
```

```
U 072152Z DEC 41
FM CNO
TO CINCPAC
COMPANAM
CINCAF
PACIFIC NORTHERN PACIFIC SOUTHERN HAWAIIAN
NAVAL COASTAL FRONTIERS
BT
UNCLAS
BY ORDER OF THE PRESIDENT EXECUTE AGAINST JAPAN
UNRESTRICTED AIR AND SUBMARINE WARFARE
BT
NNNN
```

"Although, to our knowledge, Congress has not yet declared war on Japan, I believe it safe to say that outcome is undoubtedly inevitable. I only have one short message from President Roosevelt a few hours ago simply stating to the effect, we are in this with you now. Before we proceed to the primary and sole subject of this late night gathering, 'C' do you, or anyone else for that matter, have anything else to add, to ensure the War Cabinet is as informed as humanly possible?"

Only Menzies spoke after no one else responded. "It is dawn in the South China Sea. We expect to have confirmation of the additional Japanese offensive actions in a few hours. The Japanese have executed an elaborate, highly coordinated attack on multiple British and American territories in a large portion of the Asia and Pacific region. MI6 is watching closely and using every source available to us in our efforts to clarify the situation in the Pacific as quickly as possible. In closing my comments, I would like to add that the situation in Hawaii is very confused, and understandably so. Thus, it is very difficult to appreciate the damage inflicted by the Japanese carrier air attack. Fortunately, the immediate attack has ended, although the American forces do not know if an invasion is imminent. They are assuming the invasion is imminent and taking appropriate actions. Let it suffice for now to say the Japanese appear to have achieved total surprise and the damage is quite likely to be extensive. We know from our assistant naval attaché in Honolulu that numerous very large explosions occurred. Nearly the whole anchorage is covered in heavy, black plumes of smoke. He, the attaché, believes, although he has not yet verified, that two and most likely more of the nine battleships have been sunk at their moorings."

"Dear God above," muttered Arthur Greenwood.

"What of their carriers?" asked Anthony Eden.

"We have no direct information," Menzies answered. "We have queried the attaché rather than bother the Americans at this moment of trial. We should have some indication in a few hours. We can only hope at this stage that lady luck stood by them on that score."

"Very well," Churchill said and gestured to the attendees, soliciting any other contributions. Seeing or hearing none, he continued, "then, the matter before this august body . . . I propose His Majesty's Government declare a state of war with Imperial Japan. We have been attacked directly, as have the Americans. Our mutual agreement with President Roosevelt at Placentia Bay committed us to mutual defense."

"They hardly came to our defense, when we stood alone," interjected Ernest Bevin.

"I strongly and respectfully disagree; however, we shall not debate the past," Churchill said rather sternly. "We need the Americans. As President Roosevelt so succinctly stated, we are in this together, now. That is all that matters. Do we need to debate the issue on the table?" Churchill looked to each member of the War Cabinet. Each in turn rejecting any necessity for debate. Bevin opened his mouth, as if to speak, and then closed his mouth and shook his head in the negative. "Very well. It is unanimous. I shall inform the King first thing tomorrow, well, actually, later this morning and Parliament later today. We are at war with Japan. I will also inform my friend, President Roosevelt, of our action and affirm our solidarity with them in their hour of such tragic loss." Churchill looked around the room. Several men nodded their heads in agreement, but no one spoke. "We are adjourned. Thank you for your attentiveness. Please do not stray from your communications means. I suspect we shall be reconvening on short notice and frequently, at least until the situation clarifies. That said we are adjourned."

The participants quietly and with an air of solemnity filed out of the conference room. General Ismay remained behind and waited until everyone had departed.

"Tragic events, sir," Ismay said softly.

"Yes, Pug, tragic indeed. Yet, this day I have prayed for since the war began is finally here. Tonight, this morning, I shall sleep the sleep of the saved."

Ismay smiled. "We all will. There will be no ambiguity after today."

"Quite so. Once we get things sorted out, I want to meet with President Roosevelt as soon as possible. I did not raise my intended visit to Washington with the War Cabinet this evening, but I shall do so in the next few days.

Please convey to the chiefs after your sleep . . . I will need them to go with us. I envision our first true combined chiefs conference in Washington. Our objective will be a joint war plan, now that the Americans are in this crusade."

"Yes sir. I will tend to it first thing."

"Excellent, Pug. Thank you. Now, I'm going to grab a heavy cognac and go to the roof, if you would care to join me. I want to see the city at peace, darkened as it still is, and thank the good Lord for our deliverance. I want to feel how far we have come since just one year ago."

"I would be honored to join you, Prime Minister."

The two men headed off to their late night reminiscence and contemplation of the future.

—

Monday, 8.December.1941
Oval Office
The White House
Washington, District of Columbia
United States of America
00:10 hours

Immediately after the Army airplane had landed in the early evening at Bolling Field, across the Anacostia River from the federal enclave, four staff cars had been lined up and waiting for their arrival. Bill Donovan had just enough time to make a quick stop at his corner office in the COI Headquarters, on the first floor of the East Building of the E Street Complex. As he expected, several members of his senior staff had been waiting for his arrival, apparently having been alerted by someone of his emergency return. Donovan's chosen deputy, Colonel Gonzalo Edward 'Ned' Buxton, had not hesitated for salutations and had begun briefing his boss before the office door had closed. None of them had taken the time to sit. In the 20 minutes he had available, Donovan had absorbed as much as he could of the situation. His long-time, loyal driver and new Cadillac were waiting for him as he rushed out of the headquarters building to make the hastily called emergency Cabinet meeting in the West Wing. Donovan had repeated the cycle once more after the Cabinet meeting to provide the best answers COI could, as well as offer the latest updates from the Pacific.

For the third time that evening, Bill Donovan arrived at the White House, this time just after midnight with a leather satchel containing relevant message traffic, should the President wish to see documentation. A Secret Service officer he did not recognize met him at the lobby entrance. The agent did not speak and ushered him directly to the Oval Office, as if he was running interference for the Coordinator of Information through a still bustling West Wing.

The President gestured for Donovan to join Harry Hopkins and another man sitting with his back to the door. As he stepped between the facing couches, the man stood and turned to greet the new arrival. Donovan hesitated for just a moment when he recognized CBS European Bureau Chief Edward R. Murrow.

"Great to see you, again, Bill."

"Ed," was Donovan's succinct response as the two men shook hands.

The President anticipated Donovan's hesitation. "It's OK, Bill. I want Ed to hear this. Avoid means and methods . . . for now."

Donovan remained standing and faced the President, as Murrow retook his seat. "Excuse me, Mister President, may I ask the purpose of having a renowned Press correspondent present for this discussion?"

"Yes, of course. Simple answer, because I want him here. He happened to be in town when the news broke. I am trying to finalize the content of my speech to Congress, which is where I need your assistance as well. Ed has been helping."

"Very well, sir." Donovan sat next to Hopkins. "I believe you have the condition reports from the military branches. I have nothing of substance to add. Our field agents in Southeast Asia are reporting substantial military activity on a broad front in Siam and the Kra Peninsula. The Japanese are well into a complex offensive plan to expand their domination over the Pacific region and Asian continent. We expect further attacks on the Philippines, Hong Kong, and by extension on Singapore itself. We have limited, unreliable information about the Dutch East Indies being included in their offensive plans."

"What about the Hawaiian Islands, Midway and Wake Islands? What threat do we face on the West Coast?" asked the President.

"I presume ONI briefed you on, among other topics, the sinking of the U.S.-flag, freighter *Cynthia Olson* Sunday morning, 300 miles off the West Coast."

"No, they did not," Roosevelt responded rather curtly.

"She broadcast an SOS before she went silent. According to the message, the *Olson* was attacked by a surfaced submarine's deck gun and torpedo from what appears to be the Imperial Japanese Navy I-26, based on the tower markings reported by the *Olson*'s captain."

"When did it occur and what was she carrying?"

"We have not yet correlated the times, but our first pass indicates the attack was over and done a half hour before the attack at Pearl Harbor began.

Her cargo was reported as lumber and other Army supplies. The ship had been chartered by the War Department."

"Excuse me, Mister President. The reporter in me wants to stay to hear Colonel Donovan's report; however, the employee that I am must make the night train to New York City."

"Certainly, Ed. Thank you for stopping by. Your contributions have been and will be greatly appreciated. Keep up the great work."

"Thank you, Mister President. I shall be listening when you speak to Congress and the nation later today. Good luck."

"I shall need all the luck I can garner."

Murrow stood, shook hands with Donovan, Hopkins and lastly with the President, and then left the room, closing the door behind him.

Donovan waited a few seconds, and then continued, "As far as the threat to the West Coast, clearly, the Japanese Navy is operating on our West Coast. As I'm sure you recognize, Mister President, there is a monumental difference between a submarine harassing our lines of commerce and an invasion or occupation force. The alert messages to the Pacific region forces, including the West Coast, have been issued and reinforced. Our analysis of available information indicates the threat is minimal to remote . . . at least for now. We should expect minor shore bombardment more as a terror actions, to instill fear, rather than any substantive military threat."

The President stared at Donovan as he considered the words of his Coordinator of Information. "Very well. Keep an eye on this question. While the Navy and War Departments will investigate the what, why and how of what happened yesterday, I am far more concerned about what lays ahead and fulfilling my duty as commander-in-chief under the Constitution. To that end, I shall report to Congress later today. I shall ask Congress for a declaration of war against Imperial Japan. I ask for your counsel regarding the content?"

Donovan smiled. "Thank you for the opportunity. Given what we know so far, and what we are likely to learn as the next few hours progress, the people deserve to know the extent of the Japanese operation plans. This was not just a strike at our forces on Oahu Island. They struck several British territories and forces, and will probably hit Dutch territories. The British have declared war against Japan for those attacks."

"Yes. The Prime Minister and I have exchanged messages since the attack began."

"We have sent urgent, highest priority messages to our field agents to collect relevant information as quickly as possible. I will ensure we inform you immediately as we learn more . . . and I strongly suspect there will be more until we see the full extent of their moves."

"Thank you, Bill," responded Roosevelt. "You have only been on the job for a few months, and I am most grateful for your service."

"Thank you, Mister President."

"While our attention is on the East, what do you know about any unique action by Germany or Italy in the wake of Japan's attack on us . . . and the British?"

"The Tripartite Pact of a year and a half ago implies mutual action, although it does not cover offensive operations, but rather defensive events. Our Research and Analysis Group recommends that we should expect the Germans, and collaterally the Italians, to join their Axis partner."

"Does the Navy know this?"

"Yes sir. They are preparing for anti-submarine operations in our coastal waters."

"West Coast, too?"

"Yes sir, although the West Coast is a little more problematic with the devastation at Pearl Harbor."

"Will the Germans and/or Italians attack us, as the Japanese did? And, if so, where?"

"We see no signs of offensive action by the Germans against our interests. We think the primary threat for now will be offensive submarine operations, which could begin within days."

"Very well. Now, it is nearly two in the morning. I have an important speech to deliver in a few hours. All of us need to get some sleep."

Donovan stood. "Thank you, Mister President. I truly hope and pray you are able to sleep. The COI staff is working through the night. I shall be available, as you may require. Good luck today."

"Thank you, Bill. Keep up the great work."

"We will, Mister President. Good night . . . or rather good morning."

They all chuckled at the time. Donovan departed. He did not go to his home, but rather he went back to his office. There was so much to learn and so little time to learn it.

—

Monday, 8.December.1941
RAF North Weald
Epping, Essex, England
United Kingdom
16:00 hours

The long day was nearly over. No.71 Squadron had launched early this morning, during morning twilight before dawn to reposition to RAF Manston. They flew two back-up cover sorties over the southern Channel, protecting both the ground attack and top cover squadrons during operations in France. The Germans had improved substantially in the intercept timing, shooting up several raids at their most vulnerable time. Fighter Command had adapted to using a fresh fighter squadron to scrub off any German fighters attempting to intercept the withdrawing British fighters. They had a quick turnaround for fuel after the first mission, since they had not fired a shot, and even at that they were ten minutes late making their on-station time for the second mission. The second CIRCUS support mission had been more eventful. They dove quickly and tangled with a squadron of 109's. No confirmed victories, but several hits. They had taken some hits, too; fortunately, none serious. They successfully accomplished their assigned mission. Their turn around at Manston took longer the second time, since the ground crews had to load ammunition during refueling operations, and two of their aircraft needed temporary repairs. They were released from their CIRCUS support mission. Just as they were preparing to mount up for the return to North Weald, they were given a quick reaction mission to cover a northbound convoy out of the Thames Estuary. The Germans were picked up by Chain Home making an attempt on the convoy. By the time they reached the convoy, the Germans had abandoned their attempted attack. They were instructed to remain overhead the convoy until they reached their bingo fuel level.

The thick overcast blanketing the southeast of England meant they had to execute an extended squadron penetration of the cloud layer. Thankfully, there was only moderate turbulence in the clouds. They broke out below the cloud deck at just under 2,000 feet, about five miles from the airfield and lined up well for landing. Their Spitfire Mark VB fighters landed smoothly and taxied to their parking spots.

Flying Officer Brian Drummond completed the shutdown of his aircraft's big Merlin engine and secured the battery.

Corporal Jacobs jumped up on the left wing and stood on the wing root, facing his pilot. "Welcome back, sir. Any squawks?"

"Thanks, Henry. Nope, she was perfect. I don't think I took any hits, but I know you will give her a good look-see. Should be just fuel. We did not fire a shot after the second sortie today."

"Yes sir. You know we will, sir."

Before Brian could respond, Jacobs jumped down off the wing and hurried to assist Hawking and Easton. Brian unstrapped from his parachute and seat, elected to keep all of his flight equipment on including his gloves, grabbed the solid edge of the aircraft's forward windscreen and stood up. The movement of a single Spitfire landing caught his eye, since his helmet and headset dampened external sound. He stepped out onto the left wing root, and then off the wing to the ground. He patted his fighter's empennage skin, as if to say thank you, girl. When he rounded the tail of his machine, the single Spitfire spun its tail to park – 'PR-K' – Jonathan's machine. Brian walked to the tail of Jonathan's Spitfire while his best friend shutdown the flying machine.

"What brings you to North Weald, my friend?"

"Haven't you heard?"

"We've been on missions since before dawn this morning," Brian answered.

"Then, you haven't heard?"

"Heard what?" Brian answered impatiently.

"The Japanese attacked Pearl Harbor in Hawaii."

"What!"

Just then, Joshua Forcier burst out of the Dispersal Hut. "Brian," he shouted, "the Japs attacked Pearl Harbor."

Brian turned back to Jonathan. "When?"

"Last night. Well, it was early Sunday morning in Hawaii. We were released and I asked the Skipper to fly down here. President Roosevelt is supposed to speak to a joint session of Congress in a couple of hours. I was certain you and your mates would be listening and I wanted to listen with you."

"Yeah," Frog said, "the Skipper said the same thing. He's calling Group, now."

"Do you know anything else?" Brian asked Jonathan.

"The BBC European Service said it was a surprise attack Sunday morning. Fires are continuing to burn according to the report."

"Did they invade?" asked Brian.

"There was no mention of it in the radio report," Jonathan said. "Maybe the President will offer more information in his broadcast. The radio announcer did say there was extensive damage and fatalities."

"This probably means we are in this war, now."

"That would be my guess," Jonathan added.

Dusty Langford joined them outside in the cold. "The Skipper said we are released and that the President is supposed to speak tonight at 18:30.

According to the Skipper, Wing Commander Treadle arranged for the radio guys to set up a high quality radio in the Mess for the broadcast tonight."

The pilots began departing. There was only one topic of discussion between them. Brian doffed his flight gear, and stowed his kit on the assigned peg. He stopped at Meares' office. "See you at the Mess, Skipper."

"My condolences to you and your countrymen, Brian. This is a very sad day."

"Thank you, sir. See you at the Mess."

"Count on it, Hunter."

Brian nodded, departed the commander's office, and joined Harness, Dusty and Frog for the short walk to the Mess. Their words between them were the same as the others. There was only one topic this evening.

—

Monday, 8.December.1941
House of Representatives
Capitol Building
Washington, District of Columbia
United States of America
12:30 hours

The rustling of nearly 500 Members of the Senate and House of Representatives along with an equal number of the staffs, press and spectators in the gallery came to their feet. The hastily called joint session of Congress left three score of Members too distant for attendance. President Roosevelt, assisted by his eldest son Captain of Marines James 'Jimmy' Roosevelt II, ambled to the podium. Vice President Henry Agard Wallace and Speaker of the House Samuel Taliaferro 'Sam' Rayburn of Texas stood on the chamber dais as the President entered from the side door. He grasped the podium festooned with microphones marked with small placards – CBS, NBC, MBS – straightened himself, removed a single sheet of paper from his jacket inside pocket, took a couple of deep breaths and glanced over his shoulder to signal his readiness. Applause and cheers blossomed from the standing assemblage.

"Senators and Representatives," announced Speaker Rayburn, "I have the distinguished honor of presenting the President of the United States."

More applause filled the chamber. The President waited for the rustling of his audience taking their seats and hushed. He let the silence blanket the chamber for a handful of seconds.

"Mister Vice President, Mister Speaker," Roosevelt began with a solemn but strong voice, "Members of the Senate and the House of Representatives,

"Yesterday, December seventh, 1941 – a date which will live in infamy – the United States of America was suddenly and deliberately attacked by naval and air forces of the Empire of Japan.

"The United States was at peace with that nation and, at the solicitation of Japan, was still in conversation with its Government and its Emperor, looking toward the maintenance of peace in the Pacific. Indeed, one hour after Japanese air squadrons had commenced bombing in the American island of Oahu, the Japanese Ambassador to the United States and his colleague delivered to our Secretary of State a formal reply to a recent American message. While this reply stated that it seemed useless to continue the existing diplomatic negotiations, it contained no threat or hint of war or of armed attack.

"It will be recorded that the distance of Hawaii from Japan makes it obvious that the attack was deliberately planned many days or even weeks ago. During the intervening time, the Japanese Government has deliberately sought to deceive the United States by false statements and expressions of hope for continued peace.

"The attack yesterday on the Hawaiian Islands has caused severe damage to American naval and military forces. I regret to tell you that very many American lives have been lost. In addition, American ships have been reported torpedoed on the high seas between San Francisco and Honolulu.

"Yesterday, the Japanese Government also launched an attack against Malaya.

"Last night, Japanese forces attacked Hong Kong.

"Last night, Japanese forces attacked Guam.

"Last night, Japanese forces attacked the Philippine Islands.

"Last night, the Japanese attacked Wake Island. And, this morning, the Japanese attacked Midway Island.

"Japan has, therefore, undertaken a surprise offensive extending throughout the Pacific area. The facts of yesterday and today speak for themselves. The people of the United States have already formed their opinions and well understand the implications to the very life and safety of our nation.

"As Commander-in-Chief of the Army and Navy, I have directed that all measures be taken for our defense.

"But always will our full nation remember the character of the onslaught against us."

Applause filled the House chamber and interrupted the President's speech. He took the opportunity of the interruption to take a sip of water from a small glass beside the podium.

"No matter how long it may take us to overcome this premeditated invasion, the American people in their righteous might, will win through to absolute victory."

An eruption of enthusiastic applause and cheers punctuated the President's statement.

"I believe that I interpret the will of the Congress and of the people when I assert that we will not only defend ourselves to the uttermost, but will make very certain that this form of treachery shall never again endanger us."

Again, applause and cheers amplified the President's words.

"Hostilities exist. There is no blinking at the fact that our people, our territory and our interests are in grave danger.

"With confidence in our armed forces – with the unbounding determination of our people - we will gain the inevitable triumph - so help us God."

The President shifted his weight from leg to leg as the assembly applauded his statement.

"I ask that the Congress declare that since the unprovoked and dastardly attack by Japan on Sunday, December seventh, 1941, a state of war has existed between the United States and the Japanese Empire."

Continuous cheers and vigorous applause marked the end of the President's address as the entire chamber stood to recognize the clear, demonstrative statement of what had happened yesterday.

The President's statement to Congress, to the American people, and to the world took all of seven minutes to deliver. The whole world, at least those who were listening, heard his statement. Cheers and applause continued as the President departed the House chamber to leave Congress to their work and returned to the White House. Congress promptly began the process of passing the necessary, constitutionally required legislation. The process did not take long.

—

At 16:10 [R] Eastern Standard Time on the afternoon of 8.December.1941, President Roosevelt signed the recently passed legislation titled: JOINT RESOLUTION Declaring that a state of war exists between the Imperial Government of Japan and the Government and the people of the United States and making provisions to prosecute the same [Public Law 77-I-328; Senate Joint Resolution 116; Chapter 561, page 795, Session I; 55 Statute 795]. Congress passed the declaration of war in the House of Representatives by a vote of 388 to 1, followed by the Senate with a vote of 82 to nil. The only vote against the declaration of war against Japan in both chambers of Congress that day was Representative Jeanette Pickering Rankin of Montana – the first woman elected to Congress, a Republican, and a dedicated lifelong pacifist; she would not be re-elected in 1942.

The United States of America entered World War II as a belligerent, and the second global war in a generation, or the continuation of the Great War, had been enjoined.

—

Cap Parlier

Author

—

Cap and his wife, Jeanne, moved from the Great Plains of Kansas to the warmth and diversity of Arizona. Their four children have established their families and are raising their grandchildren. Their first grandchild and granddaughter became an adult herself and graduated from university. Cap is a graduate of the U.S. Naval Academy, a retired Marine aviator, Vietnam veteran and experimental test pilot, and has finally retired from the corporate world to devote his time to his passion for writing a good story. He has numerous other projects completed and in the works including screenplays, historical novels as well as atypical novels at various stages of the creation process.

—

Interested readers may wish to visit his website at <http://www.parlier.com> for his essays and other items, or subscribe to his weekly Blog: "*Update from the Sunland.*" Cap can be reached at: cap@SaintGaudensPress.com.

—

www.ingramcontent.com/pod-product-compliance
Lightning Source LLC
Chambersburg PA
CBHW020652110726
47901CB00001B/149